Introduction to Wave Scattering, Localization, and Mesoscopic Phenomena

Introduction to Wave Scattering, Localization, and Mesoscopic Phenomena

Ping Sheng

Department of Physics
Hong Kong University of Science and Technology
Clear Water Bay, Kowloon, Hong Kong

and

Corporate Research Laboratories
Exxon Research & Engineering Co.
Clinton, New Jersey, USA

ACADEMIC PRESS

San Diego   New York   Boston   London   Sydney   Tokyo   Toronto

This book is printed on acid-free paper.

Copyright © 1995 by ACADEMIC PRESS, INC.

All Rights Reserved.
No part of this publication may be reproduced or transmitted in any form or by any means, electronic or mechanical, including photocopy, recording, or any information storage and retrieval system, without permission in writing from the publisher.

Academic Press, Inc.
A Division of Harcourt Brace & Company
525 B Street, Suite 1900, San Diego, California 92101-4495

United Kingdom Edition published by
Academic Press Limited
24-28 Oval Road, London NW1 7DX

Library of Congress Cataloging-in-Publication Data

Sheng, Ping, date.
    Introduction to wave scattering, localization, and mesoscopic phenomena / by Ping Sheng.
      p.   cm.
   Includes index.
   ISBN 0-12-639845-3
    1. Waves. 2. Scattering (Physics) 3. Localization theory.
I. Title.
QC157.S447   1995
531'.1133-dc20
                                    94-31847
                                        CIP

PRINTED IN THE UNITED STATES OF AMERICA
95 96 97 98 99 00 QW 9 8 7 6 5 4 3 2 1

Contents

Preface	ix
1. Introduction	1
References	14
2. Quantum and Classical Waves	15
2.1 Preliminaries	15
2.2 Green's Functions for Waves in a Uniform Medium	19
2.3 Waves on a Discrete Lattice	26
2.4 Lattice Green's Functions	31
2.5 Treating Continuum Problems on a Lattice	37
Problems and Solutions	40
Reference	48
3. Wave Scattering and the Effective Medium	49
3.1 An Overview of the Approach	49
3.2 Wave Scattering Formalism	51
3.3 Single Scatterer—The Lattice Case	55
3.4 Single Scatterer—The Continuum Case	58
3.5 Infinite Number of Scatterers—The Effective Medium and the Coherent Potential Approximation	66
3.6 CPA—The Anderson Model	69

	3.7 CPA—The Case of Classical Waves	73
	3.7.1 The Symmetric Microstructure	77
	3.7.2 The Dispersion Microstructure	82
	3.8 Accuracy of the CPA	84
	3.9 Extension of the CPA to the Intermediate Frequency Regime	85
	Problems and Solutions	87
	References	113

4. Diffusive Waves   115

4.1 Beyond the Effective Medium	115
4.2 Pulse Intensity Evolution in a Random Medium	116
4.3 The Bethe–Salpeter Equation and Its Solution by Moments	121
4.4 The Vertex Function	134
4.5 The Ward Identity	145
4.6 Modification of the Diffusion Constant Due to Frequency-Dependent Scattering Potentials	151
4.7 Evaluation of the Wave Diffusion Constant	153
4.8 Application: Diffusive Wave Spectroscopy	160
Problems and Solutions	171
References	175

5. The Coherent Backscattering Effect   177

5.1 Wave Diffusion versus Classical Diffusion	177
5.2 Coherence in the Backscattering Direction	178
5.3 Angular Profile of the Coherent Backscattering	181
5.4 Sample Size (Path Length) Dependence	185
Problems and Solutions	189
References	192

6. Renormalized Diffusion   193

6.1 Coherent Backscattering Effect in the Diagrammatic Representation	193
6.2 Evaluation of the Maximally Crossed Diagrams	195
6.3 Renormalized Diffusion Constant	199

Contents vii

    6.4 Sample Size and Spatial Dimensionality Dependences of Wave Diffusion 201
    6.5 Localization in One Dimension: The Herbert–Jones–Thouless Formula 203
    Problems and Solutions 211
    References 213

7. The Scaling Theory of Localization 215

    7.1 Distinguishing a Localized State from an Extended State 215
    7.2 The Scaling Hypothesis and Its Consequences 218
    7.3 Finite-Size Scaling Calculation of $\beta(\ln \gamma)$ 226
    7.4 Universality and Limitations of the Scaling Theory Results 232
    Problem and Solution 234
    References 239

8. Localized States and the Approach to Localization 241

    8.1 The Self-Consistent Theory of Localization 241
    8.2 Localization Behavior of the Anderson Model 244
    8.3 Classical Scalar Wave Localization 259
    8.4 Transport Velocity of Classical Scalar Waves 269
    8.5 The Scaling Function $\beta(\ln \gamma)$ 271
    Problems and Solutions 277
    References 281

9. Localization Phenomena in Electronic Systems 283

    9.1 Finite Temperatures and the Effect of Inelastic Scattering 283
    9.2 Temperature Dependence of the Resistance in 2D Disordered Films 284
    9.3 Magnetoresistance of Disordered Metallic Films 287
    9.4 Transport of Localized States at Finite Temperatures— Hopping Conduction 293
    Problems and Solutions 298
    References 300

10. Mesoscopic Phenomena 301

    10.1 What is "Mesoscopic"? 301
    10.2 Intensity Distribution of the Speckle Pattern 302

10.3	Correlations in the Speckle Pattern	304
10.4	Long-Range Correlation in Intensity Fluctuations	310
10.5	Landauer's Formula and Quantized Conductances	315
10.6	Characteristics of Mesoscopic Conductance	320
	Problems and Solutions	325
	References	326

Index 329

Preface

Wave behavior in disordered media is an old subject that has undergone a tremendous transformation in the past thirty years. Initiated by Anderson's seminal paper in predicting the phenomenon of wave localization, this transformation was achieved through a general proliferation of research activities in the physics of disorder and inhomogeneous materials. As a result, compared with thirty years ago there is now a new picture of wave characteristics in disordered media. The purposes of this volume are to delineate a coherent outline of this picture and to make the relevant technical materials accessible to a larger audience than those specialized in this area of research. Consistent with this intention, the presentation of both the physics and the mathematical framework is done with a minimum of assumed knowledge on the readers' part about the subject matter and the relevant theoretical techniques. In order not to interrupt the main line of presentation, however, many of the mathematical details are given at the end of each chapter, in the form of problems and solutions. The style of presentation is that of a physicist, whereby physical picture and reasoning are emphasized over mathematical rigor, and the mathematics is developed only to the extent necessary to support the derivation of results. Therefore, although Green's function is used throughout the book, the discussion of Green's function techniques is kept to a minimum. Also associated with the style of presentation is the subjective choice of the topics, which reflects this author's attempt to give not just a cartoon picture of the subject on the one hand and to avoid writing a review on the other. The readers can judge for themselves the degree of success to which this balance has been achieved.

The topics covered in this book can be categorized under four phenomena/concepts: effective medium, wave diffusion, wave localization, and intensity fluctuations and correlations. An overview and some explanation of the unconventional features of wave behavior in disordered media are given in Chapter 1. Basics of Green's functions are introduced in Chapter 2. The central themes of Chapter 3 are wave scattering and the concept of effective medium, which reconciles the physical appearance of a homogeneous medium with the ubiquitous existence of wave scattering. The limitation of the effective medium concept and what happens beyond that are discussed in some detail in Chapter 4. This chapter is perhaps the most mathematical one of the book, since it contains the derivation of the wave diffusion constant and the demonstration of Ward's identity. A new experimental measurement technique—diffusive wave spectroscopy—is also described. Chapters 2, 3, and 4 are essentially self-contained and may form an independent unit in a course on waves.

The coherent backscattering effect and its consequences, including wave localization, are treated in Chapters 5, 6, and 8. A physical description of the coherent backscattering is given in Chapter 5. The dynamical implications of the effect are described in Chapter 6, which also contains an alternative, more exact treatment of wave localization in one-dimensional systems. The full localization implications of coherent backscattering are presented in Chapter 8.

The phenomenological scaling theory of localization is presented in Chapter 7, which is relatively independent from the other chapters. The scaling theory illustrates how the essential qualitative characteristics of localization can be understood on the basis of a few crucial elements and offers a physical way to view the progression from an extended state to a localized one through the variation of sample size.

The effects of multiple scattering and localization are illustrated in Chapters 9 and 10 in two different contexts. In Chapter 9 the effects are viewed in bulk electronic systems through the temperature and magnetic-field variations of conductances. In Chapter 10 the effects are viewed through "mesoscopic" samples where the intensity fluctuations resulting from wave interference are manifest. These fluctuations are shown to contain novel correlations as well as to give rise to electrical conduction phenomena not seen in conventional bulk samples.

Writing this book was made possible by a six-month sabbatical leave from the Exxon Research and Engineering Company and a visiting membership at the Institute for Advanced Study, Princeton, New Jersey, for the 1992–1993 academic year. For the support of these institutions I owe my sincere gratitude. Special acknowledgments must be made of my colleagues Minyao Zhou and Zhao Qing Zhang, who provided me with invaluable criticisms and corrected my many mistakes during the writing process. Minyao Zhou, in particular, is responsible for the numerical calculations necessary for many of the figures, some of them quite involved. To them I give my heartfelt thanks. Appreciation is also extended to Ms. Marianne Kane, who typed most of the book and was tireless in going through the many revisions, and to Ms. Dot Fazekas for general secretarial support throughout the writing process.

Finally, I express my warm thanks to my wife, Deborah Wen, and to my daughters, Ellen and Ada, not only for their encouragement and spiritual support, but also for accepting the many hours I spent away from the family in writing.

Clear Water Bay, Hong Kong *Ping Sheng*
March, 1994

1

Introduction

Waves are everywhere around us. We rely on light and sound to sense our immediate surroundings. Radio waves and microwaves are indispensable means of communication. Water waves are responsible for the ocean's perpetually dynamic image. Quantum waves associated with electrons and atoms, while not directly visible, are important in maintaining the structure and stability of solids. With such a ubiquitous presence, wave phenomena naturally occupy a central position in our study of the physical world. Indeed, for waves in simple systems and ordered structures, an extensive literature already exists. However, for the more difficult problem of waves in disordered media, i.e., multiply scattered waves, a coherent (but by no means complete) understanding has only recently emerged, and from what is already known the picture is very different from that we normally associate with waves. In particular, the possibility that a wave can become localized in a random medium is especially intriguing because localization involves a change in the basic wave character. A localized wave has no spatial periodicity or possibility for transport and thus requires a new theoretical framework for its description and understanding. The purposes of this volume are to delineate the main features of this emerging picture of wave behavior in disordered media and to introduce the theoretical techniques for describing these features. Mesoscopic phenomena, which are the natural manifestations of wave scattering and interference effects, are also treated. A brief sketch below of the prominent random-wave characteristics serves as both an introduction to the subject and a map to what follows.

In an infinite, uniform medium, a (plane) wave may be characterized by a frequency and a direction of propagation. In contrast, a wave cannot propagate freely in a disordered medium because of the many scatterings it encounters. There are two types of scattering. One type, inelastic scattering, alters both the wave frequency and the propagation direction. Another type, elastic scattering, preserves the frequency but alters the propagation direction. This book is concerned mainly with the effect of elastic scattering. Accordingly, the term "incoherent" is defined to mean waves having different propagation directions but the same frequency.

The consequence of multiple elastic scatterings may be described in accordance with the scale of observation. There are two obvious yardsticks in the problem. One is the average size R of the inhomogeneous scatterers. If the density of scatterers is not too low, then the interscatterer separation is also on the same order as R. Another yardstick is the wavelength λ. The ratio between R and λ is an important parameter in determining the average distance of coherent propagation between two scatterings. That distance, usually called the mean free path, is the relevant length scale for separating the different regimes of wave phenomena. When $\lambda \gg R$, the scattering is weak and the mean free path is large ($\gg R$) for classical waves, i.e., electromagnetic and elastic waves. In addition, the scatterers and their placement geometry are beyond the resolution limit of the wave. Therefore, on the local scale of one to two mean free paths or less, a disordered medium appears as a *homogeneous effective medium* to the probing wave. In fact, since all matter is discrete at the atomic level, our everyday understanding of a uniform homogeneous medium reflects this effective medium concept. The same effective medium characterization holds for the quantum wave at the local scale. However, on the scale of many mean free paths, the effective medium can no longer be a valid description; even if locally the scattering is weak, over long distances the scattering effect accumulates and a wave can still be significantly randomized. When that happens, the result—*diffusive transport*—is similar to that of a classical particle undergoing random Brownian motion. The same result holds for the case of $\lambda \ll R$, except that the onset of diffusive transport occurs at a scale comparable to R, and there is no longer a valid effective medium because the local microstructure can now be clearly resolved by the wave.

The fact that wave transport can be diffusive has a prominent example in our everyday experience of heat conduction. From statistical mechanics, it is well known that heat in electrically insulating solids is carried by randomly scattered (mostly short-wavelength) elastic waves. Since heat conduction in solids is known to be governed by the diffusion equation even in the absence of inelastic scattering (Sheng and Zhou, 1991), one immediately concludes that randomly scattered elastic waves transport diffusively.

Despite such clear-cut examples, however, diffusive transport for waves still raises some important questions concerning basic principles. A basic property of the wave equation is that of causality, which means that the speed at which a wave can carry information is always finite. For the diffusion equation, however, causality is not valid because the speed of propagation for a disturbance can be infinite. In addition, directly related to causality is the phase information of a wave, which is completely absent in diffusion. Just from this simple consideration it is already clear that the diffusion description of multiply scattered waves cannot be completely accurate since, whatever its appearance, the multiply scattered wave is still a solution of the wave equation (albeit with random coefficients) and therefore must satisfy the basic properties of its solutions. Thus for wave diffusion there should be some deviation from classical diffusion where the wave character asserts itself. In the past decade, this expectation was borne out by the experimental demonstration (Tsang and Ishimaru, 1984; van Albada and Lagendijk, 1985; Wolf and Maret, 1985) of the *coherent backscattering effect*, or the weak-localization effect as it is sometimes called, which represents not only a deviation from classical diffusion, but also the precursor to wave localization.

The coherent backscattering effect means what the name implies: After a wave is multiply scattered many times, its phase coherence is preserved in the direction opposite to its incident direction (backscattering direction), but not in other directions. The reason for this behavior is fully explained in a later chapter, but we may appreciate some of its consequences here. By preserving the coherence in the backscattering direction, the probability of backscattering is enhanced through constructive interference. This leads to a decrease in the diffusion constant from its classical value, because whatever the direction of the wave, the increased backscattering tends to drag it back as if the wave medium were more "viscous" than it should be classically. The downward renormalization of the diffusion constant and the coherent backscattering effect itself have several important features. First, the decrease in the diffusion constant is proportional to the scattering strength. If the scattering is weak, the coherent backscattering effect is correspondingly weak so that it can be ignored in general. But if the backscattering is strong enough, the decrease in the diffusion constant can make it vanish, thus leading to wave *localization*.

Second, the coherent backscattering effect is fully operative only when the system is time-reversal invariant, meaning that macroscopically there should be no preferred direction of time established, for example, by a uniform average velocity of the scatterers or by the presence of a magnetic field. In the presence of effects that break time-reversal invariance, the correction to the wave diffusion constant is diminished.

Third, the coherent backscattering is fully effective only when all scatterings are elastic. In the presence of inelastic scattering, which is inevitable in real materials having various dissipation mechanisms, the coherent backscattering effect is again diminished. However, the manner in which this occurs is directly coupled to the next feature; that is, the magnitude of the coherent backscattering effect, manifested in the amount of (negative) correction to the diffusion constant, is a monotonically increasing function of the physical sample size. Where the sample size is infinite, the de facto "sample size" is set by the inelastic scattering rate, because over a distance where there are several inelastic scatterings, the coherent backscattering can no longer be operative, and that distance essentially becomes the limiting sample size. Since electron inelastic scattering is generally temperature dependent, the combination of the last two features offers a means by which to observe the coherent backscattering effect (indirectly) in electronic systems through the temperature dependence of conductivity in disordered materials.

From the point of view of classical physics, the sample size dependence of the coherent backscattering effect is revolutionary because it makes the renormalized diffusion constant no longer intensive, as would generally be expected, since diffusion constant belongs to the same class of intensive quantities as density, electrical conductivity, and temperature. Figure 1.1 shows the renormalized diffusion constant schematically as a function of

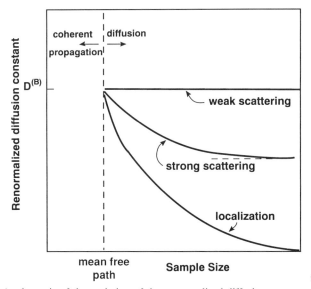

Figure 1.1 A schematic of the variation of the renormalized diffusion constant with sample size, i.e., the scale of observation. The sample size must be larger than the mean free path before diffusion can be observed. Weak scattering, strong scattering, and localization are shown. $D^{(B)}$ denotes the classical Boltzmann value of the diffusion constant.

sample size for three cases. In the weak scattering limit, the diffusion constant is independent of the sample size, as expected classically. When the scattering is strong, the diffusion constant is renormalized downward as a function of the sample size, with an asymptotic value that can be significantly less than its classical value (which is nonetheless still observable at small sample sizes). When the asymptotic value of the renormalized diffusion constant vanishes, then by definition a localized state is created. Therefore, a pulse injected into a strongly scattering medium would evolve initially as in a uniform medium, then quickly make a transition into diffusive transport, accompanied by a gradual slowdown of diffusion over time. Localization occurs when the overall diffusion is stopped.

Two comments can be made concerning the sample size dependence of the coherent backscattering effect. First, on the most elementary level, the backscattering aspect of the effect is responsible for its sample size dependence, because larger samples are more opaque than smaller samples. Thus one can expect more backscattering and less transmission for the larger samples. The coherence part of the effect then enhances the backscattering intensity from what is expected from ordinary diffusion. Through refinement, this simple viewpoint can be made quantitative, as will be seen in a later chapter. Second, the size dependence is necessary if one looks ahead to localization. When a wave is trapped by randomness, there appears a new length scale—the localization length—which naturally introduces sample size dependence into the physical transport property of the system. Let us expand on this point below by using the quantum waves associated with the electronic system as an example.

If one measures the electrical conductivity of a sample in which the electrons are localized—a disordered insulator—the usual measurement at finite temperatures is expected to yield only the conductivity of the electrons thermally activated from their localized states to some higher-energy mobile states able to carry electrical current. Now imagine a thought experiment where the temperature is lowered to absolute zero so that all electrons are in their localized states. The electrical conductivity of a bulk insulating sample at zero degree is usually regarded as zero—or too small to be measured. Nevertheless, if the sample dimension is small enough so that it becomes comparable to the localization length—a possibility which is increasingly becoming a technological reality nowadays—then even an insulator can conduct some electricity. Conduction can occur because localization prevents the wave, in this case the quantum wave associated with an electron, only from moving outside the spatial domain defined by the localization length, but the electron can still be mobile inside the domain of localization. If the edge of the localized domain is not abrupt but has an exponential tail, then the ability of a localized electron to conduct electricity would decrease exponentially as a function of the sample size when its linear dimension increases beyond the localization length. The result is the sample size dependence of the transport characteristics expected from localized states. If one did not know about the coherent backscattering effect but wanted to invent a mechanism for localization, such a mechanism would need to incorporate some kind of sample size dependence as necessitated by its desired result. From this perspective, the sample size dependence is an essential and necessary attribute for a localization mechanism.

Since localization is a major theme of this book, a brief digression on the development of the localization concept would be helpful. In the early days of solid-state physics, the recognition that electronic states in a periodic lattice form energy bands was a breakthrough that clarified a basic question about why some materials are electrically conducting and some are insulating. In the simplest version of the band picture, electrical conductors have a half-filled energy band whereas an insulator has filled bands. Mott took the conductor picture a step further and proposed that if the lattice constant of a half-filled band conductor can be continuously increased, then at some point the conductor becomes an insulator. The rationale behind this hypothesis is that when the atoms are separated far enough, they will behave as individual neutral atoms instead of as a metal in which the conduction electrons can pass freely from one atomic site to the next. The basic physics of the so-called Mott transition is the Coulomb interaction between the electrons (Mott, 1949; 1974). It does not involve disorder. In contrast, Anderson proposed in 1958 that electronic diffusion can vanish in a sufficiently random potential, in the absence of any electron–electron interaction (Anderson, 1958). This proposal, together with Mott's previous works, formed the theoretical basis for the study of the metal–insulator transition in doped semiconductors. However, it was not until the late 1970s and the early 1980s that the Anderson localization was linked to the coherent backscattering effect and explained on that basis. At about the same time, the study of Anderson localization was extended to classical waves, which offer an advantage over the disordered electronic systems where the Mott mechanism and the Anderson mechanism are inseparable: In classical waves, the Anderson localization may be studied alone, without the additional complication of wave–wave interaction. In this volume, the term "localization" denotes the phenomenon only in the sense of the Anderson mechanism.

Although the coherent backscattering effect was important to the understanding of wave multiple scattering phenomena and localization, an overview of the localization phenomenon was actually first achieved through the different perspective of a phenomenological theory, the *scaling theory of localization* (Abrahams *et al.*, 1979). The scaling theory is a scheme for interpolating between an extended wave state and a localized one. The term "scaling" is popular in physics nowadays. In the present context, it has the following connotations. First, scaling implies that the conclusions of the theory are independent of the many details of the physical model. For example, it is often the case that electron multiple scattering and localization are studied in the context of a lattice model of the solid atomic lattice. A scaling theory of localization would imply that the conclusions of the theory are independent of the type of lattice, be it simple cubic, body-centered cubic, or whatever. The conclusions are also independent of the type of random scattering, the statistics of the randomness, etc. Therefore, scaling implies broad applicability: Because the theory depends on few essential physical quantities, one hopes to obtain a better overall picture of the phenomenon without being encumbered by details. Herein lies the attraction of a scaling theory.

Second, scaling here means literally changing the scale, or physical size, of the sample under consideration. The scaling theory of localization considers how a quantity, defined as the dimensionless conductance γ, varies under a sample size change. For an electronic system, γ is simply the ordinary conductance Γ divided by the quantum unit of conductance, e^2/h, where e denotes the electronic charge and h is Planck's constant. For classical waves, γ may be expressed alternatively as the ratio of two energy scales, which are defined and motivated in the chapter on the scaling theory. How the dimensionless conductance γ varies with sample size is very simple for electrical conductors (samples in which the wave states are extended). How γ varies with sample size for disordered insulators, in which the wave states are all localized, can also be inferred from the earlier discussion on the sample size dependence of the coherent backscattering effect. To these known facts the scaling theory adds the following assumption: The rule governing the variation of γ with sample size can depend on only one parameter, which is the value of γ itself. With this seemingly innocent proposition, the scaling theory reaches some startling conclusions, the most striking of which is the dependence of the localization phenomenon on the *spatial dimensionality* of a sample. In particular, the scaling theory tells us that regardless of how weak the randomness, all waves are localized in one- or two-dimensional samples of infinite extent. For three-dimensional samples, the scaling theory predicts the possibility of coexistence for extended and localized wave states, where the extended states can exist in one or several frequency regimes and the localized states can exist in the other regimes. The frequency that separates a localized regime from a neighboring extended regime is called the "mobility edge." In the jargon of the physics community, spatial dimension two is called the "marginal dimension" for localization. In order to understand these predictions, let us first clarify the meaning of spatial dimensionality as applied to physical samples, as well as the conditions under which the predictions apply.

All physical objects must have finite cross sections or thicknesses. Therefore, there cannot be true two-dimensional (2D) or one-dimensional (1D) samples of vanishing thickness or cross section. Spatial dimensionality means that if wave propagation and scattering are allowed only in two (backward and forward) directions defined by a line, the sample is one dimensional (1D); if propagation and scattering are allowed only in directions defined by a surface, the sample is 2D; and if they are allowed in all three-dimensional space, the sample is 3D. The restriction on the direction(s) of wave propagation and scattering can be achieved by making the thickness, or cross-sectional dimension, of a sample smaller than, or comparable to, the wavelength. For example, if a wave is confined inside a sample whose thickness is smaller than the wavelength (but larger than half the wavelength), then the excitation in the thickness direction must be a standing wave. In addition, all other standing wave states are higher in frequency, with frequency increments large enough to make them inaccessible. Therefore, all scattering and propagation are confined to the planar directions, and the sample may be described as two-dimensional (2D). Another possibility of observing 2D waves is found in interfacial excitations, where the wave amplitude decays exponentially away from the interface and the wave can propagate only along the interface, as the name implies. Similar reasoning applies to the description of a one-dimensional sample.

On the basis of the scaling theory predictions, should all thin and wirelike objects be good electrical and thermal insulators? The answer depends on the conditions under which the measurements are made. The conclusions of the scaling theory are meant to apply only to samples at absolute zero temperature, so that finite-temperature effects, such as the activation of charge carriers and the inelastic scattering (that essentially limits the sample size as described before), are all absent. Therefore, the predicted 1D and 2D localization effects should not be ordinary, everyday experience. However, they should apply under laboratory conditions where the temperature of a sample is lowered to close to absolute zero and the sample dimensions are controlled to within the tolerance of being qualified as 1D or 2D.

A particularly intriguing prediction of the scaling theory is the inevitability of wave localization in 1D and 2D samples even at the weak scattering limit. To better understand this limiting case, let us first construct and analyze an "apparent" dilemma. Consider a one-dimensional sample of finite extent L. The heterogeneities in the sample are characterized by a scale R which is assumed to be smaller than the wavelength λ. To be more specific, the wave is classical and is modulated by a gentle envelope whose width is many wavelengths so that the additional frequency components are negligible. The question is: What happens when $L \to \infty$ and $\lambda \to \infty$? If $\lambda \to \infty$ first, the previous discussion on effective medium shows that the 1D random medium should always appear homogeneous to the probing wave, with no net scattering in the static limit of $\lambda/R \to \infty$. This would remain true for every finite L as $L \to \infty$. Therefore, the end result of $\lambda, L \to \infty$ should be an infinite effective medium where there is no localization. Now consider the reverse order of taking the limits. By letting $L \to \infty$ first, one can immediately invoke the prediction of the scaling theory—that all waves are localized in 1D samples of infinite extent—and this conclusion remains true for every λ as $\lambda \to \infty$. If a unique physical state is assumed for $\lambda, L \to \infty$, then the two contradicting conclusions yield a dilemma. The above description applies to a 1D sample, but similar reasoning leads to the same dilemma in 2D samples.

What can be the resolution of this dilemma? The answer is that both conclusions are correct, because even as $L, \lambda \to \infty$, the ratio L/λ still needs to be adjusted. There can indeed be different physical states depending on whether that ratio approaches 0 or ∞, which are the two possibilities probed by taking the limits in two different orders. Therefore, only the assumption about a unique $\lambda, L \to \infty$ physical state is incorrect. But what does this imply physically for wave localization? If $L \to \infty$ first, then as $\lambda \to \infty$, locally (inside the wave packet) the effective medium should be an increasingly good approximation, because scattering by classical waves diminishes as $\lambda/R \to \infty$. As the wave packet travels through the medium, the scattering is small at every instant of time. Therefore, if localization were to occur as predicted, the small scatterings at successive times *must accumulate* so that the net effect is large enough to localize. In other words, even if $R/\lambda = \varepsilon$ is small, over large travel distances, measured in terms of $1/\varepsilon$ (or even larger in 2D), there is still an order one effect, i.e., localization. If the travel distance is limited by the sample size L, then the localization limit can never be reached and one always obtains the effective medium limit instead. Therefore, the following physical picture emerges from an analysis of this apparent dilemma: In the weak scattering limit, a localized wave can exhibit propagating behavior *locally*; through increasing sample size and the effect of accumulated scatterings, the wave transport character is altered progressively from propagating to diffusion to localized. Of course, one recognizes the coherent backscattering effect in this scattering accumulation process, especially its sample size dependence. But why should this accumulation process be especially effective only in 1D and 2D? What is so magical about the spatial dimensionality two? Some insight into this question may be obtained through the special character of diffusion and its interaction with the coherent backscattering effect.

The usual way to describe diffusion is through the net distance r traveled by a random walker at the end of a time period t. The diffusion relation is described by $r^2 \propto t$, independent of the spatial dimensionality where the random walker executes its "walks." That is, in time t the random walker covers an "area" proportional to r^2. If the random walker is limited to moving on a line or a flat surface, the trace of its path would appear dense, but if the random walker is a flying particle that can freely traverse the three-dimensional space, then the trace of its path would appear to be flimsy, because an area does not go a long way toward covering a volume. While this description of diffusion may be somewhat simplistic, it can be made rigorous by adding qualifying conditions. Let us consider the path traversed by a random walker from $t = 0$ onward. As $t \to \infty$, in $d = 1, 2$ the random walker will visit the infinitesimal neighborhood of any given point with probability one, i.e., with certainty. In $d = 3$, however, the probability is infinitesimal that it would visit any given infinitesimal neighborhood. One way to appreciate that difference is by calculating the probability, at a time $t_m > 0$ onward, that a random walker will return to the neighborhood of the origin, the point where the walker started its motion at $t = 0$. According to the solution of the diffusion equation, the probability density for the walker at distance r from the origin at time t is given by $P(r, t) = (4\pi Dt)^{-d/2} \exp(-r^2/4\pi Dt)$, where d denotes the spatial dimensionality of the random walk and D the diffusion constant. The desired probability is therefore given by

$$\lim_{T \to \infty} \int_{t_m}^{T} P(0, t) dt = \lim_{T \to \infty} \int_{t_m}^{T} \frac{dt}{(4\pi Dt)^{d/2}}$$

For $d = 1, 2$ the integral diverges as $T \to \infty$, independent of t_m, implying that the random walker will certainly return to the neighborhood of the origin. This is intuitively plausible from the fact that the path of a random walker covers an "area." For $d = 3$, on the other hand, the integral is proportional to $1/\sqrt{t_m}$, so that as t_m increases, the probability decreases toward zero. Thus, in terms of this probability, there is a qualitative difference between diffusion in one and two dimensions and diffusion in three dimensions. If the probability of returning to the origin is heuristically equated with backscattering, then this special property of diffusion translates into more effective coherent backscattering in one and two dimensions, leading to localization. From this viewpoint, the marginal dimension of localization is a direct result of the exponent two in the diffusion relation $r^2 \propto t$.

Although the scaling theory gives a global view of what can happen in different spatial dimensions, the details of this explanation need to be filled in with results of calculations based on the mechanisms of diffusion plus coherent backscattering. This proves to be possible, because the correction to the diffusion constant introduced by the coherent backscattering is consistent with the scaling hypothesis. As a result, formulas for the localization length and its frequency dependence, plus explicit forms for the scaling functions, can all be derived in spatial dimensions one and two. In spatial dimension three, however, our knowledge of the subject is still incomplete. It is known that if the randomness is introduced into a periodic structure or if the system is discrete, such as an electron in a disordered lattice, then the localized states and mobility edge(s) can exist. This is due to the fact that in the absence of randomness, periodic structures yield frequency bands, and near the band edges the Bragg scattering gives rise to standing waves, i.e., waves with zero group velocity. Such wave states are, in a sense, waiting to be localized as randomness is introduced. However, for a continuous random system without any long-range periodic correlation, only the quantum wave is known to localize at low energies. The case for classical wave localization is not yet settled. One possibility for classical wave localization arises from resonant scattering, e.g., Mie resonance for electromagnetic wave scattering from small particles, which can give a much higher scattering cross section than usual and, consequently, a significant slowdown of wave diffusion. Whether the renormalized diffusion constant can have a vanishing value as sample size increases to infinity, and thereby generate a localized state, depends on some additional factors. However, the slowdown phenomenon is interesting in itself.

From elementary kinetic theory, the diffusion constant may be expressed as $D = (1/3)vl$, where v is the velocity of the diffusing "particle" and l is its elastic collision mean free path. When D decreases, either l or v, or both, may be the cause. For wave diffusion, the mean free path may be measured as the distance a plane wave can penetrate into a scattering medium before it loses its phase front. The velocity v, on the other hand, is a problem because the usual wave speed, whether the phase velocity or the group velocity, is defined for wave states that have well-defined wave vectors. Definitions of phase velocity requires a constant-phase surface perpendicular to the wave vector, and the definition of the group velocity —the wave vector derivative of frequency—requires both the wave vector and the dispersion relation between frequency and wave vector to be well defined. However, in a strongly scattering medium a wave vector cannot possibly describe a multiply scattered wave state because such a state has neither a unique direction of propagation nor a unique wavelength. What, then, is the v in the wave diffusion constant?

The v relevant to the wave diffusion is the average speed at which the wave *energy* is transported locally. In the presence of strong scattering, especially resonant scattering, this transport velocity can be significantly smaller than the free-space phase velocity for classical waves (van Albada *et al.*, 1991), but it is always identical to the group velocity for the quantum wave associated with free electrons. This striking difference between the classical and the quantum waves has its origin in the different dispersion relations for the two kinds of waves and the manner in which the randomness enters the two types of wave equations. Therefore, in approaching localization through increasingly strong scattering, the mean free path must decrease for the quantum case but both the transport speed and the mean free path can decrease for the classical waves.

Wave multiple scattering and localization can lead to various physical manifestations. For electronic systems, the direct proportionality relation between the conductivity and the diffusion constant is known as the "Einstein relation." Therefore, the observation of electronic conductivity is equivalent to observing electron diffusion. For classical phonons, heat conduction has already been mentioned as a manifestation of phonon diffusion. In dirty conductors, indirect observation of the coherent backscattering effect is contained in the measurements of anomalous temperature dependence of resistivity and the anomalous magnetoresistance, both in conducting films. However, these indirect measurements suffer from the ambiguity that a competing effect—that of electron–electron interaction—can yield similar results, so there is no way to attribute the measured behavior entirely to the coherent backscattering effect. A more direct observation of the consequences of wave multiple scattering is via measurements on *mesoscopic samples*. Here "mesoscopic" denotes a sample size regime intermediate between the molecular and the bulk. More precisely, it means a sample size whose linear dimension is smaller than a "dephasing length," $\sqrt{D\tau_{in}}$, where τ_{in} is the inelastic collision time for the wave. In a mesoscopic sample, all collisions are elastic, so the effects of coherent backscattering and interference can be manifest and not averaged out. For classical waves, the inelastic scattering rate is generally low and temperature insensitive. Wave–wave interaction is also negligible, in contrast to the interaction between electrons. Thus a mesoscopic sample in this case is a bulk sample, and the experiment can be done at room temperature. In fact, the cleanest quantitative demonstration of the coherent backscattering effect came from light scattering from bulk disordered systems. For electronic systems, on the other hand, the

Introduction

inelastic scattering length is generally small and inversely temperature dependent. Therefore, the observation of mesoscopic phenomena in electronic systems generally requires small samples whose linear dimensions are on the order of micrometers or smaller. In addition, the experiments must be performed at low temperatures to ensure that the inelastic scattering rate is sufficiently low. However, with today's microfabrication and cryogenic technologies these requirements are not difficult barriers, and the relentless push for miniaturization in the electronic industry means that the mesoscopic phenomena could very well serve as the basis for tomorrow's quantum devices.

Mesoscopic phenomena are manifest in the measurements of conventional quantities in mesoscopic samples. Electrical conductivity is an example. The definition of mesoscopic conductance is problematic because electrical conductance is always associated with dissipation, whereas in mesoscopic samples the absence of inelastic scattering means there is no dissipation. Thus, there is a need to define what one means by conductance in a mesoscopic sample. By analogy with the resistance of a tunnel junction, Landauer has proposed that the conductance of a mesoscopic sample should be proportional to the wave transmission probability (Landauer, 1957). In that case, dissipation occurs not inside the sample but in the leads connected to the sample, and through consideration of equilibrium with the leads an expression for the mesoscopic conductance can be derived.

Conductance of mesoscopic samples has many unconventional characteristics. Not only does it vary with sample size (due to the coherent backscattering effect), it can also fluctuate wildly from sample to sample, even if the samples are manufactured identically in the same batch. The conductivity can also show large fluctuations upon the application of a varying magnetic field. Moreover, as the sample size increases, these fluctuations do not decrease; i.e., the fluctuations are not averaged out as one would expect from additive noise. All of these effects are basically due to the fact that even though the waves are multiply scattered, they can still exhibit phase interference and retain some long-range memory.

The discussion thus far is a brief and qualitative guide to the contents of this volume. At this point it is perhaps also important to point out the relevant subjects that are left out. First is wave dissipation. In this volume, dissipation is treated only as a constraint, or limitation, on the effects due to elastic scattering. No attempt is made to examine either the mechanism of dissipation or the combined effect of both elastic scattering and dissipation. The second neglected topic is nonlinearity, which can couple waves of different frequencies. Since the magnitude of nonlinearity depends on the amplitude of a wave, its neglect means that we will be concerned only with small-amplitude waves, and waves of different frequencies will be treated as independent. The third neglected topic is wave–wave interaction. A well-known example in this regard is the electron–electron interaction through their electrical charges. Here, overlooking wave–wave interaction implies that the wave scattering is treated as a "single-body" problem, in contrast to the "many-body" problem where the presence of more than one electron changes the nature of the system. For electrons, interaction effects are important at low densities. At high densities, the interaction is weakened by screening, resulting in an electron dressed in a cloud of screening charges that can be treated as an independent "quasi-particle." The neglect of electron–electron interaction here implies that whenever the term electron is used in this volume, it denotes a quasi-particle in the high-density limit. For classical waves, on the other hand, the interaction between electromagnetic waves is weak, and the same is true for elastic waves of small amplitudes. Therefore, their neglect is well justified.

The three topics not covered—dissipation, nonlinearity, and interaction—are interesting and rich subjects in themselves. They have been omitted solely to attain the simplicity and coherence made possible by focusing on the subjectively chosen main line of exposition. The author, rather than trying to be inclusive and complete, intends this volume to clarify a few essential points, leaving the readers to further explore this challenging field.

References

Abrahams, E., Anderson, P. W., Licciardello, D. C., and Ramakrishnan, T. V. (1979). *Phys. Rev. Lett.* **42**, 673.
Anderson, P. W. (1958). *Phys. Rev.* **109**, 1492.
Landauer, R. (1957). *IBM J. Res. Dev.* **1**, 223.
Mott, N. F. (1949). *Proc. Phys. Soc. London, Ser. A* **62**, 416.
Mott, N. F. (1974). "Metal–Insulator Transitions," Taylor and Francis, London.
Sheng, P., and Zhou, M. (1991). *Science* **253**, 539.
Tsang, L., and Ishimaru, A. (1984). *J. Opt. Soc. Am.* **A1**, 836.
van Albada, M. P., and Lagendijk, A. (1985). *Phys. Rev. Lett.* **55**, 2692.
van Albada, M. P., van Tiggelen, B. A., Lagendijk, A., and Tip, A. (1991). *Phys. Rev. Lett.* **66**, 3132.
Wolf, P. E., and Maret, G. (1985). *Phys. Rev. Lett.* **55**, 2696.

2

Quantum and Classical Waves

2.1 Preliminaries

In a uniform medium, a wave at a fixed frequency ω has the property that a snapshot of its amplitude would display spatial periodicity. The period, or wavelength λ, is related to ω by the so-called dispersion relation, which is a basic property of the wave and the medium in which it propagates. For a quantum particle/wave, the dispersion relation reflects the definition of energy. That is, if by $E = \hbar\omega$ one denotes the total energy of the quantum particle/wave, where $\hbar$ is Planck's constant, and by $p = \hbar k$ the momentum of the particle, where $k = 2\pi/\lambda$ is the magnitude of the wave vector $\mathbf{k}$ that points at the direction of wave/particle propagation, then for a particle of mass m, E is the sum of kinetic and potential energies:

$$E = \hbar\omega = \frac{p^2}{2m} + V = \frac{\hbar^2 k^2}{2m} + V. \tag{2.1}$$

Here V denotes the potential energy, taken to be a constant for the uniform medium. Equation (2.1) represents the relation between ω and k for a quantum particle/wave. By representing the temporal oscillation of the wave by $\exp(-i\omega t)$ and the spatial periodicity by $\exp(i\mathbf{k} \cdot \mathbf{r})$, E can be obtained from the wave amplitude function $\phi = \exp[-i(\omega t - \mathbf{k} \cdot \mathbf{r})]$ by the operation of $i\hbar\,\partial/\partial t$ on ϕ, i.e.,

$$i\hbar\frac{\partial\phi}{\partial t} = E\phi. \tag{2.2}$$

Also, $p^2/2m$ can be obtained by the operation of $-(\hbar^2/2m)\nabla^2$ on ϕ, where $\nabla^2 = \partial^2/\partial x^2 + \partial^2/\partial y^2 + \partial^2/\partial z^2$ is the Laplacian operator, i.e.,

$$-\frac{\hbar^2}{2m}\nabla^2\phi = \frac{\hbar^2 k^2}{2m}\phi. \qquad (2.3)$$

Another way of stating Eq. (2.3) is that $\exp(i\mathbf{k}\cdot\mathbf{r})$ is the eigenfunction of the Laplacian operator with the eigenvalue $-k^2$. From Eqs. (2.1), (2.2), and (2.3), one thus infers the Schrödinger's equation,

$$i\hbar\frac{\partial\phi}{\partial t} + \frac{\hbar^2}{2m}\nabla^2\phi - V\phi = 0, \qquad (2.4)$$

from the dispersion relation.

In contrast to the quantum wave, the classical electromagnetic and elastic waves have the dispersion relation

$$\omega = v|\mathbf{k}|. \qquad (2.5)$$

One way to get rid of the absolute value sign is to rewrite Eq. (2.5) as

$$\omega^2 = v^2 k^2. \qquad (2.6)$$

By using the same time and space derivative operations on the wave amplitude function $\exp[-i(\omega t - \mathbf{k}\cdot\mathbf{r})]$, Eq. (2.6) implies the following scalar wave equation:

$$\frac{\partial^2\phi}{\partial t^2} - v^2\nabla^2\phi = 0. \qquad (2.7)$$

The scalar wave equation differs from the electromagnetic and elastic wave equations in that the true classical waves have *vector character*, i.e., different polarizations. For the electromagnetic wave, there are two polarizations transverse to $\mathbf{k}$, and for the elastic wave there is in addition a longitudinal polarization parallel to $\mathbf{k}$. In a uniform, isotropic medium, classical waves of different polarizations are each described by a scalar wave equation, with a velocity v which can be different for the elastic longitudinal wave and the elastic transverse (shear) waves because they rely on different material properties for their propagation. Since different polarizations are decoupled from each other in a uniform, isotropic medium, the scalar wave equation is thus accurate for each polarization component of the classical waves. This is not true in a disordered medium. Because of the coupling between the different polarization components at a scattering interface (between materials of different wave properties),

2.1 Preliminaries

classical waves with different polarizations can be interconverted into each other. As a result, the scalar wave equation is no longer accurate in a disordered medium. However, in spite of such complexities arising from the polarization consideration, many of the basic classical wave characteristics are still qualitatively captured by the scalar wave equation. In particular, for the aspects of wave transport and localization considered in this volume, the scalar wave equation is adequate in illustrating the basic physics involved. For simplicity, the scalar wave would serve as our model for the description of classical waves.

An important difference between the Schrödinger (quantum) wave equation and the scalar (classical) wave equation is the order of the time derivative. Whereas it is first order in the quantum case, it is second order in the classical case. This difference arises from the different dispersion relations for the two types of waves and implies that, whereas for the classical wave the speed of propagation is a constant of the wave medium, for a free quantum particle ($V = 0$) the wave speed, defined as dE/dp for the group velocity, is dependent on its momentum (just like a classical particle). Quantum waves of different frequencies thus travel with different speeds. This important difference between quantum and classical waves is reflected in many aspects of their behavior in disordered media.

For waves of a given energy or frequency, the quantum and classical wave equations can be put in the same form:

$$\nabla^2 \phi + \kappa^2 \phi = 0, \tag{2.8}$$

where

$$\kappa^2 = \frac{2m(E-V)}{\hbar^2} \tag{2.9}$$

for the quantum case, and

$$\kappa^2 = \frac{\omega^2}{v^2} \tag{2.10}$$

for the classical case. Here a striking difference between the quantum wave and the classical wave is already apparent, because whereas $E - V$ can be negative in Eq. (2.9), ω^2/v^2 is always positive. Therefore, the classical wave has a quantum correspondence only when $E > V$.

In general, κ^2 is a function of $\mathbf{r}$, and it can be written as

$$\kappa^2 = \kappa_0^2 - \sigma(\mathbf{r}), \tag{2.11}$$

where κ_0^2 represents the uniform background and $\sigma(\mathbf{r})$ represents the deviation from it. In the quantum case, if V is a constant, it can be absorbed into E as a redefinition of that quantity. Therefore, it is always possible to write

$$\kappa_0^2 = \frac{2m\omega}{\hbar} \qquad (2.12)$$

and

$$\sigma(\mathbf{r}) = \frac{2mV(\mathbf{r})}{\hbar^2} \qquad (2.13)$$

for the quantum wave. In the classical case, if v_0 denotes the wave speed in the uniform medium, such as the speed of light in vacuum, and $v = v_0/\sqrt{\epsilon}$ is the speed inside a scattering object, then ϵ is generally known as the dielectric constant, where $\epsilon = n^2$, n being the index of refraction. By expressing $v^2 = v_0^2/\epsilon(\mathbf{r})$, where $\epsilon = 1$ in the background, then

$$\kappa_0^2 = \frac{\omega^2}{v_0^2}, \qquad (2.14)$$

and

$$\sigma(\mathbf{r}) = [1 - \epsilon(\mathbf{r})]\frac{\omega^2}{v_0^2}. \qquad (2.15)$$

The quantity $\sigma(\mathbf{r})$ is seen to depend on ω for the classical wave case but is independent of ω for the Schrödinger wave. More precisely, the deviation from a uniform medium occurs *additively* with frequency for the quantum case, but *multiplicatively* with frequency for the classical case. One immediate result is that as $\omega \to 0$, classical waves always scatter weakly because $\sigma(\mathbf{r})$ vanishes in that limit. The same is not true for the quantum case.

In the next section the behavior of quantum and classical waves in a uniform medium is examined under different spatial dimensionalities. The meaning of spatial dimensionality for a physical sample has been discussed in the last chapter. Whereas real 2D and 1D samples must have physical thicknesses and finite cross sections, in the subsequent theoretical development 2D and 1D are treated as meaning true mathematical surfaces and lines. This is, of course, an approximation to reality, but one which is accurate for the essential physics of the lower dimensionalities.

2.2 Green's Functions for Waves in a Uniform Medium

Green's function is one of the most useful tools in physics because of the physical information it can provide about the system described by the differential equation. Simply put, Green's function in the present context is just the solution to the wave equation with a specific initial source and a boundary condition at infinity. One might ask why that is more useful than just any solution $\phi(t, \mathbf{r})$ to the homogeneous wave equation. The answer could be illustrated by the following analogy.

Suppose there are N mechanical players in a game whose rules, while well defined, are not known a priori to an outside observer. The game involves the exchange of some currency between the players (such as in Monopoly, except here the players have no free will), which we just call "money." The purpose of the observer is to figure out the game rules and predict what is going to happen. The observer can record the money each player has at every instant and how it varies with time. The recorded quantity, $M(t, i)$, where i is the index for the players, may be compared to the wave function $\phi(t, \mathbf{r})$, since both give successive snapshots of the state of affairs. However, an intelligent observer might want to ask the question: What is the interrelationship between the amounts of money of any two players, if any? This important information cannot be obtained by simply looking at each player individually. Instead, one has to look at the *correlation* between pairs of players as a function of time separation, i.e., the correlation between $M(t, A)$ and $M(t + \Delta t, B)$. If the observer has the power to do a controlled experiment on the players, this correlation may be obtained by (1) setting $M = 0$ for all players, (2) giving an amount M_0 to A at $t = 0$, and (3) watching how much of M_0 is transferred to B at a later time t. This is precisely the sort of information that a Green's function would provide. By making A and B the same player, the Green's function can also monitor the information contained in the wave function. Therefore it is more general. In the case of the wave equation, the precise knowledge of the Green's function is sufficient to figure out all the rules of the system; i.e., it is equivalent to an exact solution of the problem. In the absence of exact solutions, which are often difficult beyond simple or periodic systems, Green's function also provides an approach to the solution of the wave equation that is amenable to systematic perturbation analysis. Moreover, the meaning of these analyses can be interpreted physically. Therein lies the attractiveness of the Green's function approach.

The specific initial condition for the Green's function is that at $t = 0$, the system is excited by a pulse localized at $\mathbf{r} = \mathbf{r}'$; i.e., the right-hand sides of Eqs. (2.4) and (2.7) are replaced by $\delta(t)\delta(\mathbf{r} - \mathbf{r}')$, where $\mathbf{r}'$ denotes the

---

in Cartesian coordinates and

$$\int_0^{2\pi} d\varphi \int_0^{\pi} d\theta \int_0^{\infty} \frac{\sin\theta \, k^2 \, dk}{(2\pi)^3}$$

in spherical coordinates. Substitution of Eqs. (2.18) and (2.19a) into Eq. (2.17) gives

$$\frac{L^d}{N} \int \frac{d\mathbf{k}}{(2\pi)^d} \exp[i\mathbf{k} \cdot (\mathbf{r} - \mathbf{r}')][\kappa_0^2(\omega) - k^2] G_0(\omega, \mathbf{k})$$

$$= \int \frac{d\mathbf{k}}{(2\pi)^d} \exp[i\mathbf{k} \cdot (\mathbf{r} - \mathbf{r}')]. \tag{2.20}$$

Equating the amplitudes of each $\mathbf{k}$ component on both sides yields

$$\left[\kappa_0^2(\omega) - k^2\right] G_0(\omega, \mathbf{k}) = \frac{N}{L^d}, \tag{2.21}$$

or

$$G_0(\omega, \mathbf{k}) = \frac{1}{\kappa_0^2(\omega) - k^2} \frac{N}{L^d}. \tag{2.22}$$

The Green's function $G_0(\omega, \mathbf{k})$ gives the response of the homogeneous medium to a source excitation that has frequency ω and spatial frequency $\mathbf{k}$. It is noted that if the source $\mathbf{k}$ and ω coincide with the wave dispersion relation of the medium, i.e., $k^2 = \kappa_0^2(\omega)$, then G_0 diverges. In analogy with the response of a forced oscillator which has its own resonance frequencies, this divergent response is expected, since the dispersion relation essentially gives the condition of natural resonance in a medium. On the other hand, when the source frequency and $\mathbf{k}$ do not coincide with a natural mode of the system, Eq. (2.22) indicates that it can still excite many modes just by its proximity to the resonance condition, as in the case of a forced oscillator.

The divergence in $G_0(\omega, \mathbf{k})$ would cause a problem in the transformation back into the spatial domain. To fix this problem, it is generally the practice to add a small imaginary constant to the denominator, then to take the limit as the constant approaches zero. This simple procedure is illustrated in the solution to Problem 2.1 at the end of this chapter. The source position. The reason for such a source is that it can excite the natural resonances (or eigenfunctions) of *all* temporal and spatial frequencies in the system (since the source contains all temporal and spatial frequencies), so that the subsequent development may contain information about them. A useful approach for analyzing the Green's function is through its frequency components. The left-hand side of the wave equation for the frequency component ω is given by Eq. (2.8), and the same frequency component on the right-hand side is simply $\delta(\mathbf{r} - \mathbf{r}')$ because $\delta(t)$ gives 1 as the amplitude for every frequency component. Equating both sides yields

$$(\nabla^2 + \kappa^2) G(\omega, \mathbf{r}, \mathbf{r}') = \delta(\mathbf{r} - \mathbf{r}'), \quad (2.16)$$

where $G(\omega, \mathbf{r}, \mathbf{r}')$ denotes the Green's function in the frequency and spatial domain and $\mathbf{r}'$ is used to mark the source point. The source point position is in general an important parameter for the Green's function if the system is inhomogeneous, since the response to a point source can depend on its position. In addition to Eq. (2.16), a unique determination of $G(\omega, \mathbf{r}, \mathbf{r}')$ also requires the boundary condition that $G(\omega, \mathbf{r}, \mathbf{r}') \to 0$ as $|\mathbf{r} - \mathbf{r}'| \to \infty$, as well as the physical condition of causality; i.e., cause always precedes effect. In this chapter, only the Green's function for the uniform medium is considered. The effects induced by $\sigma(\mathbf{r})$ are treated in the subsequent chapters.

For the uniform medium case of $\kappa = \kappa_0$, the response of the system is independent of the source position, and the Green's function depends only on the relative separation $\mathbf{r} - \mathbf{r}'$ between the source and the detector. Therefore,

$$(\nabla^2 + \kappa_0^2) G_0(\omega, \mathbf{r} - \mathbf{r}') = \delta(\mathbf{r} - \mathbf{r}'). \quad (2.17)$$

A simple solution approach to Eq. (2.17) is to first solve for the spatial frequency component $G_0(\omega, \mathbf{k})$ by writing

$$G_0(\omega, \mathbf{r} - \mathbf{r}') = \frac{L^d}{N} \int \frac{d\mathbf{k}}{(2\pi)^d} \exp[i\mathbf{k} \cdot (\mathbf{r} - \mathbf{r}')] G_0(\omega, \mathbf{k}) \quad (2.18)$$

and

$$\delta(\mathbf{r} - \mathbf{r}') = \int \frac{d\mathbf{k}}{(2\pi)^d} \exp[i\mathbf{k} \cdot (\mathbf{r} - \mathbf{r}')], \quad (2.19a)$$

where d denotes the spatial dimensionality, L^d is the sample volume [$(2\pi)^d L^{-d}$ may be regarded as the smallest volume unit $(\delta k)^d$ in the wave vector space], and N is the number of states, or atoms, in the sample. $L^d/N = (\delta r)^d$ is the smallest volume unit in the system. The notation here is based on viewing the continuum as the limit of infinitesimal discretization, so that both the continuum case and the lattice case may be treated on the same footing. The presence of N in Eq. (2.18) is due to the fact that a plane wave has magnitude $\sqrt{N}$. The convention in this book is to associate a factor $1/N$ with every $\mathbf{k}$-integration (summation) to compensate for that fact. To be consistent, $\delta(\mathbf{k} - \mathbf{k}')$ as defined by

$$\int d\mathbf{k}\, \delta(\mathbf{k} - \mathbf{k}') = 1$$

has the following Fourier representation:

$$\delta(\mathbf{k} - \mathbf{k}') = \frac{1}{(2\pi)^d} \int d\mathbf{r} \exp[i(\mathbf{k} - \mathbf{k}') \cdot \mathbf{r}]. \quad (2.19b)$$

In Eq. (2.19a)

$$\int \frac{d\mathbf{k}}{(2\pi)^d}$$

denotes integration over all $\mathbf{k}$ vectors. For $d = 1$, that means

$$\int_{-\infty}^{\infty} \frac{dk}{2\pi}.$$

For $d = 2$, that means

$$\int_{-\infty}^{\infty} \int_{-\infty}^{\infty} \frac{dk_x\, dk_y}{(2\pi)^2}$$

in Cartesian coordinates and

$$\int_{0}^{2\pi} d\theta \int_{0}^{\infty} \frac{k\, dk}{(2\pi)^2}$$

in circular (cylindrical) coordinates. For $d = 3$, that means

$$\int_{-\infty}^{\infty} \int_{-\infty}^{\infty} \int_{-\infty}^{\infty} \frac{dk_x\, dk_y\, dk_z}{(2\pi)^3}$$

2.2 Green's Functions for Waves in a Uniform Medium

result is

$$\frac{L^d}{N} G_0^\pm(\omega, \mathbf{k}) = \lim_{\eta \to 0} \frac{1}{\kappa_0^2(\omega) - k^2 \pm i\eta}$$

$$= P \frac{1}{\kappa_0^2(\omega) - k^2} \mp i\pi \delta\left[\kappa_0^2(\omega) - k^2\right]. \quad (2.23)$$

Here the sign of the imaginary constant is important in determining the behavior of the Green's function in real space, as will be seen later. The symbol P means taking the principal value of $[\kappa_0^2(\omega) - k^2]^{-1}$ as that quantity diverges, which is defined in the solution to Problem 2.1. The imaginary part of the Green's function, on the other hand, is interesting because the delta function picks out exactly the modes of the system. If one wants to count the number of modes per unit frequency range and per unit volume L^d/N, i.e., to get the density of states $\rho_0(\omega)$, it can be achieved by integrating the delta function over all possible free-space eigenstates that are characterized by the $\mathbf{k}$ vectors, i.e.,

$$\rho_0(\omega) = \frac{d\kappa_0^2(\omega)}{d\omega} \int \frac{d\mathbf{k}}{(2\pi)^d} \delta\left[\kappa_0^2(\omega) - k^2\right]. \quad (2.24)$$

Here the factor

$$\frac{d\kappa_0^2(\omega)}{d\omega} = \begin{cases} \dfrac{2m}{\hbar} & \text{quantum} \\ \dfrac{2\omega}{v_0^2} & \text{classical} \end{cases} \quad (2.25)$$

accounts for the fact that the delta function counts the states in units of κ_0^2, so it is required for conversion to states per unit frequency. It is shown in Problem 2.2 that $\rho_0(\omega)$ is given by

$$\rho_0(\omega) = \begin{cases} \dfrac{m^{3/2}}{\sqrt{2}\,\hbar^{3/2}\pi^2} \sqrt{\omega} & \text{3D} \\ \dfrac{m}{2\pi\hbar} & \text{2D} \\ \dfrac{1}{\pi}\sqrt{\dfrac{m}{2\hbar}}\dfrac{1}{\sqrt{\omega}} & \text{1D} \end{cases} \quad (2.26)$$

for the quantum case, and

$$\rho_0(\omega) = \begin{cases} \dfrac{\omega^2}{2\pi^2 v_0^3} & \text{3D} \\ \dfrac{\omega}{2\pi v_0^2} & \text{2D} \\ \dfrac{1}{\pi v_0} & \text{1D} \end{cases} \quad (2.27)$$

for the classical wave case. Here ω is defined to be positive.

In Eq. (2.24), the integral may be reexpressed as

$$\int \frac{d\mathbf{k}}{(2\pi)^d} \delta[\kappa_0^2(\omega) - k^2] = -\frac{1}{\pi} \frac{L^d}{N} \int \frac{d\mathbf{k}}{(2\pi)^d} \operatorname{Im} G_0^+(\omega, \mathbf{k})$$

$$= -\frac{1}{\pi} \operatorname{Im} G_0^+(\omega, \mathbf{r} = \mathbf{r}'), \quad (2.28)$$

where Im means taking the imaginary part of G^+. The last equality in Eq. (2.28) follows because at $\mathbf{r} = \mathbf{r}'$, the phase factor $\exp[i\mathbf{k} \cdot (\mathbf{r} - \mathbf{r}')]$ in the Fourier transform of $G_0^+(\omega, \mathbf{k})$ becomes 1, so the real and imaginary parts of $G_0^+(\omega, \mathbf{r} = \mathbf{r}')$ are just the $\mathbf{k}$-integrals of the real and imaginary parts of $G_0^+(\omega, \mathbf{k})$. Substitution of Eq. (2.28) into Eq. (2.24) gives

$$\rho_0(\omega) = -\frac{d\kappa_0^2(\omega)}{d\omega} \frac{1}{\pi} \operatorname{Im} G_0^+(\omega, \mathbf{r} = \mathbf{r}'). \quad (2.29)$$

The reason for expressing the density of states in terms of the imaginary part of the Green's function (in the ω and $\mathbf{r}$ representation) is the general validity of the formula not only for waves in a uniform medium but also for waves in random media. In solution to Problem 2.3, it is shown that

$$\rho(\omega, \mathbf{r}) = -\frac{d\kappa_0^2(\omega)}{d\omega} \frac{1}{\pi} \operatorname{Im} G^+(\omega, \mathbf{r} = \mathbf{r}') \quad (2.30)$$

gives the "local density of states" at $\mathbf{r}$ in the general case. The fact that the density of states can always be obtained this way is due to the Green's function's general property that it diverges at the resonant modes of the system, and as a result its imaginary part offers a way of picking them out as seen from Eq. (2.23). Another general property of the Green's function is that $G^+(\omega, \mathbf{r}, \mathbf{r}')$ and $G^-(\omega, \mathbf{r}', \mathbf{r})$ are complex conjugates of each other

2.2 Green's Functions for Waves in a Uniform Medium

(note the interchange of $\mathbf{r}$ and $\mathbf{r}'$), and G_0^+, G_0^- are complex conjugates in both the real space and the momentum space. These and other facts about the Green's function are made clear in the solution to Problem 2.3.

The Green's function in the $(\omega, \mathbf{k})$ domain is noted to have the same expression in different spatial dimensionalities. However, $G_0^\pm(\omega, \mathbf{r} - \mathbf{r}')$ is different in different spatial dimensions. This can be seen by the Fourier transformation of $G_0^\pm(\omega, \mathbf{k})$. In 3D, we have

$$G_0^\pm(\omega, \mathbf{r} - \mathbf{r}') = \lim_{\eta \to 0} \frac{1}{4\pi^2} \frac{L^d}{N}$$

$$\times \int_0^\infty \frac{k^2 \, dk}{\kappa_0^2 - k^2 \pm i\eta} \int_0^\pi d(\cos \theta) \exp(ik|\mathbf{r} - \mathbf{r}'| \cos \theta) \frac{N}{L^d}$$

$$= \lim_{\eta \to 0} \frac{1}{4\pi^2 |\mathbf{r} - \mathbf{r}'|} \frac{1}{i} \int_{-\infty}^\infty \frac{k \exp(ik|\mathbf{r} - \mathbf{r}'|)}{\kappa_0^2 - k^2 \pm i\eta} \, dk$$

$$= -\frac{\exp[\pm i(\kappa_0 \pm i\eta)|\mathbf{r} - \mathbf{r}'|]}{4\pi |\mathbf{r} - \mathbf{r}'|} \qquad (2.31)$$

through contour integration in the upper complex k plane (so that the integrand vanishes in the limit of $k \to +i\infty$). $G_0^\pm(\omega, \mathbf{r} - \mathbf{r}')$ is seen to correspond to outgoing ($+$) and incoming ($-$) spherical waves. The sign of $i\eta$ determines whether the wave is outgoing or incoming because G_0 must vanish as $|\mathbf{r} - \mathbf{r}'| \to \infty$, so that $\kappa_0 + i\eta$ must be outgoing whereas $\kappa_0 - i\eta$ must be incoming. In 2D, similar contour integration and comparison with the integral representation of the Hankel function yield $G_0^\pm$ as outgoing and incoming cylindrical waves:

$$G_0^\pm(\omega, \mathbf{r} - \mathbf{r}') = \mp \frac{i}{4} H_0^{(1,2)}(\kappa_0 |\mathbf{r} - \mathbf{r}'|), \qquad (2.32)$$

where $H_0^{(1,2)}$ denotes the zeroth-order Hankel function of the first (outgoing wave) kind and the second (incoming wave) kind. In 1D, $G_0^\pm$ is given by

$$G_0^\pm(\omega, \mathbf{r} - \mathbf{r}') = \mp \frac{i}{2\kappa_0} \exp(\pm i\kappa_0 |\mathbf{r} - \mathbf{r}'|). \qquad (2.33)$$

It is easy to check that the imaginary part of $G_0^\pm(\omega, \mathbf{r} = \mathbf{r}')$, when it is substituted into Eq. (2.29), gives the correct density of states. A further Fourier transform with respect to frequency ω would yield Green's function in the $(t, \mathbf{r})$ domain. Since κ_0^2 is linear in ω for quantum waves and quadratic in ω for classical waves, the Green's functions for the two types of waves are expected to differ in the $(t, \mathbf{r})$ domain (Economou, 1983). However, they share the same causal characteristics: In the time domain $G_0^+(t, |\mathbf{r} - \mathbf{r}'|) = 0$ for $t < 0$ and $G_0^-(t, |\mathbf{r} - \mathbf{r}'|) = 0$ for $t > 0$. This fact is easy to see in the case of 3D classical waves, where the Fourier transform of the numerator of (2.31) gives $\delta(|\mathbf{r} - \mathbf{r}'|/v_0 \mp t)$. Thus, G_0^+ is usually denoted the retarded (with respect to $t = 0$) Green's function, whereas G_0^- is denoted the advanced Green's function.

2.3 Waves on a Discrete Lattice

In the case of electrons in solids, the interaction of an electron with the solid lattice is known to play a dominant role in the electron's resulting behavior. It is therefore desirable that the lattice effect be taken into account first, before examining any effect due to disorder. Basic to the consideration of the lattice effect is the lattice potential. If the potential V at each site of an ordered, perfect lattice is given by a Coulomb or some form of attractive potential, then the solution of the Schrödinger's equation for each site, considered in isolation from other sites, would result in many atomic energy levels. On adding to this problem the interaction between the sites, i.e., the effect of the lattice structure, the combined state of affair can become very complicated indeed. However, in many electronic problems of physical interest only one atomic state is important, e.g., the valence electronic state. If the energy separations between the valence state and other atomic states are large compared to the interaction energy between the sites, then the problem can be simplified by solution in two stages. The first stage considers only the atomic-level problem for each site, in isolation from other sites. The next stage considers the effect of lattice structure on each atomic level, without worrying about the intermixing of the atomic levels. This approach is the basis of the "tight-binding approximation" in which the lattice-structure effect is treated as a perturbation on the single-site atomic energy levels. In what follows, the description of the quantum Schrödinger wave is based on this tight binding approach, where the one relevant atomic level is assumed to be identified, and our attention is therefore focused on the effect of the lattice structure plus the randomness.

The question before us is: If $\phi(t, \mathbf{l})$ stands for the wave amplitude of the relevant atomic level at lattice site $\mathbf{l}$ and time t, then what is the effect of the lattice on its frequency component $\phi(\omega, \mathbf{l})$? To answer this question, let us start with the wave form in a perfect lattice, $\exp(-i\omega t + i\mathbf{k} \cdot \mathbf{l})$, and ask what general spatial operator would yield $\exp(i\mathbf{k} \cdot \mathbf{l})$ as its eigenfunc-

2.3 Waves on a Discrete Lattice

tion. Since in continuum $\exp(i\mathbf{k} \cdot \mathbf{r})$ is the eigenfunction of the Laplacian operator ∇^2, the obvious choice in the present case is the Laplacian defined on the lattice, i.e., a discrete Laplacian $\Delta^{(2)}$. In 1D, where $l = \cdots, -2, -1, 0, 1, 2, \cdots$ and the lattice constant is denoted by a, the operation of the discrete Laplacian is defined by

$$\Delta^{(2)}\phi(\omega, l) = \frac{1}{a^2}[\phi(\omega, l+1) + \phi(\omega, l-1) - 2\phi(\omega, l)]. \quad (2.34)$$

Equation (2.34) expresses the operation of taking the discrete second derivative of ϕ. For a simple square lattice in 2D and a simple cubic lattice in 3D, the operation of $\Delta^{(2)}$ is a direct extension from 1D:

$$\Delta^{(2)}\phi(\omega, \mathbf{l}) = \frac{1}{a^2}\left[-2d\phi(\omega, \mathbf{l}) + \sum_{\mathbf{n}} \phi(\omega, \mathbf{l} + \mathbf{n})\right], \quad (2.35)$$

where $\mathbf{n}$ denotes the vectors pointing from the lattice site $\mathbf{l}$ to its nearest neighbors. The sum therefore has four terms in 2D and six terms in 3D. By substituting $\exp(-i\omega t + i\mathbf{k} \cdot \mathbf{l})$ for $\phi(\omega, \mathbf{l})$ and similar forms for $\phi(\omega, \mathbf{l} + \mathbf{n})$ into Eqn (2.35), we get

$$\Delta^{(2)}\phi(\omega, \mathbf{l}) = -e(\mathbf{k})\phi(\omega, \mathbf{l}), \quad (2.36)$$

where $e(\mathbf{k})$ is given by

$$e(\mathbf{k}) = \frac{2}{a^2} \sum_{\alpha=1}^{d} (1 - \cos k_\alpha a). \quad (2.37)$$

Here $k_{1,2,3}$ means $k_{x,y,z}$. When $k_\alpha a \to 0$, i.e., in the long-wavelength limit, $1 - \cos k_\alpha a \cong k_\alpha^2 a^2/2$, and

$$e(\mathbf{k}) \cong \sum_{\alpha=1}^{d} k_\alpha^2 = k^2, \quad (2.38)$$

which corresponds precisely to the negative of the eigenvalue for the Laplacian. Since $\hbar^2 k^2/2m$ is the kinetic energy of a quantum wave/particle, $\hbar^2 e(\mathbf{k})/2m^*$ may thus be identified as the similar quantity on a lattice. Here m^* is the "effective mass," to be defined below.

Equation (2.37) indicates significant differences between lattice waves and waves in continuum. In the first place, $e(\mathbf{k})$ is periodic in $\mathbf{k}$. This is due to the fact that lattice has a minimum length scale, a, which means that there cannot be a wave with wavelength less than $2a$, or a $\mathbf{k}$ with $|k_\alpha| \geq \pi/a$. When $|k_\alpha|$ is formally extended beyond π/a, it becomes equivalent to a state with $|k_\alpha| \leq \pi/a$. A related second difference is that $e(\mathbf{k})$ has a maximum, defined by $4d/a^2$. The eigenvalues of $-\Delta^{(2)}$ thus form a band, extending from 0 to $4d/a^2$. At the maximum, which occurs at $|k_\alpha| = \pi/a$, the dispersion relation gives a group velocity $de(\mathbf{k})/dk_\alpha = 0$ in the direction of k_α, which indicates a standing wave caused by Bragg scattering from the lattice structure. At frequencies beyond the maximum so that $e(\mathbf{k}) > 4d/a^2$, i.e., outside the band, it is clear that the only way for Eq. (2.37) to be satisfied is by switching k_α from real to imaginary so that $\cos(k_\alpha a)$ becomes $\cosh(k_\alpha a)$. However, a "state" which is associated with an imaginary k must decay exponentially toward infinity. These waves are called evanescent waves and are analogous to what happens in waveguides below the cutoff frequency. The third difference is that on a lattice, the "mass" of a quantum wave/particle is not necessarily the same as that in the continuum. Since in the long-wavelength limit the dispersion relation of the lattice wave has the same form as that in the continuum, the *effective mass* m^* on a lattice may be defined by the curvature of the kinetic energy expression around $\mathbf{k} = 0$:

$$\frac{1}{m^*} = \frac{1}{\hbar^2} \frac{d^2}{dk^2} (\text{kinetic energy})_{\mathbf{k}=0}. \qquad (2.39)$$

The kinetic energy is dependent on the strength of coupling between the neighboring sites. If the interaction between neighboring sites is weak, the electron tends to be bound at a given site, and the transfer between the sites would occur slowly, thus implying a large m^*. The effective mass m^* may therefore be regarded as a parameter for measuring the lattice effect on the electron.

The total energy E for a quantum wave on a lattice is

$$E = \frac{\hbar^2}{2m^*} e(\mathbf{k}) + \epsilon_0 - \frac{\hbar^2}{2m^*}\left(\frac{2d}{a^2}\right), \qquad (2.40)$$

where ϵ_0 is the energy of the atomic level, defined relative to potential $V = 0$. The fact that ϵ_0 always defines the center of the energy band accounts for the last term, which is half of the energy bandwidth. Equation (2.40) relates $\mathbf{k}$ and $E = \hbar\omega$ for the quantum Schrödinger wave on a lattice. Just as in the case of waves in continuum, the corresponding wave equation may be inferred from the dispersion relation. By noting that $e(\mathbf{k})$ is obtained as the eigenvalue of $-\Delta^{(2)}$, it is easy to see that the lattice wave equation is

$$(\beta \Delta^{(2)} + \kappa_0^2)\phi(\omega, \mathbf{l}) = 0, \qquad (2.41)$$

2.3 Waves on a Discrete Lattice

where κ_0^2 is now renormalized with respect to $\epsilon_0 - (\hbar^2 d/m^* a^2)$:

$$\kappa_0^2 = \frac{2m(E - \epsilon_0)}{\hbar^2} + \frac{2\beta d}{a^2}, \qquad (2.42)$$

with

$$\beta = \frac{m}{m^*}. \qquad (2.43)$$

So far we have discussed only quantum waves on a lattice. However, Eq. (2.41) is equally applicable to waves in a classical spring-and-mass lattice. This can be seen by letting $\beta = 1$ and $\kappa_0^2 = \omega^2/v_0^2$, where $v_0^2 = Ka^2/m$ is the wave speed in the long-wavelength limit, K being the spring constant of the springs linking identical nearest-neighbor particles of mass m. Equation (2.41) then becomes

$$Ka^2 \Delta^{(2)} \phi(\omega, \mathbf{l}) = m\omega^2 \phi(\omega, \mathbf{l}). \qquad (2.44)$$

The left-hand side of Eq. (2.44) is the negative of the force on the particle situated at $\mathbf{l}$, if ϕ denotes the displacement of the particle away from its equilibrium position. For example, in 1D it is

$$K\{[\phi(\omega, l+1) - \phi(\omega, l)] - [\phi(\omega, l) - \phi(\omega, l-1)]\},$$

which gives the force on the particle at $\mathbf{l}$ due to the net contraction/extension of the springs on its two sides. The right-hand side of Eq. (2.44) is the negative mass times acceleration of the particle (at frequency ω). Therefore this is precisely the dynamical equation for the vibrational modes of an elastic lattice. One thus concludes that Eq. (2.41) is the general form for both the quantum and the classical waves on a lattice.

An alternative way of expressing Eq. (2.41) is to use a vector $\boldsymbol{\phi}$ to express the wave function, where each component of the vector corresponds to the value of the wave function at a site. The operation $-\beta \Delta^{(2)}$ can then be expressed by a matrix $\mathbf{D}$ whose diagonal elements are

$$(\mathbf{D})_{\mathbf{l},\mathbf{l}} = \frac{2\beta d}{a^2} \qquad (2.45)$$

and whose off-diagonal elements are

$$(\mathbf{D})_{\mathbf{l},\mathbf{l}+\mathbf{n}} = -\frac{\beta}{a^2} \qquad (2.46)$$

and zero otherwise. A succinct way of writing the matrix is by using Dirac's bra and ket notation, where $|\phi\rangle = \boldsymbol{\phi}$ means the wave function expressed as a column vector and $\langle\phi|$ means the same wave function but expressed as a row vector. $\langle\phi|\phi'\rangle$ means the inner product (dot product between two vectors), and $|\phi'\rangle\langle\phi|$ means the outer product. In Dirac's notation,

$$\mathbf{D} = \sum_{\mathbf{l}} \frac{2\beta d}{a^2}|\mathbf{l}\rangle\langle\mathbf{l}| + \sum_{\mathbf{l},\mathbf{n}}\left(-\frac{\beta}{a^2}\right)|\mathbf{l}\rangle\langle\mathbf{l}+\mathbf{n}|, \qquad (2.47)$$

where the second sum is over all the nearest neighbors of $\mathbf{l}$, for all $\mathbf{l}$, and $|\mathbf{l}\rangle$ means a column vector which has all zero components except for the component $\mathbf{l}$, which is 1. The same holds for the row vector $\langle\mathbf{l}|$. The outer product $|\mathbf{l}\rangle\langle\mathbf{l}+\mathbf{n}|$ is a matrix whose elements are all zero except the one at the row corresponding to $\mathbf{l}$ and the column corresponding to $\mathbf{l}+\mathbf{n}$, which has the value 1. $\mathbf{D}$ is expressed in Eq. (2.47) as the sum of such matrices. Equation (2.41) can then be written as

$$\mathbf{D}|\phi\rangle = \kappa_0^2 |\phi\rangle. \qquad (2.48)$$

There are two ways to introduce randomness into the lattice problem—either via random variation in the atomic energy level or via random variation in β. If $\Delta\epsilon_\mathbf{l}$ denotes the random deviation (from ϵ_0) of the atomic energy level at site $\mathbf{l}$, then $\Delta\epsilon_\mathbf{l}$ randomizes only the diagonal matrix elements. For the phonon case, the same effect is obtained by varying the mass of the particle at each site by $\Delta m_\mathbf{l}$. When there is randomness, κ_0^2 should be replaced by $\kappa_0^2 - \sigma(\mathbf{l})$ in Eq. (2.41), where

$$\sigma(\mathbf{l}) = \begin{cases} \dfrac{2m\,\Delta\epsilon_\mathbf{l}}{\hbar^2} & \text{quantum} \\ \dfrac{\omega^2}{v_0^2}\dfrac{\Delta m_\mathbf{l}}{m} & \text{classical}. \end{cases} \qquad (2.49)$$

Since one is free to define ϵ_0 and m as the mean values of the respective quantities, $\Delta\epsilon_\mathbf{l}$ and $\Delta m_\mathbf{l}$ may be regarded as having zero mean:

$$\langle\sigma(\mathbf{l})\rangle_c = 0, \qquad (2.50)$$

where $\langle\ \rangle_c$ means configurational averaging, which is equivalent to averaging over all sites in the present case. By defining a new matrix $\mathbf{M}$ such that the diagonal elements consist of only the deviations, i.e.,

$$\mathbf{M} = \sum_{\mathbf{l}} \sigma(\mathbf{l})|\mathbf{l}\rangle\langle\mathbf{l}| + \sum_{\mathbf{l},\mathbf{n}}\left(-\frac{\beta}{a^2}\right)|\mathbf{l}\rangle\langle\mathbf{l}+\mathbf{n}|, \qquad (2.51)$$

2.4 Lattice Green's Functions

the equation

$$[\beta \Delta^{(2)} + \kappa_0^2 - \sigma(\mathbf{l})]\phi(\omega, \mathbf{l}) = 0 \tag{2.52}$$

may be written as

$$\mathbf{M}|\phi\rangle = q_0^2|\phi\rangle, \tag{2.53}$$

where

$$q_0^2 = \kappa_0^2 - \frac{2\beta d}{a^2}, \tag{2.54a}$$

with

$$\kappa_0^2 = \frac{2m(E - \epsilon_0)}{\hbar^2} + \frac{2\beta d}{a^2}$$

in the quantum case and

$$\kappa_0^2 = \frac{\omega^2}{v_0^2}, \tag{2.54b}$$

$\beta = 1$ in the phonon case. Equations (2.51) and (2.53), together with the condition (2.50), are known as the *Anderson model*. Here the disorder is denoted "diagonal randomness."

The second way to introduce randomness into the model is by varying β. That would randomize the off-diagonal elements as well as the diagonal elements of **D**. However, for the quantum case $q_0 = 2m(E - \epsilon_0)/\hbar^2$, therefore β does not enter into the diagonal elements of **M**. Randomizing β thus constitutes an "off-diagonal randomness" model. For the phonon problem, on the other hand, varying β would randomize both the diagonal and off-diagonal elements in a correlated manner. In what follows, only the case of diagonal randomness is pursued because the randomness in β is known to have an effect on the scattering and localization behavior qualitatively similar to that in the diagonal randomness case. Studying one is therefore sufficient to illustrate the physics involved.

2.4 Lattice Green's Functions

The Green's function for the perfect lattice is defined by the solution to the following equation:

$$(\beta \Delta^{(2)} + \kappa_0^2)G_0(\omega, \mathbf{l} - \mathbf{l}') = \frac{1}{a^d}\delta_{\mathbf{l},\mathbf{l}'}, \tag{2.55}$$

where $\delta_{l,l'}$ is the Kronecker delta defined by

$$\delta_{l,l'} = \begin{cases} 1 & \text{if } l = l' \\ 0 & \text{otherwise}. \end{cases} \quad (2.56)$$

To solve for $G_0(\omega, l - l')$, we use the same approach as in the continuum case by first solving for the spatial frequency component $G_0(\omega, \mathbf{k})$, related to $G_0(\omega, l - l')$ by

$$G_0(\omega, l - l') = \frac{L^d}{N} \int_{1BZ} \frac{d\mathbf{k}}{(2\pi)^d} \exp[i\mathbf{k} \cdot (l - l')] G_0(\omega, \mathbf{k}). \quad (2.57)$$

The Kronecker delta may also be expressed as

$$\frac{1}{a^d} \delta_{l,l'} = \int_{1BZ} \frac{d\mathbf{k}}{(2\pi)^d} \exp[i\mathbf{k} \cdot (l - l')]. \quad (2.58)$$

Here the notation 1BZ means restricting the values of $|k_x|, |k_y|, |k_z| \leq \pi/a$, which is the first Brillouin zone for the simple cubic lattice in 3D and the simple square lattice in 2D. These are the only lattices considered throughout the succeeding chapters. By substituting Eqs. (2.57) and (2.58) into Eq. (2.55), the operation of $\Delta^{(2)}$ on $\exp[i\mathbf{k}(l - l')]$ is seen to yield $-e(\mathbf{k})$ as its eigenvalue. Therefore, by equating the left- and right-hand sides of each $\mathbf{k}$ component, one gets

$$\left[\kappa_0^2 - \beta e(\mathbf{k})\right] G_0(\omega, \mathbf{k}) = \frac{N}{L^d} = \frac{1}{a^d} \quad (2.59)$$

or

$$G_0(\omega, \mathbf{k}) = \frac{1}{\kappa_0^2 - \beta e(\mathbf{k})} \frac{1}{a^d}. \quad (2.60)$$

Comparison with the continuum case, Eq. (2.22), shows the only difference to be the replacement of k^2 by $\beta e(\mathbf{k})$, and L^d/N by a^d. In 1D, the calculation of $G_0(\omega, l - l')$ from $G_0(\omega, \mathbf{k})$ may be performed analytically. By substituting Eq. (2.60) into Eq. (2.57) and using the definition of $e(\mathbf{k})$ [Eq. (2.37)], we get

$$G_0^{\pm}(\omega, l - l') = a \int_{-\pi/a}^{\pi/a} \frac{dk}{2\pi} \frac{\exp[ik(l - l')]}{\kappa_0^2 - 2\beta(1 - \cos ka)/a^2 \pm i\eta} \frac{1}{a}$$

$$= \frac{a}{2\pi\beta} \int_{-\pi}^{\pi} d(ka)$$

$$\times \frac{\exp[ika|l - l'|/a]}{2(\kappa_0^2 a^2/2\beta - 1) + [\exp(ika) + \exp(-ika)]}, \quad (2.61)$$

2.4 Lattice Green's Functions

where $\pm i\eta$ is dropped in the second line of Eq. (2.61), but the constant $(\kappa_0^2 a^2/2\beta - 1)$ should be remembered to have a $\pm i\eta$ imaginary part. Also, since the integral is invariant with respect to the interchange of l and l' (because the ka integral is from $-\pi$ to $+\pi$), it must be a function of $|l - l'|$. By changing to the variable $z = \exp(ika)$ and noting that $d(ka) = dz/iz$, the definite integral may be converted into a contour integral around the unit circle in the complex z-plane:

$$G_0^\pm(\omega, l - l') = \frac{a}{2\pi i \beta} \oint dz \, \frac{z^{|l-l'|/a}}{z^2 + 2uz + 1}, \qquad (2.62)$$

where $u = (\kappa_0^2 a^2/2\beta - 1) \pm i\eta$. The denominator of the integrand has two roots,

$$z_1 = -u + \sqrt{u^2 - 1}, \qquad (2.63)$$

$$z_2 = -u - \sqrt{u^2 - 1}. \qquad (2.64)$$

For $|u| \leq 1$, $|z_1| = |z_2| = 1$, and the contour integration has to rely on $\pm i\eta$ to give

$$G_0^\pm(\omega, l - l') = \frac{\mp ai}{2\beta\sqrt{1 - u^2}} \left(-u \pm i\sqrt{1 - u^2}\right)^{|l-l'|/a}. \qquad (2.65a)$$

This form of $G_0^\pm(\omega, l - l')$ may be put in a more illuminating form by defining a k_0 as the solution to the equation $\kappa_0^2 = \beta e(k_0)$. Then from the definition of u and $e(k_0)$ one gets $-u = \cos k_0 a$, so that

$$G_0^\pm(\omega, l - l') = \frac{\mp ai}{2\beta \sin k_0 a} \exp(\pm ik_0|l - l'|)$$

$$= \frac{\mp i}{(\partial \beta e(\mathbf{k})/\partial k)_{k_0}} \exp(\pm ik_0|l - l'|). \qquad (2.65b)$$

Comparison with Eq. (2.33) shows remarkable similarity. The only difference is that the preexponential factor is now modified. The continuum form of the preexponential factor is recovered in the limit of $k_0 \to 0$ and $\beta = 1$.

When $u > 1$, then $|z_1| < 1$ and $|z_2| > 1$, which means only the pole at $z = z_1$ is inside the contour. But if $u < -1$, the reverse is true. In either case, $\pm i\eta$ does not play a role any more, and

$$G_0(\omega, l - l') = \frac{\pm a}{2\beta\sqrt{u^2 - 1}} \left(-u \pm \sqrt{u^2 - 1}\right)^{|l-l'|/a}, \qquad (2.66)$$

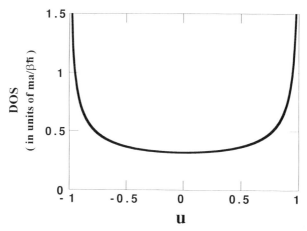

Figure 2.1 The electron density of states for a 1D lattice, normalized to per site per frequency unit $\beta\hbar/ma^2$. Here $u = (\hbar\omega - \epsilon_0)/(\beta\hbar^2/ma^2)$. The integrated density of states over the whole band is 1. The solid curve is described by the function $(\pi\sqrt{1-u^2})^{-1}$.

where the $+$ and $-$ signs correspond to the cases $u > 1$ and $u < -1$, respectively. Since the right-hand side of Eq. (2.66) is a real number, it follows that the density of states is zero for frequencies that give $|u| > 1$. In this case $G_0(\omega, l - l')$ is also noted to decay geometrically (exponentially) as a function of $|l - l'|$.

To calculate the density of states from Eqs. (2.29), (2.65), and (2.66), the substitution $u = ma^2(E - \epsilon_0)/\beta\hbar^2$ is used for the quantum case to obtain

$$\rho_0(\omega) = -\frac{1}{\pi}\frac{d\kappa_0^2}{d\omega}\,\text{Im}\,G_0^+(\omega, \mathbf{l} = \mathbf{l}')$$

$$= \begin{cases} \dfrac{ma}{\pi\hbar\beta}\dfrac{1}{\sqrt{1 - (\hbar\omega - \epsilon_0)^2/(\beta\hbar^2/ma^2)^2}} & |\hbar\omega - \epsilon_0| \leq \dfrac{\beta\hbar^2}{ma^2} \\ 0 & |\hbar\omega - \epsilon_0| > \dfrac{\beta\hbar^2}{ma^2} \end{cases}$$

(2.67)

This is shown in Figure 2.1. Near the lower band edge $\rho_0(\omega)$ is noted to have the same divergence as in a uniform medium. For the phonon case, $u = (m\omega^2/2K) - 1$, $\beta = 1$, so that

$$\rho_0(\omega) = \begin{cases} \dfrac{2}{\pi a}\dfrac{1}{\sqrt{(4K/m) - \omega^2}} & \omega \leq 2\sqrt{K/m} \\ 0 & \omega > 2\sqrt{K/m} \end{cases}$$

(2.68)

2.4 Lattice Green's Functions

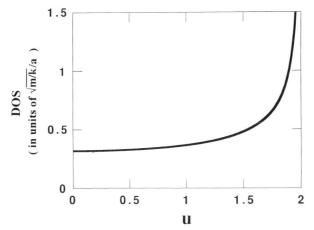

Figure 2.2 The phonon density of states on a 1D lattice, normalized to per site per frequency unit $\sqrt{K/m}$. Here $u = \omega/\sqrt{K/m}$. The integrated density of states over the whole band is 1. The solid curve is described by the function $(\pi\sqrt{1-(u/2)^2})^{-1}$.

As $\omega \to 0$, $\rho_0(\omega)$ for the continuum case is recovered exactly (by using the identity $v_0 = a\sqrt{K/m}$), as expected. At the upper band edge the qualitative behavior is similar to that of the quantum case. This is shown in Figure 2.2. For the sake of uniformity in notation the variable u is redefined here to mean the dimensionless frequency $\omega/\sqrt{K/m}$.

In 2D, the general Green's function can no longer be calculated analytically. This is because a constant value of $e(\mathbf{k})$ does not correspond to a fixed $|\mathbf{k}|$, i.e., the Brillouin zone is not circular, and as a result analytic calculation becomes difficult. However, the Green's function for $\mathbf{l} = \mathbf{l}'$, and therefore the density of states, can still be explicitly evaluated in terms of elliptic integrals. Details of the evaluation are given in the solution to Problem 2.4. The answer is

$$\rho_0(\omega) = \frac{1}{2\pi^2\beta} \frac{d\kappa_0^2(\omega)}{d\omega} \mathbf{K}\left[\sqrt{1 - \left(\frac{\kappa_0^2 a^2 - 4\beta}{4\beta}\right)^2}\right] \qquad (2.69)$$

for $|(\kappa_0^2 a^2 - 4\beta)/4\beta| < 1$, and zero otherwise. Here **K** denotes the complete elliptic integral of the first kind. Written out explicitly, the 2D quantum case gives

$$\rho_0(\omega) = \frac{m}{\pi^2\beta\hbar} \mathbf{K}\left[\sqrt{1 - \left(\frac{m(\hbar\omega - \epsilon_0)a^2}{2\beta\hbar^2}\right)^2}\right] \qquad (2.70)$$

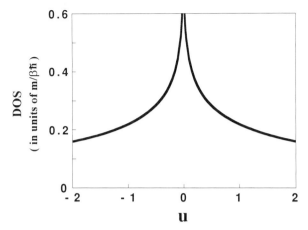

Figure 2.3 The electronic density of states for the 2D square lattice, normalized to per site per frequency unit $\beta\hbar/ma^2$. Here $u = (\hbar\omega - \epsilon_0)/(\beta\hbar^2/ma^2)$. The density of states integrates to 1. The solid curve is given by $(1/\pi^2)\mathbf{K}[\sqrt{1-(u/2)^2}]$. The divergence at the band center is logarithmic in nature.

for $|\hbar\omega - \epsilon_0| \leq 2\beta\hbar^2/ma^2(|u| < 2)$, and zero otherwise. This is shown in Figure 2.3. At the band edge, where the argument of $\mathbf{K}$ vanishes, the value of $\mathbf{K}(0) = \pi/2$, and the continuum density of states is recovered if $\beta = 1$. It is also interesting to note that at the center of the band $\mathbf{K}(1)$ has a divergence that is logarithmic in nature. For the phonon case,

$$\rho_0(\omega) = \frac{\omega}{\pi^2 v_0^2} \mathbf{K}\left[2\left(\frac{\omega}{\omega_0}\right)\sqrt{1 - \left(\frac{\omega}{\omega_0}\right)^2}\right], \quad (2.71)$$

for $0 \leq \omega \leq \omega_0 = 2\sqrt{2}\sqrt{K/m}$, and zero otherwise. This is shown in Figure 2.4. Again, at ω close to zero the continuum density of states is recovered.

In 3D, it is no longer possible to have an analytic density of states expression for the simple cubic lattice, even for the $\mathbf{l} = \mathbf{l}'$ case. However, in the solution to Problem 2.5 it is shown that the density of states can nevertheless be expressed as an integral which has to be evaluated numerically:

$$\rho_0(\omega) = -\frac{1}{\pi}\frac{d\kappa_0^2}{d\omega}\text{Im}\, G_0^+(\omega, \mathbf{l} = \mathbf{l}'), \quad (2.72)$$

2.5 Treating Continuum Problems on a Lattice

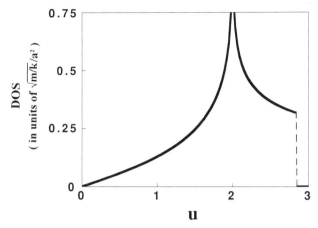

Figure 2.4 The phonon density of states on a 2D square lattice, normalized to per site per frequency unit $\sqrt{K/m}$, plotted as a function of $u = \omega/\sqrt{K/m}$. The density of states integrates to 1, and the upper band edge is at $\sqrt{8}\sqrt{K/m}$. The solid curve is given by $(u/\pi^2)\mathbf{K}[(u/\sqrt{2})\sqrt{1-(u^2/8)}]$. The divergence in the density of states is logarithmic in nature.

where

$$\text{Im}\, G_0^{\pm}(\omega, \mathbf{l} = \mathbf{l}') = \frac{1}{2\pi^2 a\beta}\, \text{Im} \int_0^{\pi} d\theta \mu^{\pm}(\theta)\mathbf{K}[\mu^{\pm}(\theta)], \quad (2.73)$$

$$\mu^{\pm}(\theta) = \frac{4\beta}{|\kappa_0^2 a^2 - 6\beta + 2\beta\cos\theta| \pm i\eta}. \quad (2.74)$$

In Figures 2.5 and 2.6 the densities of states for the quantum case and the phonon case are plotted, respectively. The 3D density of states differs from those of 1D and 2D in that it has no divergences.

2.5 Treating Continuum Problems on a Lattice

In the prior discussion it has been noted that the continuum limit can be achieved on a lattice at the $k \to 0$ limit. Therefore, one can always use the lattice to simulate waves in continuum by making sure that the lattice spacing is much smaller than the wavelength so that the $k \to 0$ limit is well

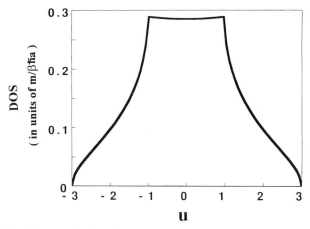

Figure 2.5 The electronic density of states for a 3D simple cubic lattice, normalized to per site per frequency unit $\beta\hbar/ma^2$. Here $u = (\hbar\omega - \epsilon_0)/(\beta\hbar^2/ma^2)$. The area under the curve integrates to 1. The solid curve is given by the function

$$-\frac{1}{\pi^3}\,\text{Im}\int_0^\pi d\theta\, \mu^+(u,\theta)\mathbf{K}[\mu^+(u,\theta)],$$

where

$$\mu^+(u,\theta) = \frac{2}{|u + \cos\theta| + i\eta}.$$

approximated. Technically, however, there is a limit to the usefulness of the lattice approach in its application to wave scattering problems. One potential complication has already been mentioned, namely that the internal degrees of freedom at each lattice site, e.g., the atomic energy levels, may intermix. If there are inhomogeneities which have many internal degrees of freedom and which intermix with the lattice effect, then the simplicity of the lattice model would be lost. Another technical complication can arise when the inhomogeneities are *correlated* from site to site. Thus, if the lattice approach is used to model light scattering from spherical particles whose diameters are on the order of the light wavelength, it is necessary to subdivide the interior of each particle into many lattice sites so that the wave behavior may be modeled correctly. However, in that case the inhomogeneous randomness must be correlated over the size of each sphere. Technically, correlation would present difficulty in a lattice model, which is best suited to treat uncorrelated randomness that can vary independently from site to site. Even numerically, 3D problems

2.5 Treating Continuum Problems on a Lattice

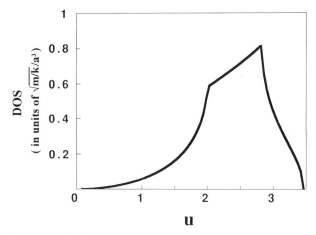

Figure 2.6 The phonon density of states on a 3D simple cubic lattice, normalized to per site per frequency unit $\sqrt{K/m}$. Here $u = \omega/\sqrt{K/m}$. The integrated area under the curve is 1. The upper band edge is at $\omega = \sqrt{12}\sqrt{K/m}$. The solid curve here is given by

$$-\frac{u}{\pi^2}\,\text{Im}\int_0^\pi d\theta\, \mu^+(u,\theta)\mathbf{K}[\mu^+(u,\theta)],$$

where

$$\mu^+(u,\theta) = \frac{4}{|u^2 - 6 + 2\cos\theta| + i\eta}.$$

with correlated randomness are still technically challenging. This is one of the reasons that the problem of light localization is not yet settled in 3D. However, the viewpoint—that the waves in continuum may be modeled by waves in a correlated lattice—is useful in illuminating one source of difference between the scattering and localization behavior of classical waves and that of an electron on a lattice: the existence of resonances for scatterers in continuum and the inert, point-particle nature of the single-site scattering impurity in the lattice case. The equivalence between the continuum model and the correlated lattice model tells us that the scatterer resonances may be viewed as a consequence of correlation between the scattering sites on a lattice.

In what follows, the quantum case, pertaining to electrons in solids, will be treated by the lattice model, since this is fairly realistic. For classical waves the scalar wave equation with continuous spatial variables will be used as our model. Although the basic Green's function formalism is common to the two cases, actual implementation can involve different approaches. In the case of short-wavelength phonons, it is already seen that the problem is qualitatively identical to that of the electron. Therefore it will not be covered separately.

Problems and Solutions

2.1. Show that

$$\lim_{\eta \to 0} \frac{1}{x \pm i\eta} = P\left(\frac{1}{x}\right) \mp i\pi \delta(x).$$

By multiplying $x \mp i\eta$ to the numerator and denominator, one gets

$$\frac{1}{x \pm i\eta} = \frac{x \mp i\eta}{x^2 + \eta^2} = \frac{x}{x^2 + \eta^2} \mp i \frac{\eta}{x^2 + \eta^2}. \qquad (P2.1)$$

For the real part, taking the principal value means that in performing the integral over x one should first do the integral with a finite η, and then let η approach zero. That is,

$$P \int \frac{dx}{x} = \lim_{\eta \to 0} \int \frac{x}{x^2 + \eta^2} dx. \qquad (P2.2)$$

It is clear that if the integration limits are symmetrical about $x = 0$, the result would be zero since $x/(x^2 + \eta^2)$ is an odd function of x. When the limits are not symmetrical about 0, the result would depend on the degree of asymmetry but would never diverge.

For the imaginary part, the integral over x is given by

$$\int_{-\infty}^{\infty} \frac{\eta}{x^2 + \eta^2} dx = \int_{-\pi/2}^{\pi/2} \frac{\eta^2 \sec^2 \theta \, d\theta}{\eta^2 (1 + \tan^2 \theta)} = \pi, \qquad (P2.3)$$

where the change of variable $x = \eta \tan \theta$ is used. What should be noted is that the final answer is independent of η. Therefore, when η approaches zero, $\eta/(x^2 + \eta^2)$ becomes a narrower and taller Lorentzian, but its area, π, remains constant. In the limit of $\eta \to 0$, that Lorenzian becomes a delta function, with a factor π to account for its integral value.

It should be noted here that a delta function always has the dimension that is the inverse of its argument's dimension. For example, $\delta(k)$ has the dimension of length because k has the dimension of [length]$^{-1}$. The reason is that

$$\int_{-\infty}^{\infty} \delta(k)\, dk = 1. \tag{P2.4}$$

Since the right-hand side is dimensionless, $\delta(k)$ must have the dimension of length in order to cancel that of k.

2.2 Derive the continuum density of states in 1D, 2D, and 3D uniform media for both the quantum wave and the classical scalar wave.

Starting from Eq. (2.24), the main task is to evaluate the integral

$$\int \frac{d\mathbf{k}}{(2\pi)^d}\, \delta\!\left[\kappa_0^2(\omega) - k^2\right].$$

In 1D, we have

$$\int_{-\infty}^{\infty} \frac{dk}{2\pi}\, \delta\!\left[\kappa_0^2(\omega) - k^2\right] = \frac{2}{2\pi} \int_0^{\infty} d(k^2) \cdot \frac{dk}{d(k^2)}\, \delta\!\left[\kappa_0^2(\omega) - k^2\right]. \tag{P2.5}$$

The integral immediately gives $[2\pi\kappa_0(\omega)]^{-1}$. In 2D, the integral may be written as

$$\frac{1}{4\pi} \int_0^{\infty} d(k^2)\, \delta\!\left[\kappa_0^2(\omega) - k^2\right] = (4\pi)^{-1}. \tag{P2.6}$$

In 3D, the integral is

$$\frac{1}{8\pi^3} \int_0^{\infty} 2\pi k\, d(k^2)\, \delta\!\left[\kappa_0^2(\omega) - k^2\right] = \frac{\kappa_0(\omega)}{4\pi^2}. \tag{P2.7}$$

By substituting these expressions into Eq. (2.24) and using Eq. (2.25) plus the respective $\kappa_0(\omega)$ expressions for the quantum and the classical cases, Eqs. (2.26) and (2.27) are obtained.

2.3 Derive a general expression of (local) density of states for the scalar wave in an inhomogeneous medium. Express the diagonal Green's function in terms of the density of states.

The Green's function for the general scalar wave equation is given by Eq. (2.16):

$$(\nabla^2 - \sigma(\mathbf{r}) + \kappa_0^2) G(\omega, \mathbf{r}, \mathbf{r}') = \delta(\mathbf{r} - \mathbf{r}'). \tag{P2.8}$$

Although this equation is not solvable in general, one can nevertheless formally define the eigenfunctions of the operators $\nabla^2 - \sigma(\mathbf{r})$ as $\phi_n(\mathbf{r})$, with eigenvalues $-d_n$:

$$[\nabla^2 - \sigma(\mathbf{r})]\phi_n(\mathbf{r}) = -d_n \phi_n(\mathbf{r}). \tag{P2.9}$$

Here n is the index for the eigenfunctions and their corresponding eigenvalues. It is noted that for the operator ∇^2, $\phi_n(\mathbf{r})$ is just $\exp(i\mathbf{k} \cdot \mathbf{r})$, with the eigenvalue $-k^2$. A formal solution of Eq. (P2.8) may be obtained by first expressing $G(\omega, \mathbf{r}, \mathbf{r}')$ and $\delta(\mathbf{r}, \mathbf{r}')$ as

$$G(\omega, \mathbf{r}, \mathbf{r}') = \frac{1}{N} \sum_n G(\omega, d_n) \phi_n(\mathbf{r}) \phi_n^*(\mathbf{r}'), \tag{P2.10}$$

$$\delta(\mathbf{r} - \mathbf{r}') = (1/L^d) \sum_n \phi_n(\mathbf{r}) \phi_n^*(\mathbf{r}'), \tag{P2.11}$$

where the summation over n is seen to correspond to summation over $\mathbf{k}$ in the free-space case, i.e.,

$$\sum_{\mathbf{k}} = L^d \int \frac{dk}{(2\pi)^d}.$$

Equation (P2.11) is the definition of the completeness of the eigenfunctions, which is guaranteed for the operator $\nabla^2 - \sigma(\mathbf{r})$ if $\sigma(\mathbf{r})$ is real. By substituting Eqs. (P2.10) and (P2.11) into Eq. (P2.8) and equating the left-hand and right-hand sides for each component n, it is seen that

$$\frac{L^d}{N} G^{\pm}(\omega, d_n) = \lim_{\eta \to 0} \frac{1}{\kappa_0^2(\omega) - d_n \pm i\eta}$$

$$= P \frac{1}{\kappa_0^2(\omega) - d_n} \mp i\pi \delta[\kappa_0^2(\omega) - d_n]. \tag{P2.12}$$

The imaginary part of $G(\omega, \mathbf{r} = \mathbf{r}')$ is given by

$$\frac{L^d}{N} \operatorname{Im} G^+(\omega, \mathbf{r} = \mathbf{r}') = -\frac{\pi}{N} \sum_n \delta[\kappa_0^2(\omega) - d_n] |\phi_n(\mathbf{r})|^2. \tag{P2.13}$$

If one defines a local density of states as

$$\rho(\omega, \mathbf{r}) = \frac{d\kappa_0^2(\omega)}{d\omega} \frac{1}{L^d} \sum_n \delta[\kappa_0^2(\omega) - d_n] |\phi_n(\mathbf{r})|^2, \tag{P2.14a}$$

then

$$\rho(\omega, \mathbf{r}) = -\frac{1}{\pi} \frac{d\kappa_0^2(\omega)}{d\omega} \operatorname{Im} G^+(\omega, \mathbf{r} = \mathbf{r}'). \qquad \text{(P2.14b)}$$

The volume-averaged density of states is then

$$\begin{aligned}
\bar{\rho}(\omega) &= \frac{1}{L^d} \int \rho(\omega, \mathbf{r}) \, d\mathbf{r} \\
&= \frac{d\kappa_0^2(\omega)}{d\omega} \frac{1}{L^d} \sum_n \delta[\kappa_0^2(\omega) - d_n] \frac{1}{L^d} \int |\phi_n(\mathbf{r})|^2 \, d\mathbf{r} \\
&= \frac{d\kappa_0^2(\omega)}{d\omega} \frac{1}{L^d} \sum_n \delta[\kappa_0^2(\omega) - d_n].
\end{aligned} \qquad \text{(P2.15)}$$

For a uniform medium, $|\phi_n(\mathbf{r})|^2 = |e^{i\mathbf{k}\cdot\mathbf{r}}|^2 = 1$, so there is no distinction between the local density of states and the volume-averaged density of states. We also note here that from Eqs. (P2.12) and (P2.10),

$$G^+(\omega, \mathbf{r}, \mathbf{r}') = [G^-(\omega, \mathbf{r}', \mathbf{r})]^*,$$

where * means complex conjugation.

There is an important and general relation between the local density of states and the diagonal component of the Green's function which can easily be shown as follows. From Eqs. (P2.10), (P2.12), and (P2.14a), we have

$$\begin{aligned}
G^+(\omega, \mathbf{r} = \mathbf{r}') &= \frac{1}{L^d} \sum_n \frac{|\phi_n(\mathbf{r})|^2}{\kappa_0^2(\omega) - d_n + i\eta} \\
&= \frac{1}{L^d} \sum_n \int d\kappa_1^2 \frac{\delta[\kappa_1^2 - d_n]|\phi_n(\mathbf{r})|^2}{\kappa_0^2(\omega) - \kappa_1^2 + i\eta} \\
&= \int d\omega' \frac{\rho(\omega', \mathbf{r})}{\kappa_0^2(\omega) - \kappa_0^2(\omega') + i\eta},
\end{aligned} \qquad \text{(P2.16)}$$

where we have set $\kappa_1^2 = \kappa_0^2(\omega')$. The real part of G^+ may be obtained by taking the principal part of the integral. Equation (P2.16) is also known as the Kramers–Kronig relation. Its physical basis lies in the causal nature of the retarded Green's function; i.e., response always follows the excitation. Translated into the frequency domain, causality means G^+ is always analytic in the upper half of the complex ω plane. This is clear from Eq.

(P2.12), where the poles for $G^+(\omega, d_n)$ occur at the lower half of the complex ω plane. The relation (P2.16) is valid for both the continuum and the lattice Green's functions.

2.4 Derive an expression for the density of states on a 2D square lattice.
From Eqs. (2.37), (2.57), and (2.60), one can write for the 2D square lattice

$$G_0^\pm(\omega, \mathbf{l} = \mathbf{l}') = \frac{L^2}{N} \int\int_{-\pi/a}^{\pi/a} \frac{dk_x \, dk_y}{(2\pi)^2}$$
$$\times \frac{1}{\kappa_0^2 - (2\beta/a^2)[(1 - \cos k_x a) + (1 - \cos k_y a)] \pm i\eta} \frac{1}{a^2}$$
$$= \int_{-\pi}^{\pi} \frac{d\theta_x}{2\pi} \int_{-\pi}^{\pi} \frac{d\theta_y}{2\pi} \frac{1}{(\kappa_0^2 a^2 - 4\beta) + 2\beta(\cos\theta_x + \cos\theta_y)}.$$
(P2.17)

In the last expression $\theta_{x,y} = k_{x,y} a$, and the $\pm i\eta$ is absorbed into κ_0^2. By a change of variables

$$\theta_1 = \frac{\theta_x + \theta_y}{2}, \quad \theta_2 = \frac{\theta_x - \theta_y}{2},$$

one can write

$$G_0^\pm(\omega, \mathbf{l} = \mathbf{l}') = \frac{1}{\pi^2} \int_0^\pi d\theta_1 \int_0^{\pi - \theta_1} d\theta_2 \left| \frac{\partial(\theta_x, \theta_y)}{\partial(\theta_1, \theta_2)} \right|$$
$$\times \frac{1}{(\kappa_0^2 a^2 - 4\beta) + 4\beta \cos\theta_1 \cos\theta_2}. \quad \text{(P2.18)}$$

In obtaining Eq. (P2.18) the identity

$$\cos\theta_x + \cos\theta_y = \cos(\theta_1 + \theta_2) + \cos(\theta_1 - \theta_2) = 2\cos\theta_1 \cos\theta_2$$

has been utilized. The notation $|\partial(\theta_x, \theta_y)/\partial(\theta_1, \theta_2)|$ denotes the Jacobian of transformation, and in this case it has the value 2. Since the integrand is an even function of θ_1 and θ_2, it is written as an integral over only the first quadrant, with a factor 4 multiplied. The limits of integration come from the fact that the domain of integration is modified under the transformation as shown in Figure P2.1. The shaded region denoted 1 is the region of integration in Eq. (P2.18). It is easy to show that the integral over the region denoted 2 is identical to the integral over region 1 under a transformation of variables $\theta_1' = \pi - \theta_1$ and $\theta_2' = \pi - \theta_2$. Therefore, Eq.

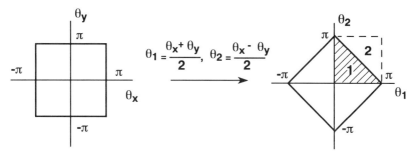

Figure P2.1 The integration domain for the variables θ_x, θ_y is a square $[-\pi, \pi] \times [-\pi, \pi]$. The domain after the transformation into θ_1 and θ_2 is a smaller tilted square. The 1 and 2 denote the two triangular regions in the first quadrant referred to in the text.

(P2.18) can be rewritten as

$$G_0^\pm(\omega, \mathbf{l} = \mathbf{l}') = \frac{1}{\pi} \int_0^\pi d\theta_1 \left[\frac{1}{2\pi} \int_{-\pi}^\pi d\theta_2 \frac{1}{(\kappa_0^2 a^2 - 4\beta) + 4\beta \cos\theta_1 \cos\theta_2} \right].$$

(P2.19)

The integral inside the bracket may be transformed into a contour integral just as shown in the 1D case, Eqs. (2.61) and (2.62). The result is, for $|\kappa_0^2 a^2 - 4\beta| > 4\beta \cos\theta_1$,

$$[\ ] = \frac{\operatorname{sgn}(\kappa_0^2 a^2 - 4\beta)}{\left[(\kappa_0^2 a^2 - 4\beta)^2 - (4\beta)^2 \cos^2\theta_1\right]^{1/2}}$$

$$= \frac{1}{(\kappa_0^2 a^2 - 4\beta)(1 - h^2 \cos^2\theta_1)^{1/2}},$$

(P2.20)

where

$$h = \frac{4\beta}{\kappa_0^2 a^2 - 4\beta}.$$

(P2.21)

On the other hand, if $|\kappa_0^2 a^2 - 4\beta| < 4\beta \cos\theta_1$, then

$$[\ ] = \frac{\mp i}{|\kappa_0^2 a^2 - 4\beta|(h^2 \cos^2\theta_1 - 1)^{1/2}}.$$

(P2.22)

The real part of Eq. (P2.19) is therefore

$$G_0^{\pm}(\omega, \mathbf{l} = \mathbf{l'}) = \frac{2}{\pi(\kappa_0^2 a^2 - 4\beta)} \int_{u(h)}^{\pi/2} \frac{d\theta}{(1 - h^2 \cos^2 \theta)^{1/2}}, \quad (P2.23)$$

with the condition that $|h \cos \theta| < 1$, where $u(h) = 0$ if $|h| \leq 1$, and $u(h) = \cos^{-1}(1/|h|)$ if $|h| > 1$. Alternatively, the imaginary part of the Green's function is given by

$$G_0^{\pm}(\omega, \mathbf{l} = \mathbf{l'}) = \frac{\mp 2i}{\pi|\kappa_0^2 a^2 - 4\beta|} \int_0^{u(h)} \frac{d\theta}{(h^2 \cos^2 \theta - 1)^{1/2}}, \quad (P2.24)$$

with the condition that $|h| > 1$ and $|h \cos \theta| > 1$.

To evaluate (P2.23) and (P2.24) explicitly, the easiest case is $|h| \leq 1$, which yields

$$G_0(\omega, \mathbf{l} = \mathbf{l'}) = \frac{2}{\pi(\kappa_0^2 a^2 - 4\beta)} \mathbf{K}\left(\frac{4\beta}{\kappa_0^2 a^2 - 4\beta}\right) \quad (P2.25)$$

for $|\kappa_0^2 a^2 - 4\beta| > 4\beta$, i.e., outside the energy band. The symbol $\mathbf{K}$ stands for the complete elliptic integral of the first kind, defined by Eq. (P2.34). In this case the Green's function is completely real, and the density of states is therefore zero. If $|h| \geq 1$, i.e., inside the energy band, the real part of $G_0(\omega, \mathbf{l} = \mathbf{l'})$ can be written from Eq. (P2.23) as

$$\mathrm{Re}[G_0(\omega, \mathbf{l} = \mathbf{l'})] = \frac{2}{\pi(\kappa_0^2 a^2 - 4\beta)} \int_{\cos^{-1}(1/|h|)}^{\pi/2} \frac{d\theta}{(1 - h^2 \cos^2 \theta)^{1/2}}$$

$$= \frac{\mathrm{sgn}(\kappa_0^2 a^2 - 4\beta)}{2\pi\beta} \int_0^{\pi/2} \frac{d\phi}{(1 - h^{-2} \cos^2 \phi)^{1/2}}$$

$$= \frac{\mathrm{sgn}(\kappa_0^2 a^2 - 4\beta)}{2\pi\beta} \mathbf{K}\left(\frac{\kappa_0^2 a^2 - 4\beta}{4\beta}\right), \quad (P2.26)$$

for $|\kappa_0^2 a^2 - 4\beta| \leq 4\beta$. The change of variable that accomplished the task in Eq. (P2.26) is $\cos \phi = |h| \cos \theta$. For the imaginary part inside the energy

Problems and Solutions

band, one starts with Eq. (P2.24), from which one obtains

$$
\begin{aligned}
\text{Im}\left[G_0^\pm(\omega, \mathbf{l} = \mathbf{l}')\right] &= \mp \frac{1}{2\pi\beta} \int_0^{\cos^{-1}(1/|h|)} \frac{d\theta}{(\cos^2\theta - h^{-2})^{1/2}} \\
&= \mp \frac{1}{2\pi\beta} \int_0^{\pi/2} \frac{d\phi}{[1 - (1 - h^{-2})\sin^2\phi]^{1/2}} \\
&= \mp \frac{1}{2\pi\beta} \mathbf{K}\left[\sqrt{1 - \left(\frac{\kappa_0^2 a^2 - 4\beta}{4\beta}\right)^2}\right] \quad (P2.27)
\end{aligned}
$$

for $|\kappa_0^2 a^2 - 4\beta| \leq 4\beta$. The transformation that does the trick in Eq. (P2.27) is $\sin\phi = (1 - h^{-2})^{1/2} \sin\theta$. By a further transformation $\phi' = \pi/2 - \phi$, one obtains the last line of Eq. (P2.27). The density of states is given by

$$
\begin{aligned}
\rho_0(\omega) &= -\frac{1}{\pi} \frac{d\kappa_0^2(\omega)}{d\omega} \text{Im}[G^+(\omega, \mathbf{l} = \mathbf{l}')] \\
&= \frac{1}{2\pi^2\beta} \frac{d\kappa_0^2(\omega)}{d\omega} \mathbf{K}\left[\sqrt{1 - \left(\frac{\kappa_0^2 a^2 - 4\beta}{4\beta}\right)^2}\right]. \quad (P2.28)
\end{aligned}
$$

for $|\kappa_0^2 a^2 - 4\beta| < 4\beta$, and zero otherwise.

2.5 Derive an expression for the diagonal elements of the Green's function on a 3D simple cubic lattice.

The diagonal elements of Green's function are defined as those with $\mathbf{l} = \mathbf{l}'$, or

$$
\begin{aligned}
G_0^\pm(\omega, \mathbf{l} = \mathbf{l}') &= \iiint_{-\pi/a}^{\pi/a} \frac{dk_x dk_y dk_y}{(2\pi)^3} \\
&\quad \times \frac{1}{\kappa_0^2 - (2\beta/a^2)(3 - \cos k_x a - \cos k_y a - \cos k_z a) \pm i\eta} \\
&= \frac{1}{2\pi a} \int_{-\pi}^{\pi} d\theta_z \left[\int_{-\pi}^{\pi} \frac{d\theta_y}{2\pi} \int_{-\pi}^{\pi} \frac{d\theta_x}{2\pi} \right. \\
&\quad \times \left. \frac{1}{(\kappa_0^2 a^2 - 6\beta + 2\beta\cos\theta_z) + 2\beta(\cos\theta_x + \cos\theta_y)}\right], \\
&\quad (P2.29)
\end{aligned}
$$

where $\pm i\eta$ is absorbed into κ_0^2. From the previous problem, the double integral inside the square brackets can be performed just as in the 2D square lattice case, with the result that

$$G_0(\omega, \mathbf{l} = \mathbf{l}') = \frac{1}{2\pi^2 a \beta} \int_0^\pi d\theta \, \mu(\theta) \mathbf{K}[\mu(\theta)], \qquad \text{(P2.30)}$$

with

$$\mu(\theta) = \frac{4\beta}{\kappa_0^2 a^2 - 6\beta + 2\beta \cos \theta} \qquad \text{(P2.31)}$$

when the integrand is real. When the integrand is imaginary, one has to remember the $\pm i\eta$ in κ_0^2, and moreover, the Green's function is now given by

$$G_0^\pm(\omega, \mathbf{l} = \mathbf{l}') = \frac{1}{2\pi^2 a \beta} \int_0^\pi d\theta \, \mu^\pm(\theta) \mathbf{K}[\mu^\pm(\theta)], \qquad \text{(P2.32)}$$

where $\mu^\pm(\theta)$ is given by

$$\mu^\pm(\theta) = \frac{4\beta}{|\kappa_0^2 a^2 - 6\beta + 2\beta \cos \theta| \pm i\eta}. \qquad \text{(P2.33)}$$

In both cases $\mathbf{K}(\mu)$ stands for

$$\mathbf{K}(\mu) = \int_0^{\pi/2} \frac{d\phi}{(1 - \mu^2 \cos^2 \phi)^{1/2}}. \qquad \text{(P2.34)}$$

Reference

Economou, E. N. (1979). "Green's Function in Quantum Physics," Springer-Verlag, Berlin.

3

Wave Scattering and the Effective Medium

3.1 An Overview of the Approach

This chapter begins the consideration of waves in random media. Before delving into more detailed considerations, however, a discussion of the overall approach would be helpful in pointing out the rationale of the subsequent development.

In a random medium, a complete solution of the wave equation is represented by the knowledge of $G(\omega, \mathbf{r}, \mathbf{r}')$ for all values of ω, $\mathbf{r}$, $\mathbf{r}'$ in the presence of disorder $\sigma(\mathbf{r})$. Complications arise because the accurate solution of the wave equation is not generally possible in the presence of $\sigma(\mathbf{r})$. Moreover, there is also the problem of how to extract the desired information from $G(\omega, \mathbf{r}, \mathbf{r}')$ even if it were known. Here a comparison with the classical random walk would be illuminating. Suppose the position of a random walker is known at every instant of time, given by $\mathbf{r}(t)$. This is analogous to the knowledge of the Green's function for the wave equation and represents the complete solution of the random walk problem in principle. But the diffusive behavior, which results from the statistical character of $\mathbf{r}(t)$, is not directly evident from $\mathbf{r}(t)$. What is necessary to bring out the diffusive behavior is the evaluation of the moments of $\mathbf{r}(t)$—i.e. $\langle \mathbf{r} \rangle_c$, $\langle \mathbf{r}^2 \rangle_c$, etc.—where the angular brackets with the subscript c denote averaging over different configurations [of the random perturbations $\sigma(\mathbf{r})$ in the wave scattering case, and different random walk trajectories in the random walker case]. Similarly, for the wave problem the first objective of our approach is the approximate evaluation of $\langle G \rangle_c$. Since $\langle G \rangle_c$ contains much less detailed information than G, its calculation is simpler than that of G. For example, whereas G depends on the source position $\mathbf{r}'$, $\langle G \rangle_c$ can depend only on the source–detector separation

$\mathbf{r} - \mathbf{r}'$ because after configurational averaging, $\langle G \rangle_c$ can no longer depend on any particular $\sigma(\mathbf{r})$ [since an averaged quantity cannot depend on the averaged variable, and in this case independence from $\sigma(\mathbf{r})$ means all spatial positions are equivalent], and only $\mathbf{r} - \mathbf{r}'$ plus the statistical properties of the $\sigma(\mathbf{r})$ ensemble are relevant. The knowledge of $\langle G \rangle_c$ can tell us the following. First, it gives the wave propagation characteristics in an averaged sense. This is what is meant by the "effective medium" as seen by a wave. Second, it gives the spatial scale beyond which this effective medium description can no longer be valid. Mathematically, the second point is manifest in the exponential decay of $\langle G \rangle_c$ as a function of $|\mathbf{r} - \mathbf{r}'|$. However, it should be emphasized that the decay here does not indicate wave localization; just as in random walk $\langle \mathbf{r} \rangle_c = 0$ does not imply that the random walker is localized at the origin. Rather, the exponential decay of $\langle G \rangle_c$ means that the wave *coherence*, in the sense of a unique wave propagation direction and phase relation, is lost. This decay length is defined as the *mean free path*. For the random walk counterpart, $\langle \mathbf{r} \rangle_c = 0$ essentially means the absence of ballistic motion.

Whereas $\langle G \rangle_c$ gives us the character of the coherent part of the wave propagation in a random medium, the transport dynamics beyond the scale of mean free path are contained in the second moment, $\langle GG^* \rangle_c$, just as in the random walk case the diffusion dynamics emerge from the time dependence of $\langle r^2 \rangle_c$. Therefore the next objective in our approach is the evaluation of $\langle GG^* \rangle_c$. The outcome of this calculation will show that at the long-time, long-propagation-distance limit, intensity transport is indeed diffusive in character, and the attendant diffusion constant may be explicitly calculated. It is in the context of diffusive transport that the coherent backscattering effect may be demonstrated as a correction to the diffusion constant. Our exposition will give a global view of the possible consequences of coherent backscattering through a description of the scaling theory of localization, which is then substantiated by showing consistency between the assumption of the scaling theory and the calculations involving $\langle GG^* \rangle_c$.

Since the intensity transport behavior is given by $\langle GG^* \rangle_c$, fluctuations in the intensity transport, which are observable in mesoscopic samples, have to be calculated from $\langle GG^*GG^* \rangle_c$. The fourth moment represents the intensity–intensity correlation and contains information concerning the long-range memory and phase interference effects that may be retrieved from intensity (or current) fluctuations, e.g., speckle patterns formed by light after passing through a random medium. Together, these moments of G are indicative of the special statistical character of wave transport in random media. Their evaluation will be the main technical task for this and subsequent chapters.

Wave transport characteristics obtained from the moments of G necessarily differ from those in a single configuration. Nevertheless, these moments can give a general sense of the single-configuration characteristics because in most physical situations the *ergodic hypothesis* is valid, which means the configurationally averaged behavior may be equated to the infinite-time average of the single-configuration behavior. However, the detailed spatial and temporal variations of the wave field in a single configuration can never be retrieved from the moments of G. For such information there is no substitute for an exact solution to the problem.

3.2 Wave Scattering Formalism

A useful property of the Green's function is that it can solve inhomogeneous differential equations of the form

$$(\nabla^2 + \kappa_0^2)\phi(\omega, \mathbf{r}) = f(\mathbf{r}), \qquad (3.1)$$

where $f(\mathbf{r})$ is an inhomogeneous source term with arbitrary spatial dependence. For example, if instead of $\delta(t)\delta(\mathbf{r} - \mathbf{r}')$ the excitation source is of the form $\delta(t)f(\mathbf{r})$, then Eq. (3.1) would result. We would like to show that

$$\phi(\omega, \mathbf{r}) = \phi_0(\omega, \mathbf{r}) + \int G_0(\omega, \mathbf{r} - \mathbf{r}')f(\mathbf{r}')\,d\mathbf{r}' \qquad (3.2)$$

is the solution to Eq. (3.1). Here ϕ_0 represents the solution to the homogeneous equation [i.e., right-hand side of Eq. (3.1) equals zero]. By substituting Eq. (3.2) into Eq. (3.1), it is seen that since the operation of $(\nabla^2 + \kappa_0^2)$ is on the $\mathbf{r}$ variable of $G(\omega, \mathbf{r}, \mathbf{r}')$, it can be performed inside the integral. This result gives

$$\int \delta(\mathbf{r} - \mathbf{r}')f(\mathbf{r}')\,d\mathbf{r}' = f(\mathbf{r}), \qquad (3.3)$$

thus recovering the right-hand side of Eq. (3.1).

Let us now write the Green's function equation in the presence of $\sigma(\mathbf{r})$ as

$$(\nabla^2 + \kappa_0^2)G(\omega, \mathbf{r}, \mathbf{r}') = \delta(\mathbf{r} - \mathbf{r}') + \sigma(\mathbf{r})G(\omega, \mathbf{r}, \mathbf{r}'). \qquad (3.4)$$

By comparison with Eqs. (3.1) and (3.2), one can write

$$G(\omega, \mathbf{r}, \mathbf{r}') = \int d\mathbf{r}_1 \, G_0(\omega, \mathbf{r} - \mathbf{r}_1)[\delta(\mathbf{r}_1 - \mathbf{r}') + \sigma(\mathbf{r}_1)G(\omega, \mathbf{r}_1, \mathbf{r}')]$$

$$= G_0(\omega, \mathbf{r} - \mathbf{r}') + \int d\mathbf{r}_1 \, G_0(\omega, \mathbf{r} - \mathbf{r}_1)\sigma(\mathbf{r}_1)G(\omega, \mathbf{r}_1, \mathbf{r}'),$$

(3.5)

where the homogeneous term ϕ_0 is eliminated by the boundary condition that G must vanish as $|\mathbf{r} - \mathbf{r}'| \to \infty$. At this point it would be convenient to introduce the operator notation to rewrite Eq. (3.5). That is, if the space is regarded as finely discretized so that every location $\mathbf{r}$ is associated with an index number and a small volume $(\delta r)^d$, where δr is the linear size of the discretized unit, e.g., the atomic unit cell, then the Green's function $G(\omega, \mathbf{r}, \mathbf{r}')$ has two indices, one associated with r and the other one associated with r'; i.e., G may be regarded as a matrix. Similarly, $G_0(\omega, \mathbf{r} - \mathbf{r}_1)$ is also a matrix, and $\sigma(\mathbf{r}_1)$ is a diagonal matrix. By writing the second term on the right-hand side of Eq. (3.5) as

$$\sum_{\mathbf{r}_1} \sum_{\mathbf{r}_2} (\delta r)^{2d} G_0(\omega, \mathbf{r} - \mathbf{r}_1) \sigma(\mathbf{r}_1) \left[\delta_{\mathbf{r}_1, \mathbf{r}_2}/(\delta r)^d \right] G(\omega, \mathbf{r}_2, \mathbf{r}'),$$

where the quantity in square brackets is a discrete representation of $\delta(\mathbf{r}_1 - \mathbf{r}_2)$, one may regard the double integral as a double summation. The whole term is thus equivalent to the multiplication of three matrices:

$$\mathbf{G}_0 \mathbf{V} \mathbf{G},$$

where the elements of the **V** matrix are given by

$$(\mathbf{V})_{\mathbf{r}_1, \mathbf{r}_2} = (\delta r)^{2d} \sigma(\mathbf{r}_1) \delta_{\mathbf{r}_1, \mathbf{r}_2}/(\delta r)^d$$

$$= (\delta r)^d \sigma(\mathbf{r}_1) \delta_{\mathbf{r}_1, \mathbf{r}_2}.$$

V is called the impurity potential operator, and in our convention it has the dimension [length]$^{d-2}$. Alternatively, one can use Dirac's bra and ket notation as described in the last chapter to write the matrix multiplication as

$$\sum_{\mathbf{r}_1} \sum_{\mathbf{r}_2} \langle \mathbf{r} | \mathbf{G}_0 | \mathbf{r}_1 \rangle \langle \mathbf{r}_1 | \mathbf{V} | \mathbf{r}_2 \rangle \langle \mathbf{r}_2 | \mathbf{G} | \mathbf{r}' \rangle.$$

3.2 Wave Scattering Formalism

Here $\langle \mathbf{r}|\mathbf{G}_0|\mathbf{r}_1\rangle = G_0(\omega, \mathbf{r} - \mathbf{r}_1)$, $\langle \mathbf{r}_2|\mathbf{G}|\mathbf{r}'\rangle = G(\omega, \mathbf{r}_2, \mathbf{r}')$, and

$$\sum_{\mathbf{r}} |\mathbf{r}\rangle\langle\mathbf{r}| = \mathbf{I},$$

the identity matrix. This should be clear since $|\mathbf{r}\rangle\langle\mathbf{r}|$ means a matrix with only one nonzero element (with value 1) at the diagonal position corresponding to row and column associated with $\mathbf{r}$. The summation of such matrices fills the diagonal elements of the resulting matrix with 1, which is the identity matrix. In this notation $\langle\mathbf{r}|\mathbf{r}'\rangle = \delta_{\mathbf{r},\mathbf{r}'}$, and $\delta(\mathbf{r},\mathbf{r}') = \delta_{\mathbf{r},\mathbf{r}'}/(\delta r)^d$. Either way, Eq. (3.5) may be expressed succinctly as

$$\mathbf{G} = \mathbf{G}_0 + \mathbf{G}_0 \mathbf{V} \mathbf{G}, \tag{3.6}$$

where the symbols stand for operators (e.g., matrices), and their ordering is important and cannot be altered at will. In the operator notation, Eq. (3.6) is valid regardless of whether $\mathbf{G}$, $\mathbf{G}_0$, and $\mathbf{V}$ are in the $\mathbf{r}$ domain representation or in the $\mathbf{k}$ domain representation. The manipulation of the formalism thus simplifies.

An alternative way to express Eq. (3.6) is obtained by iterating on $\mathbf{G}$:

$$\mathbf{G} = \mathbf{G}_0 + \mathbf{G}_0 \mathbf{V}[\mathbf{G}_0 + \mathbf{G}_0 \mathbf{V}(\mathbf{G}_0 + \mathbf{G}_0 \mathbf{V} \cdots]$$
$$= \mathbf{G}_0 + \mathbf{G}_0 \mathbf{V} \mathbf{G}_0 + \mathbf{G}_0 \mathbf{V} \mathbf{G}_0 \mathbf{V} \mathbf{G}_0 + \cdots$$

or

$$\mathbf{G} = \mathbf{G}_0 + \mathbf{G}_0 \mathbf{T} \mathbf{G}_0, \tag{3.7}$$

where

$$\mathbf{T} = \mathbf{V} + \mathbf{V}\mathbf{G}_0\mathbf{V} + \mathbf{V}\mathbf{G}_0\mathbf{V}\mathbf{G}_0\mathbf{V} + \cdots$$
$$= \mathbf{V}(\mathbf{I} - \mathbf{G}_0\mathbf{V})^{-1} = (\mathbf{I} - \mathbf{V}\mathbf{G}_0)^{-1}\mathbf{V} \tag{3.8}$$

is called the T matrix, or the scattering matrix. In Eq. (3.8) it is seen that if each $\mathbf{V}$ represents one scattering, then $\mathbf{T}$ includes all the multiple scatterings. The formal summation of the operator series may be expressed as the inverse of $(\mathbf{I} - \mathbf{G}_0\mathbf{V})$ (just as in the summation of the power series). In the form of the inverses, Eq. (3.6) can be put in another popular form which may be derived as follows. From Eq. (3.6) we have

$$(\mathbf{I} - \mathbf{G}_0\mathbf{V})\mathbf{G} = \mathbf{G}_0. \tag{3.9}$$

Taking the inverse of both sides and then right-multiplying by $(\mathbf{I} - \mathbf{G}_0\mathbf{V})$ give

$$\mathbf{G}^{-1} = \mathbf{G}_0^{-1}(\mathbf{I} - \mathbf{G}_0\mathbf{V})$$

or
$$\mathbf{G}^{-1} = \mathbf{G}_0^{-1} - \mathbf{V}. \tag{3.10}$$

Knowledge of **T** is sufficient to completely solve the general wave equation

$$(\nabla^2 + \kappa_0^2 - \sigma(\mathbf{r}))\phi(\omega, \mathbf{r}) = 0$$

in terms of the uniform-medium solutions $\phi_0(\omega, \mathbf{r})$ and $G_0(\omega, \mathbf{r} - \mathbf{r}')$. This can be seen by rewriting the equation as

$$(\nabla^2 + \kappa_0^2)\phi(\omega, \mathbf{r}) = \sigma(\mathbf{r})\phi(\omega, \mathbf{r}).$$

Then from Eq. (3.2) the solution may be written in terms of the bra and ket notation as

$$|\phi\rangle = |\phi_0\rangle + \mathbf{G}_0\mathbf{V}|\phi\rangle = |\phi_0\rangle + \mathbf{G}_0\mathbf{V}|\phi_0\rangle + \mathbf{G}_0\mathbf{V}\mathbf{G}_0\mathbf{V}|\phi_0\rangle + \cdots$$
$$= |\phi_0\rangle + \mathbf{G}_0^+\mathbf{T}^+|\phi_0\rangle. \tag{3.11}$$

The + superscripts on $\mathbf{G}_0$ and **T** in the last line of Eq. (3.11) are meant to select the physical solution branch where the scattering from an inhomogeneity is represented by an outgoing wave (from the inhomogeneity) rather than by an incoming wave, which would be selected by $\mathbf{G}_0^-\mathbf{T}^-$.

From Eq. (3.7), the configurationally averaged Green's function is given by

$$\langle \mathbf{G} \rangle_c = \mathbf{G}_0 + \mathbf{G}_0 \langle \mathbf{T} \rangle_c \mathbf{G}_0, \tag{3.12}$$

where $\mathbf{G}_0$ is independent of $\sigma(\mathbf{r})$ and is therefore not affected by the average. From the **T** representation of Eq. (3.8), we have

$$\langle \mathbf{T} \rangle_c = \langle \mathbf{V}(\mathbf{I} - \mathbf{G}_0\mathbf{V})^{-1} \rangle_c. \tag{3.13}$$

It is noted that $\langle \mathbf{T} \rangle_c$ contains all the higher-order correlations of **V**, such as $\langle \mathbf{V}\mathbf{G}_0\mathbf{V} \rangle_c$ and $\langle \mathbf{V}\mathbf{G}_0\mathbf{V}\mathbf{G}_0\mathbf{V} \rangle_c$. Since in the real space representation $\langle \mathbf{G} \rangle_c$ depends only on the spatial separation between the source and the receiver $\mathbf{r} - \mathbf{r}'$, its Fourier transform is a function of one **k** only, just as for $\mathbf{G}_0$. The general validity of this statement is shown in Problem 3.1 at the end of this chapter. In Problem 3.2 it is shown that a convolution integral in real space, as symbolized by $\mathbf{G}_0 \langle \mathbf{T} \rangle_c \mathbf{G}_0$, means simple multiplication of the transformed quantities in **k** space. Therefore from Eq. (3.12) it is thus clear that in the **k** representation, $\langle \mathbf{T} \rangle_c$ is also a function of one **k** only, which implies a dependence on only the separation $\mathbf{r} - \mathbf{r}'$ in real space.

3.3 Single Scatterer—The Lattice Case

One can define a Σ operator as

$$\langle \mathbf{G}\rangle_c^{-1} = \mathbf{G}_0^{-1} - \Sigma \frac{L^d}{N}. \tag{3.14}$$

In the **k** representation, since both $\langle \mathbf{G}\rangle_c$ and $\mathbf{G}_0^{-1}$ are functions of one **k** only, Σ is also a function of one **k** only, which means a dependence on $\mathbf{r} - \mathbf{r}'$ in real space, just as for $\langle \mathbf{T}\rangle_c$. From Eqs. (3.12) and (3.14), Σ is related to $\langle \mathbf{T}\rangle_c$ by

$$\Sigma = \frac{N}{L^d}\langle \mathbf{T}\rangle_c(\mathbf{I} + \langle \mathbf{T}\rangle_c\mathbf{G}_0)^{-1}. \tag{3.15}$$

Σ is denoted the self-energy operator, and Eq. (3.14) is known as the Dyson equation. In the **k** representation, Eqs. (3.14) and (3.15) are simple algebraic equations because all the relevant operators are diagonal matrices. Comparison of Eq. (3.14) with Eq. (3.10) shows that the self-energy Σ is a very different object from the operator **V** of the exact Green's function in a fixed configuration. In real space, whereas the latter is just the perturbation $\sigma(\mathbf{r})$, Σ represents a nonlocal operator as can be seen from the equation satisfied by $\langle \mathbf{G}\rangle_c$. Since in the **k**-domain we have

$$\left[\kappa_0^2(\omega) - k^2 - \Sigma(\omega, \mathbf{k})\right]\langle G\rangle_c(\omega, \mathbf{k}) = \frac{N}{L^d},$$

it follows that in real space, $-k^2$ means ∇^2, and multiplication in the **k**-domain means convolution in real space (see solution to Problem 3.2), so that

$$\left[\kappa_0^2(\omega) + \nabla^2\right]\langle G\rangle_c(\omega, \mathbf{r} - \mathbf{r}') - L^{-d}\int \Sigma(\omega, \mathbf{r} - \mathbf{r}_1)\langle G\rangle_c(\omega, \mathbf{r}_1 - \mathbf{r}')\, d\mathbf{r}_1$$

$$= \delta(\mathbf{r} - \mathbf{r}').$$

This equation reduces to the ordinary wave equation only if $\Sigma(\omega, \mathbf{r} - \mathbf{r}_1) = \Sigma'(\omega)L^d\,\delta(\mathbf{r} - \mathbf{r}_1)$. This turns out to be possible when the effective medium description is valid, as will be seen later.

3.3 Single Scatterer—The Lattice Case

Consider a single scatterer embedded in a uniform lattice. The Green's function of the system is given by

$$\mathbf{g} = \mathbf{G}_0 + \mathbf{G}_0 \mathbf{t} \mathbf{G}_0, \tag{3.16}$$

where the lowercase **g** and **t** are used to denote the Green's function and the scattering matrix in the presence of a single scatterer. In this case, an explicit expression for **t** can be obtained.

In the lattice model, a single scatterer may be expressed by a deviation $\Delta\epsilon$ of the site energy at $\mathbf{l}_0$. In the bra and ket notation, the operator **V** is given by

$$\mathbf{V} = \mathbf{v} = \sigma_0 a^d |\mathbf{l}_0\rangle\langle\mathbf{l}_0|, \quad (3.17)$$

$$\sigma_0 = \frac{2m\,\Delta\epsilon}{\hbar^2}, \quad (3.18)$$

where **v** denotes the perturbation due to a single scatterer and a is the lattice constant. The operator **v** may be thought of as a matrix with only one nonzero element, which is at the diagonal position of column $\mathbf{l}_0$ and row $\mathbf{l}_0$. From Eq. (3.8), one gets

$$\begin{aligned}\mathbf{t} &= (\sigma_0 a^d)|\mathbf{l}_0\rangle\langle\mathbf{l}_0| + (\sigma_0 a^d)^2 |\mathbf{l}_0\rangle\langle\mathbf{l}_0|\mathbf{G}_0|\mathbf{l}_0\rangle\langle\mathbf{l}_0| + \cdots \\ &= (\sigma_0 a^d)|\mathbf{l}_0\rangle\langle\mathbf{l}_0|\Big\{1 + \sigma_0 a^d G_0(\omega, \mathbf{l} = \mathbf{l}') \\ &\quad + \big[\sigma_0 a^d G_0(\omega, \mathbf{l} = \mathbf{l}')\big]^2 + \cdots\Big\} \\ &= |\mathbf{l}_0\rangle\langle\mathbf{l}_0|\frac{\sigma_0 a^d}{1 - \sigma_0 a^d G_0(\omega, \mathbf{l} = \mathbf{l}')}. \end{aligned} \quad (3.19)$$

Knowledge of **t** explicitly solves the Green's function $g(\omega, \mathbf{l}, \mathbf{l}')$ for the single-scatterer case:

$$\begin{aligned}g(\omega, \mathbf{l}, \mathbf{l}') &= G_0(\omega, \mathbf{l} - \mathbf{l}') + G_0(\omega, \mathbf{l} - \mathbf{l}_0)G_0(\omega, \mathbf{l}_0 - \mathbf{l}') \\ &\quad \times \frac{a^d \sigma_0}{1 - \sigma_0 a^d G_0(\omega, \mathbf{l} = \mathbf{l}')}. \end{aligned} \quad (3.20)$$

From Eq. (3.11), the wave function for the one-scatterer case may also be similarly expressed as

$$\begin{aligned}\phi(\omega, \mathbf{l}) &= \exp(i\mathbf{k}_0 \cdot \mathbf{l}) \\ &\quad + \left[\frac{a^2 \sigma_0 \exp(i\mathbf{k}_0 \cdot \mathbf{l}_0)}{1 - \sigma_0 a^d G_0^+(\omega, \mathbf{l} = \mathbf{l}')}\right] a^{d-2} G_0^+(\omega, \mathbf{l} - \mathbf{l}_0),\end{aligned} \quad (3.21)$$

where the notation $\mathbf{k}_0$ is used to denote a wave vector which satisfies the relation $\kappa_0^2(\omega) = \beta e(\mathbf{k}_0)$. Equation (3.21) has the following physical inter-

3.3 Single Scatterer—The Lattice Case

pretation. The difference between the uniform medium and the single-scatterer solution is represented by the second term on the right-hand side of Eq. (3.21). That term shows the single scatterer acting as a point source whose strength is given by the expression inside the square brackets. That strength is essentially the probability amplitude whereby the wave in the uniform medium is diverted into a wave emanating isotropically from the scatterer, i.e., the scattering probability. It is interesting that the scattering probability amplitude has a denominator which may vanish when

$$\frac{1}{a^d \sigma_0} = G_0^+(\omega, \mathbf{l} = \mathbf{l}').$$

This condition can never be satisfied for ω values inside the energy band because $G_0^+(\omega, \mathbf{l} = \mathbf{l}')$ has a nonzero imaginary part which gives the density of states inside the band. Outside the band, however, the condition can indeed be satisfied.

The denominator of the scattering cross section arises physically from the scattered wave that is multiply scattered back by the lattice. Since the same denominator is in the expression for g, Eq. (3.20), its zero(s) thus represents the existence of resonance mode(s) of the system. Such state(s) must have an amplitude that is localized around $\mathbf{l}_0$, because if one looks at Eq. (3.21), the spatial dependence of the new state is determined by $G_0^+(\omega, \mathbf{l} - \mathbf{l}_0)$, and since for ω outside the energy band G_0^+ must decay exponentially, as discussed in the last chapter, the new resonance mode created by the single "impurity" is thus a state bound to the neighborhood of the impurity site. Whereas in 1D and 2D the creation of such bound state(s) always accompanies the presence of an impurity, in 3D the strength of the impurity perturbation must exceed a critical threshold, i.e.,

$$|\sigma_0 a^2| \geq \frac{1}{0.2519} = 3.97, \tag{3.22}$$

before a bound state can be generated (see Figure 3.3).

In the solution to Problem 3.3, an expression for the total scattering across section O is derived; given by

$$O = \frac{-(\sigma_0 a^d)^2 \operatorname{Im} G_0^+(\omega, \mathbf{l} = \mathbf{l}')}{\bar{k}_0 |1 - \sigma_0 a^d G_0^+(\omega, \mathbf{l} = \mathbf{l}')|^2}, \tag{3.23}$$

where the definition of $\bar{k}_0$ and the relation between O and the imaginary part of $\mathbf{t}^+$, known as the optical theorem, are given in the same problem solution. In Figures 3.1, 3.2, and 3.3, the real part of $G_0^+(\omega, \mathbf{l} = \mathbf{l}')$ is

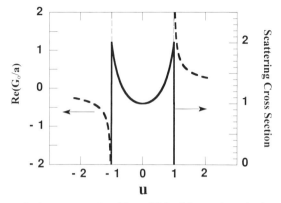

Figure 3.1 The scattering cross section (the solid line) for an impurity in a 1D lattice. The impurity strength is $\sigma_0 a^2 = 2$. The real part of the lattice Green's function $G_0 a^{-1}$ is shown by the dashed lines, with $\beta = 1$. The solid line is given by the function

$$\frac{-4\,\mathrm{Im}\,y(u)}{\bar{k}_0 a|1 - 2y(u)|^2},$$

where

$$y(u) = \frac{1}{2\sqrt{u^2 - 1}},$$

and $\bar{k}_0$ is given by Eq. (P3.18) with $d = 1$. The dashed lines are given by $\mathrm{sgn}(u)y(u)$, for $|u| > 1$. Here $u = (\hbar\omega - \epsilon_0)/(\beta\hbar^2/ma^2)$. The resonant state outside the band (at $G_0 a^{-1} = 0.5$) is too far removed from the band edge for any appreciable effect on the scattering cross section.

shown for $d = 1$, 2, and 3, respectively, together with plots of the scattering cross sections for fixed values of $\sigma_0 a^2$. An interesting observation is that although the resonant scattering cannot occur inside the band, the scattering cross section can be affected if the resonant impurity state is close to the band edge. This is clearly seen in Figures 3.2 and 3.3, where a resonant impurity state near the upper band edge enhances the scattering cross section just inside the band edge.

3.4 Single Scatterer—The Continuum Case

In the classical scalar wave case, the single scatterer is taken to be a sphere of radius R (a circle in 2D, a line in 1D) with dielectric constant

3.4 Single Scatterer—The Continuum Case

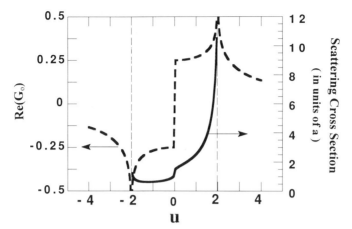

Figure 3.2 The scattering cross section (in units of a) for an impurity in a 2D square lattice (shown by the solid line). The impurity strength is $\sigma_0 a^2 = 3$. The real part of the lattice Green's function is shown by the dashed line, with $\beta = 1$. The solid line is given by the function

$$\frac{-9 \operatorname{Im} y(u)}{\bar{k}_0 a |1 - 3y(u)|^2},$$

where $\bar{k}_0$ is given by Eq. (P3.18) with $d = 2$ and

$$y(u) = \frac{1}{2\pi} \left\{ \operatorname{sgn}(u) \mathsf{K}\left(\frac{u}{2}\right) - i \mathsf{K}\left[\sqrt{1 - \left(\frac{u}{2}\right)^2}\right] \right\}.$$

The dashed line is given by the function

$$\frac{1}{\pi u} \mathsf{K}\left(\frac{2}{u}\right)$$

for $|u| > 2$ and

$$\frac{\operatorname{sgn}(u)}{2\pi} \mathsf{K}\left(\frac{u}{2}\right)$$

for $|u| < 2$. Here $u = (\hbar\omega - \epsilon_0)/(\beta\hbar^2/ma^2)$, and $\mathsf{K}(u)$ denotes the complete elliptic integral of the first kind. Since the resonant impurity state (at $\operatorname{Re} G_0 = 1/3$) is close to the upper band edge, the scattering cross section is seen to be pulled up in its vicinity.

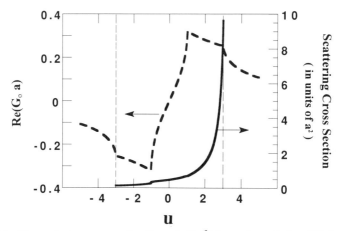

Figure 3.3 The scattering cross section (in units if a^2) for an impurity in a 3D simple cubic lattice (shown by the solid line). The impurity strength is $\sigma_0 a^2 = 5$. The real part of the lattice Green's function is shown by the dashed line, with $\beta = 1$. The solid line is given by the function

$$\frac{-25 \, \text{Im} \, y(u)}{\overline{k}_0 a |1 - 5y(u)|^2},$$

where $y = a\beta G_0^+(\omega, \mathbf{l} = \mathbf{l}')$, with G_0^+ given by Eqs. (P2.30)–(P2.33). The dashed line is given by $\text{Re}[y(u)]$. Here $u = (\hbar\omega - \epsilon_0)/(\beta\hbar^2/ma^2)$. The resonant impurity state (at Re $G_0 = 0.2$), is close to the upper band edge. As a result, the scattering cross section is enhanced in its vicinity.

$\epsilon \neq 1$. Then

$$\sigma(\mathbf{r}) = (1 - \epsilon)\frac{\omega^2}{v_0^2} \qquad r \leq R$$

$$= 0 \qquad \qquad \text{otherwise} \qquad (3.24)$$

In the bra and ket notation, the impurity operator **v** may be expressed as

$$\mathbf{v} = \sum_{\mathbf{r}} (\Delta r)^d \sigma(\mathbf{r}) |\mathbf{r}\rangle\langle\mathbf{r}|. \qquad (3.25)$$

In contrast to the lattice case, **v** is now a diagonal matrix with nonzero elements for all $|\mathbf{r}| \leq R$. The **t** operator is thus

$$\mathbf{t} = \sum_{\mathbf{r}} (\Delta r)^d \sigma(\mathbf{r}) |\mathbf{r}\rangle\langle\mathbf{r}|$$

$$+ \sum_{\mathbf{r}} \sum_{\mathbf{r}_1} (\Delta r)^{2d} \sigma(\mathbf{r}) \sigma(\mathbf{r}_1) |\mathbf{r}\rangle\langle\mathbf{r}|\mathbf{G}_0|\mathbf{r}_1\rangle\langle\mathbf{r}_1| + \cdots . \qquad (3.26)$$

3.4 Single Scatterer—The Continuum Case

Except for the first term, the rest of the series consists of full matrices, since each $|\mathbf{r}\rangle\langle\mathbf{r}_1|$ represents a nondiagonal entry at column $\mathbf{r}_1$ and row $\mathbf{r}$. The explicit form for $\mathbf{t}$ is thus difficult to obtain by using the series or the inversion of the matrix $(\mathbf{I} - \mathbf{G}_0\mathbf{v})$. However, it is observed that physically, the multiple scattering here is caused by the boundary of the scatterer with the uniform medium. The net effect of summing the series in Eq. (3.26) must therefore be equivalent to the solution of a boundary value problem, which is easily manageable. With this idea in mind, we write the wave function for the one-scatterer case by using Eq. (3.11):

$$\phi(\omega,\mathbf{r}) = \exp(i\boldsymbol{\kappa}_0 \cdot \mathbf{r}) + \sum_{\mathbf{r}_1}\sum_{\mathbf{r}_2} \langle\mathbf{r}|G_0^+|\mathbf{r}_1\rangle\langle\mathbf{r}_1|\mathbf{t}^+|\mathbf{r}_2\rangle\exp(i\boldsymbol{\kappa}_0 \cdot \mathbf{r}_2), \quad (3.27)$$

where $\boldsymbol{\kappa}_0$ denotes a wave vector with magnitude κ_0. In 3D, $G_0^+(\omega,\mathbf{r}-\mathbf{r}_1) = -\exp(i\kappa_0|\mathbf{r}-\mathbf{r}_1|)/4\pi|\mathbf{r}-\mathbf{r}_1|$ [Eq. (2.31)]. Since $\langle\mathbf{r}_1|\mathbf{t}^+|\mathbf{r}_2\rangle$ is nonzero only for $r_1, r_2 < R$, for $r \gg R$, $|\mathbf{r}-\mathbf{r}_1|$ may be expanded as

$$|\mathbf{r}-\mathbf{r}_1| = (r^2 + r_1^2 - 2rr_1\cos\theta)^{1/2} \cong r(1 - r_1\cos\theta/r) = r - r_1\cos\theta,$$

where θ is the angle between $\mathbf{r}_1$ and $\mathbf{r}$. That means for Eq. (3.27),

$$\phi(\omega,\mathbf{r}) \cong \exp(i\boldsymbol{\kappa}_0 \cdot \mathbf{r})$$

$$-\frac{\exp(i\kappa_0 r)}{4\pi r}\sum_{\mathbf{r}_1}\sum_{\mathbf{r}_2} \exp(-i\boldsymbol{\kappa}'_0 \cdot \mathbf{r}_1)\langle\mathbf{r}_1|\mathbf{t}^+|\mathbf{r}_2\rangle \exp(i\boldsymbol{\kappa}_0 \cdot \mathbf{r}_2)$$

$$= \exp(i\boldsymbol{\kappa}_0 \cdot \mathbf{r}) - \frac{\exp(i\kappa_0 r)}{4\pi r}\langle\boldsymbol{\kappa}'_0|\mathbf{t}^+|\boldsymbol{\kappa}_0\rangle, \qquad |\mathbf{r}| \gg R \quad (3.28)$$

where $\boldsymbol{\kappa}'_0$ is noted to have the direction of $\mathbf{r}$ and is thus different from $\boldsymbol{\kappa}_0$ in general. Here we have made the identification of

$$\langle\mathbf{r}_2|\boldsymbol{\kappa}_0\rangle = \exp(i\boldsymbol{\kappa}_0 \cdot \mathbf{r}_2), \quad (3.29)$$

and similarly for $\langle\boldsymbol{\kappa}'_0|\mathbf{r}_1\rangle = \exp(-i\boldsymbol{\kappa}'_0 \cdot \mathbf{r}_1)$.

A brief digression on the convention to be followed is helpful at this point. Under the conventions of this book, each $|\mathbf{k}\rangle$ or $\langle\mathbf{k}|$ has magnitude $\sqrt{N}$, so that $\langle\mathbf{k}'|\mathbf{k}\rangle = N\delta_{\mathbf{k},\mathbf{k}'}$ because

$$\langle\mathbf{k}'|\mathbf{k}\rangle = \sum_{\mathbf{r}}\langle\mathbf{k}'|\mathbf{r}\rangle\langle\mathbf{r}|\mathbf{k}\rangle = \frac{1}{(\delta r)^d}\int d\mathbf{r}\exp[i(\mathbf{k}-\mathbf{k}')\cdot\mathbf{r}]$$

$$= \frac{(2\pi)^d}{(\delta r)^d}\delta(\mathbf{k}-\mathbf{k}') = \delta_{\mathbf{k},\mathbf{k}'}\frac{L^d}{(\delta r)^d} = N\delta_{\mathbf{k},\mathbf{k}'},$$

where $\delta(\mathbf{k} - \mathbf{k}') = \delta_{\mathbf{k},\mathbf{k}'}/(\delta k)^d = L^d \delta_{\mathbf{k},\mathbf{k}'}/(2\pi)^d$ since $(\delta k)^d = (2\pi)^d L^{-d}$ as defined in the last chapter. Also,

$$\frac{1}{N} \sum_{\mathbf{k}} |\mathbf{k}\rangle\langle\mathbf{k}| = \mathbf{I},$$

and $G_0(\omega, \mathbf{k}) = \langle \mathbf{k}|\mathbf{G}_0|\mathbf{k}\rangle$, $\langle \mathbf{G}\rangle_c(\omega, \mathbf{k}) = \langle \mathbf{k}|\langle \mathbf{G}\rangle_c|\mathbf{k}\rangle$, and $G(\omega, \mathbf{k}, \mathbf{k}') = \langle \mathbf{k}'|\mathbf{G}|\mathbf{k}\rangle$.

In the literature, it is popular to define a scattering amplitude $f(\boldsymbol{\kappa}_0', \boldsymbol{\kappa}_0)$ as

$$\phi(\omega, \mathbf{r}) \xrightarrow[|\mathbf{r}|\to\infty]{} \exp(i\boldsymbol{\kappa}_0 \cdot \mathbf{r}) + f(\boldsymbol{\kappa}_0', \boldsymbol{\kappa}_0) \frac{\exp(i\kappa_0 r)}{r^{(d-1)/2}}. \quad (3.30)$$

Comparison with Eq, (3.28) gives, for $d = 3$,

$$f(\boldsymbol{\kappa}_0', \boldsymbol{\kappa}_0) = -\langle \boldsymbol{\kappa}_0'|\mathbf{t}^+|\boldsymbol{\kappa}_0\rangle/4\pi. \quad (3.31)$$

In 2D, $G_0^+(\omega, \mathbf{r} - \mathbf{r}_1)$ is given by $-iH_0^{(1)}(\kappa_0|\mathbf{r} - \mathbf{r}_1|)/4$. At $|\mathbf{r}| \gg R$, $H_0^{(1)}$ may be expanded to give

$$G_0^+(\omega, \mathbf{r} - \mathbf{r}_1) \cong -\frac{i}{4}\sqrt{\frac{2}{\pi\kappa_0}} \frac{\exp(i\kappa_0|\mathbf{r} - \mathbf{r}_1|)}{\sqrt{|\mathbf{r} - \mathbf{r}_1|}} \exp\left(\frac{-i\pi}{4}\right)$$

$$\cong -i\frac{1}{4}\sqrt{\frac{2}{\pi\kappa_0}} \frac{\exp[i(\kappa_0 r - \pi/4)]}{\sqrt{r}} \exp(i\boldsymbol{\kappa}_0' \cdot \mathbf{r}_1). \quad (3.32)$$

By substituting Eq. (3.32) into Eq. (3.27), one gets for the 2D one-scatterer wave function

$$\phi(\omega, \mathbf{r}) \xrightarrow[|\mathbf{r}|\to\infty]{} \exp(i\boldsymbol{\kappa}_0 \cdot \mathbf{r}) - \frac{i}{4}\sqrt{\frac{2}{\pi\kappa_0}} \langle \boldsymbol{\kappa}_0'|\mathbf{t}^+|\boldsymbol{\kappa}_0\rangle \frac{\exp[i(\kappa_0 r - \pi/4)]}{\sqrt{r}},$$

$$(3.33\text{a})$$

so that in 2D,

$$f(\boldsymbol{\kappa}_0', \boldsymbol{\kappa}_0) = -\frac{i}{4}\sqrt{\frac{2}{\pi\kappa_0}} \langle \boldsymbol{\kappa}_0'|\mathbf{t}^+|\boldsymbol{\kappa}_0\rangle \exp\left(-i\frac{\pi}{4}\right). \quad (3.33\text{b})$$

3.4 Single Scatterer—The Continuum Case

In 1D, $G_0^+(\omega, x - x_1) = -i\exp(i\kappa_0|x - x_1|)/2\kappa_0$. Expansion of $|x - x_1|$ as $|x| \to \infty$ gives

$$\phi(\omega, x) \xrightarrow[|x| \to \infty]{} \exp(i\kappa_0 x) - \frac{i}{2\kappa_0}\langle \kappa_0'|\mathbf{t}^+|\kappa_0\rangle \exp(i\kappa_0|x|), \quad (3.34a)$$

and

$$f(\kappa_0', \kappa_0) = -\frac{i}{2\kappa_0}\langle \kappa_0'|\mathbf{t}^+|\kappa_0\rangle. \quad (3.34b)$$

It should be noted that $\mathbf{t}^+$ has the dimension of $[\text{length}]^{d-2}$. Therefore comparison with Eq. (3.30) shows that the scattering amplitude has the dimension of length in 3D, the dimension of $[\text{length}]^{1/2}$ in 2D, and is dimensionless in 1D.

From Eqs. (3.28), (3.33), and (3.34), it is clear that the matrix elements of $\mathbf{t}^+$ in the $\mathbf{k}$ representation may be obtained from the solution of the single-scatterer wavefunction, as shown in the solutions to Problems 3.4, 3.5, and 3.6 at the end of this chapter, for a spherical scatterer of radius R and $v \neq v_0$. The answer in 3D is

$$f(\kappa_0', \kappa_0) = -\frac{i}{\kappa_0}\sum_{n=0}^{\infty} D_n P_n(\cos\theta)(2n + 1), \quad (3.35)$$

where $P_n(\cos\theta)$ is the nth Legendre polynomial, θ is the angle between κ_0' and κ_0, n is the angular momentum index, and

$$D_n = \frac{\kappa_0 j_n(\kappa R) j_n'(\kappa_0 R) - \kappa j_n'(\kappa R) j_n(\kappa_0 R)}{\kappa j_n'(\kappa R) h_n(\kappa_0 R) - \kappa_0 h_n'(\kappa_0 R) j_n(\kappa R)} \quad (3.36)$$

results from the solution of the boundary value problem. Here j_n and h_n denote the nth-order spherical Bessel function and spherical Hankel function of the first kind, respectively; prime means derivative with respect to the argument of the function; and $\kappa = \omega/v$, whereas $\kappa_0 = \omega/v_0$. In the limit of $\kappa R, \kappa_0 R \to 0$, f is dominated by D_0, whose leading order real and imaginary parts have the form

$$D_0 \cong -\frac{i}{3}\left[1 - \left(\frac{\kappa}{\kappa_0}\right)^2\right](\kappa_0 R)^3 - \frac{1}{9}\left[1 - \left(\frac{\kappa}{\kappa_0}\right)^2\right]^2(\kappa_0 R)^6 + \cdots. \quad (3.37)$$

As a result,

$$\langle \kappa'_0|t^+|\kappa_0\rangle = -4\pi f(\kappa'_0,\kappa_0) \cong \frac{4\pi}{3}\left[1-\left(\frac{\kappa}{\kappa_0}\right)^2\right]\frac{(\kappa_0 R)^3}{\kappa_0} - i\frac{4\pi}{9}$$
$$\times \left[1-\left(\frac{\kappa}{\kappa_0}\right)^2\right]^2 \frac{(\kappa_0 R)^6}{\kappa_0} + \cdots. \quad (3.38)$$

It is noted that in the $\kappa_0 R \to 0$ limit, the sphere would appear as a point scatterer relative to the wave, and the scattering is consequently isotropic, just as in the lattice case for a single impurity. Since κ and κ_0 are both proportional to ω, the scattering amplitude is noted to be proportional to ω^2 in the leading order. The scattering cross section, which is just $|f|^2$, is thus proportional to ω^4 in the low-frequency limit. This dependence is well known as the signature of *Rayleigh scattering*. Also, Eq. (3.38) tells us that the imaginary part of f is related to the total scattering cross section O, which is just the angular integral of the leading part of $|f|^2$. In the present case, since the scattering is isotropic in the $\omega \to 0$ limit, we have

$$O = \frac{4\pi}{9}\left[1-\left(\frac{\kappa}{\kappa_0}\right)^2\right]^2 \frac{(\kappa_0 R)^6}{\kappa_0^2} = \frac{4\pi}{\kappa_0} \operatorname{Im} f(\kappa_0,\kappa). \quad (3.39)$$

Alternatively, Eq. (3.39) may be expressed in a more general form as

$$O = -\frac{1}{\kappa_0}\operatorname{Im}\langle\kappa_0|t^+|\kappa_0\rangle. \quad (3.40)$$

Equation (3.39) or (3.40) is known as the *optical theorem*. It relates the total scattering cross section with the imaginary part of the *forward* scattering amplitude, i.e., $\kappa'_0 = \kappa_0$. The forward direction is special because heuristically, what is lost from the original wave has to be accounted for by scattering. The optical theorem is just a precise statement of that fact, as shown in the solution to Problem 3.7. As Eqs. (3.38) and (3.39) have demonstrated the validity of the optical theorem in the low-frequency limit, in Problems 3.4 and 3.5 it is shown that in 3D and 2D the optical theorem is valid for every angular momentum channel in the spherical scatterer case, and in 1D the theorem is trivially valid as shown below.

3.4 Single Scatterer—The Continuum Case

For 2D the solution of $\langle \kappa'_0 | t^+ | \kappa_0 \rangle$ is given in the solution to Problem 3.5. The answer is

$$\langle \kappa'_0 | t^+ | \kappa_0 \rangle = 4i \sum_{n=-\infty}^{\infty} D_n \exp(in\theta), \tag{3.41}$$

with

$$D_n = \frac{\kappa J'_n(\kappa R) J_n(\kappa_0 R) - \kappa_0 J_n(\kappa R) J'_n(\kappa_0 R)}{\kappa_0 H'_n(\kappa_0 R) J_n(\kappa R) - \kappa H_n(\kappa_0 R) J'_n(\kappa R)}. \tag{3.42}$$

Here H_n stands for the Hankel function of the first kind, of order n, and J_n stands for the Bessel function of order n. At low frequencies, D_0 dominates, and its leading terms in the expansion are given by

$$D_0 \cong -i \frac{\pi}{4} \left[1 - \left(\frac{\kappa}{\kappa_0}\right)^2\right](\kappa_0 R)^2 - \frac{\pi^2}{16}\left[1 - \left(\frac{\kappa}{\kappa_0}\right)^2\right]^2 (\kappa_0 R)^4 + \cdots. \tag{3.43}$$

The t-matrix element to the leading order is thus given by

$$\langle \kappa'_0 | t^+ | \kappa_0 \rangle \cong \pi \left[1 - \left(\frac{\kappa}{\kappa_0}\right)^2\right](\kappa_0 R)^2 - i\frac{\pi^2}{4}\left[1 - \left(\frac{\kappa}{\kappa_0}\right)^2\right]^2 (\kappa_0 R)^4 + \cdots. \tag{3.44}$$

Again, the scattering is seen to be isotropic because in the $\kappa_0 R \to 0$ limit the scatterer appears as a point impurity to the incident wave. The scattering amplitude f to the leading order may be obtained by comparison with Eq. (3.33):

$$f(\kappa'_0, \kappa_0) \cong -i\frac{\sqrt{2\pi}}{4} \frac{(\kappa_0 R)^2}{\sqrt{\kappa_0}}\left[1 - \left(\frac{\kappa}{\kappa_0}\right)^2\right] \exp\left(-i\frac{\pi}{4}\right). \tag{3.45}$$

The total scattering cross section, given by $|f|^2$ integrated over all angles, is now

$$O = \frac{\pi^2}{4} \frac{(\kappa_0 R)^4}{\kappa_0}\left[1 - \left(\frac{\kappa}{\kappa_0}\right)^2\right]^2 = -\frac{1}{\kappa_0} \text{Im} \langle \kappa_0 | t^+ | \kappa_0 \rangle. \tag{3.46}$$

The optical theorem is seen to be satisfied, although its popular version, Eq. (3.39), is not satisfied here unless one gives a different definition for f in 2D. However, in the form of Eq. (3.40) it is always valid. Equation (3.46) also gives ω^3 as the frequency dependence for the 2D Rayleigh scattering cross section in the $\omega \to 0$ limit.

In 1D, the solution to the scattering problem can be written down explicitly as shown in the solution to Problem 3.6. Since there are only two scattering directions, backward and forward, the answer may be expressed in terms of reflection and transmission coefficients ρ and τ:

$$\langle -\boldsymbol{\kappa}_0|\mathbf{t}^+|\boldsymbol{\kappa}_0\rangle = i2\kappa_0\rho, \tag{3.47}$$

$$\langle \boldsymbol{\kappa}_0|\mathbf{t}^+|\boldsymbol{\kappa}_0\rangle = i2\kappa_0(\tau - 1), \tag{3.48}$$

where ρ and τ are given by the following formulae:

$$\rho = -i\frac{\left[1 - (\kappa/\kappa_0)^2\right]\sin 2\kappa R}{2(\kappa/\kappa_0)\cos 2\kappa R - i\left[1 + (\kappa/\kappa_0)^2\right]\sin 2\kappa R}\exp(-2i\kappa_0 R), \tag{3.49}$$

$$\tau = \frac{2(\kappa/\kappa_0)}{2(\kappa/\kappa_0)\cos 2\kappa R - i\left[1 + (\kappa/\kappa_0)^2\right]\sin 2\kappa R}\exp(-2i\kappa_0 R). \tag{3.50}$$

It is easily verified that $|\rho|^2 + |\tau|^2 = 1$, and the optical theorem is just a reflection of the energy conservation in this case because the total scattering cross section is given by $O = |\tau - 1|^2 + |\rho|^2 = 2(1 - \operatorname{Re}\tau)$, whereas $-\operatorname{Im}\langle\boldsymbol{\kappa}_0|\mathbf{t}^+|\boldsymbol{\kappa}_0\rangle/\kappa_0$ is simply $-\operatorname{Im} 2i(\tau - 1) = 2(1 - \operatorname{Re}\tau)$ and is therefore identical to the total scattering cross section. Also, at low frequencies the Rayleigh scattering cross section is noted to be proportional to ω^2. The results in 1D, 2D, and 3D show that the frequency dependence of the Rayleigh scattering is ω^{d+1} in general. Rayleigh scattering is special to classical waves, and its frequency dependence is a reflection of the classical wave dispersion relation and the fact that $\sigma(\mathbf{r}) \propto \omega^2$.

3.5 Infinite Number of Scatterers—The Effective Medium and the Coherent Potential Approximation

When there are infinitely many scatterers, the T-matrix and the exact Green's function are impossible to obtain accurately. However, it is noted

3.5 Infinite Number of Scatterers

that in the wave vector representation, the averaged Green's function is given by

$$\langle G \rangle_c(\omega, \mathbf{k}) = \frac{1}{\kappa_0^2(\omega) - \beta e(\mathbf{k}) - \Sigma(\omega, \mathbf{k})} \frac{N}{L^d}, \quad (3.51)$$

where $\beta e(\mathbf{k}) = k^2$ for classical scalar waves in continuum. This form of $\langle G \rangle_c$ follows from Eq. (3.14) and (2.22). It is clear from Eq. (3.51) that all nontrivial information about $\langle G \rangle_c$ is contained in $\Sigma(\omega, \mathbf{k})$. If Σ turns out to be *independent of* $\mathbf{k}$ in some frequency regime, then the effect of $\Sigma(\omega)$ is just to renormalize $\kappa_0^2(\omega)$, the uniform medium property. This renormalized medium is called the *effective medium*.

The meaning of self-energy's $\mathbf{k}$ independence is as follows. In general, $\mathbf{k}$ dependence reflects correlated spatial structures. For example, consider the function $F(\mathbf{r}) = 1$ for $|\mathbf{r}| \le R$ and zero otherwise. In the Fourier-transformed $\mathbf{k}$ domain, $F(k) = 4\pi N(\sin kR - kR \cos kR)/(k^3 L^d)$. The $\mathbf{k}$ dependence of $F(k)$ is the reflection of a the finite sphere in real space. However, as $kR \to 0$ it is easy to verify that $F(k)$ becomes $\mathbf{k}$ independent, since relative to λ the sphere becomes a point, and a point has no length scale and no geometric structure to speak of. Indeed, the results for single scatterer in the preceding section show no $\mathbf{k}$ dependence in the t-matrix elements at the $k_0 R \to 0$ limit. If in addition the pointlike scatterers are situated randomly in space with no correlation between them, there is indeed a possibility that Σ could be $\mathbf{k}$ independent.

Suppose the effective medium exists; then some of its properties can be easily inferred. First, Σ is likely to be frequency dependent, but it may not be the same frequency dependence as $\kappa_0^2(\omega)$ for a uniform medium. For example, classical wave scattering from point scatterers gives rise to the Rayleigh frequency dependence of ω^{d+1} in the low-frequency limit. Therefore, if randomly situated point scatterers can yield an effective medium, its frequency dependence should see a slight deviation from $\kappa_0^2 \sim \omega^2$ at higher frequencies (ω^2 would dominate over ω^{d+1} at low frequencies for 2D and 3D). Second, $\Sigma(\omega)$ is in general a complex number. Its imaginary part must be negative so that the imaginary part of $\langle G^+ \rangle_c$ is negative. This is because $-\text{Im}\langle G^+ \rangle_c / \pi$ is related to the density of states for the effective medium, which must be positive or zero. If the combination $\kappa_0^2(\omega) - \Sigma(\omega)$ is viewed as a new $[\kappa_e^*(\omega)]^2$ for the effective medium, then κ_e^* should have a positive imaginary part. For example, in 3D the real space Green's function of a continuous effective medium would have the form

$$\langle G^+ \rangle_c = -\frac{\exp(i\kappa_e^* |\mathbf{r} - \mathbf{r}'|)}{4\pi |\mathbf{r} - \mathbf{r}'|}, \quad (3.52)$$

just as in a uniform medium [see Eq. (2.31)]. If κ_e^* has a positive imaginary part, $\langle G^+ \rangle_c$ decays exponentially. The meaning of this decay may be inferred from the optical theorem because the imaginary part of $\Sigma(\omega)$ is directly related to the imaginary part of the averaged T matrix, which is related to the total scattering cross section. Therefore, the spatial decay of $\langle G^+ \rangle_c$ is a result of the scattering loss from the propagating wave. Because all scatterings considered are elastic, what is lost is the unique propagation direction and phase relation, i.e., the coherence, of a plane wave. The inverse of Im κ_e^* gives a length for such decay, which is generally denoted as twice the mean free path l. The factor of 2 enters because l is defined as the *intensity* decay length for the coherent wave component. Inside the scale of a mean free path a propagating wave is expected to maintain its coherent behavior, but the dynamics beyond the scale of l have to be probed through the evaluation of $\langle GG^* \rangle_c$, to be examined in the next chapter.

If the solution for an effective medium exists, i.e., $\Sigma(\omega, \mathbf{k}) = \Sigma(\omega)$, an efficient approach to find such a solution is given by the coherent-potential approximation (CPA) (Lax, 1951). The central idea of CPA is the simple observation that *relative* to the renormalized uniform medium with $[\kappa_e^*(\omega)]^2 = \kappa_0^2(\omega) - \Sigma(\omega)$, $\langle \mathbf{G} \rangle_c$ is just a uniform-medium Green's function, given by $G_e(\kappa_e^*)$ in the k-representation. Since $\langle \mathbf{G} \rangle_c^{-1} = \mathbf{G}_e^{-1} - \overline{\Sigma} L^d/N$, that means $\overline{\Sigma}$ *must vanish in the renormalized effective medium*. From Eq. (3.15), $\overline{\Sigma} = 0$ implies

$$\langle \overline{\mathbf{T}} \rangle_c = 0 \qquad (3.53)$$

in the effective medium. Here we have used an overbar to distinguish those self-energy and scattering matrix elements defined *relative to the effective medium*. In terms of $\mathbf{G}_e$, the exact Green's function is given by

$$\mathbf{G} = \mathbf{G}_e + \mathbf{G}_e \overline{\mathbf{T}} \mathbf{G}_e. \qquad (3.54)$$

The condition of Eq. (3.53) means that in the effective medium, there is *no scattering on average*. However, scattering still exists, since $\langle \overline{\mathbf{T}}\overline{\mathbf{T}} \rangle_c \neq 0$ in general. This point is important for understanding the origin of the diffusive wave transport behavior.

A theory would not be very useful if it could be expressed formally but its consequences could not be explicitly evaluated. In Eq. (3.53), the $\overline{\mathbf{T}}$ matrix cannot be calculated in general for a disordered medium. However, the beauty of the CPA approach is that precisely in the $\langle \overline{\mathbf{T}} \rangle_c \to 0$ limit, the theory becomes calculable because $\langle \overline{\mathbf{T}} \rangle_c \to 0$ implies weak overall scattering. One way this may be achieved is for each individual scatterer to scatter weakly in the effective medium. Then the overall $\overline{\mathbf{T}}$ matrix can be written approximately as the sum of single scatterings from individual scatterers:

$$\bar{\mathbf{T}} \cong \sum_i \bar{\mathbf{t}}_i, \qquad (3.55)$$

where i is the index for the scatterers. That is, when each scattering is weak, the multiple scattering terms can be ignored because they represent small numbers raised to some power > 1. The accuracy of this approximation will be examined in a later section. With this approximation, the CPA condition becomes

$$\langle \bar{\mathbf{T}} \rangle_c \cong \sum_i \langle \bar{\mathbf{t}}_i \rangle_c = 0. \qquad (3.56)$$

The implications of the CPA condition are elucidated in the following implementations.

3.6 CPA—The Anderson Model

In the Anderson model, the randomness is induced by random deviations of the atomic energy level at site l, denoted by $\Delta\epsilon_l$, with $\langle \Delta\epsilon_l \rangle = 0$, $\Delta\epsilon_l$ and $\Delta\epsilon_m$ independent for $l \neq m$, i.e., $\langle \Delta\epsilon_l \Delta\epsilon_m \rangle = 0$, and $\Delta\epsilon_l$ has a flat distribution with width W:

$$D(\Delta\epsilon_l) = \begin{cases} \dfrac{1}{W} & -W/2 \leq \Delta\epsilon_l \leq W/2 \\ 0 & \text{otherwise}. \end{cases} \qquad (3.57)$$

Here $\Delta\epsilon_l$'s are all defined relative to the uniform medium. The equation defining the Green's function is

$$(\beta\Delta^{(2)} + \kappa_0^2(\omega) - \sigma(\mathbf{l}))G^+(\omega, \mathbf{l}, \mathbf{l}') = \frac{1}{a^d}\delta_{\mathbf{l},\mathbf{l}'}, \qquad (3.58)$$

where $\sigma(\mathbf{l}) = 2m\Delta\epsilon_l/\hbar^2$. To the leading order, the **T** operator in this case is just the addition of all the single-scatter **t**'s, each one given by Eq. (3.19):

$$\mathbf{T}^+ \cong \sum_{\mathbf{l}} \mathbf{t}_{\mathbf{l}}^+ = \sum_{\mathbf{l}} \frac{\sigma(\mathbf{l})a^d}{1 - \sigma(\mathbf{l})a^d G_0^+(\omega, \mathbf{l} = \mathbf{l}')} |\mathbf{l}\rangle\langle\mathbf{l}|. \qquad (3.59)$$

The operation of configurational averaging $\mathbf{t}_{\mathbf{l}}^+$ involves fixing the site $\mathbf{l}$ and averaging over the $\mathbf{t}^+$ matrix that appears on that site for all configurations. But that is equivalent to averaging over all possible values of $\sigma(\mathbf{l})$, i.e.,

$$\langle \mathbf{t}_\mathbf{l}^+ \rangle_c = \left[\int d\sigma \, D(\sigma) \frac{\sigma a^d}{1 - \sigma a^d G_0^+(\omega, \mathbf{l} = \mathbf{l}')} \right] |\mathbf{l}\rangle\langle\mathbf{l}|, \quad (3.60a)$$

where the label on $\sigma(\mathbf{l})$ is dropped, and $D(\sigma)$ denotes the distribution of the σ values. It becomes obvious that the value in the square bracket should be identical for every site $\mathbf{l}$ after configurational averaging. Therefore

$$\langle \mathbf{T}^+ \rangle_c \cong \left[\int d\sigma \, D(\sigma) \frac{\sigma a^d}{1 - \sigma a^d G_0^+(\omega, \mathbf{l} = \mathbf{l}')} \right] \mathbf{I}, \quad (3.60b)$$

where $\mathbf{I} = \sum_\mathbf{l} |\mathbf{l}\rangle\langle\mathbf{l}|$ is the identity operator. In accordance with the CPA recipe, we next try to make $\langle \overline{\mathbf{T}}^+ \rangle_c$ vanish through shifting $\kappa_0^2(\omega)$ by an amount Σ^+. Since the shift has to be performed *without altering the original problem* as given by Eq. (3.58), the shift of κ_0^2 must be accompanied by a similar shift in $\sigma(\mathbf{l})$, which means that the scattering potential also has to be renormalized relative to the effective medium. That is, the shift is simply the operation of rewriting the wave equation (3.58) as

$$\left[\beta \Delta^{(2)} + (\kappa_0^2(\omega) - \Sigma^+) - (\sigma(\mathbf{l}) - \Sigma^+) \right] G(\omega, \mathbf{l}, \mathbf{l}) = \frac{1}{a^d} \delta_{\mathbf{l}, \mathbf{l}'}.$$

With this shift, the CPA condition is given by (noting the ω dependence of G^+ is always implicitly defined by $\kappa_0^2(\omega)$):

$$\int d\sigma \, D(\sigma) \frac{\sigma - \Sigma^+}{1 - (\sigma - \Sigma^+) a^d G_0^+(\kappa_0^2 - \Sigma^+, \mathbf{l} = \mathbf{l}')} = 0. \quad (3.61)$$

To express this condition in a more useful form, it is noted that the integrand may be written as

$$\left[-1 + \frac{(a^d G_e^+)^{-1}}{(a^d G_e^+)^{-1} - (\sigma - \Sigma^+)} \right] \frac{1}{(G_e^+) a^d},$$

which means

$$1 = \int d\sigma \frac{D(\sigma)}{1 - (\sigma - \Sigma^+) a^d G_e^+}, \quad (3.62)$$

3.6 CPA—The Anderson Model

where $G_e^+ = G_0^+(\kappa_0^2 - \Sigma, \mathbf{l} = \mathbf{l}')$. From Eqs. (3.61) and (3.62) the CPA condition can be rewritten as

$$\int d\sigma\, D(\sigma) \frac{\sigma}{1 - (\sigma - \Sigma^+)a^d G_0^+(\kappa_0^2 - \Sigma^+, \mathbf{l} = \mathbf{l}')} = \Sigma^+. \quad (3.63)$$

This is the equation one can use to calculate Σ^+ numerically by the method of iterative substitution; i.e., pick a value of Σ^+, calculate the left-hand side, and use that value as the new Σ^+ to calculate the left-hand side until consistency is achieved. Usually a few iterations suffice in practice. For the flat distribution, Eq. (3.63) becomes

$$\frac{1}{2\sigma_w} \int_{-\sigma_w}^{\sigma_w} d\sigma \frac{\sigma}{1 - (\sigma - \Sigma^+)a^d G_0^+(\kappa_0^2 - \Sigma^+, \mathbf{l} = \mathbf{l}')} = \Sigma^+, \quad (3.64a)$$

where

$$2\sigma_w = \frac{2mW}{\hbar^2}$$

is the total width of the distribution. The spatial dimensionality of the problem is contained in the form of the Green's function $G_0^+(\kappa_0^2 - \Sigma^+, \mathbf{l} = \mathbf{l}')$. Since G_0^+ is complex, the solution Σ^+ is in general a complex number. The amount of randomness in the model is governed by σ_w. When $\sigma_w = 0$, $\Sigma^+ = 0$ necessarily, and the uniform medium is recovered.

For the 3D CPA, the evaluation of the Green's function is perhaps the most time-consuming part of the numerical calculation. A useful and efficient approach to the calculation of the 3D Green's function is to compile a numerical table of the density of states for the ordered lattice and store it in computer memory. The diagonal Green's function can then be obtained from Eq. (P2.16), which involves only one integration. That is,

$$G_0^+(\kappa_0^2 - \Sigma^+, \mathbf{l} = \mathbf{l}') = \int \frac{\rho_0(\omega')}{\kappa_0^2 - \Sigma^+ - \kappa_0^2(\omega')} d\omega'. \quad (3.64b)$$

The integration is over a finite domain of ω', since $\rho_0(\omega')$ is nonzero only inside the band. Moreover, the imaginary part of Σ^+ regularizes the singularity that appears in the integrand, thus making the numerical calculation quick and accurate.

Once Σ^+ is obtained, the mean free path l can also be evaluated because $-\text{Im}\,\Sigma^+$ is proportional to a scattering time. However, in the lattice model the expression for l is not as simple as that for the continuum case, where G_e^+ can be explicitly written down as in Eq. (3.52).

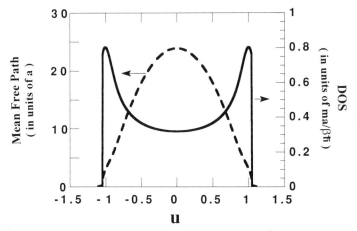

Figure 3.4 The density of states per site per unit frequency $\hbar/ma^2$ (solid line) and the mean free path l/a (dashed line) for a 1D lattice with site randomness $\sigma_w a^2 = 0.5$. β is taken to be 1. The density of states should be compared with Figure 2.1 for the perfect lattice case. Here the dashed line is given by Eq. (4.101), with $d = 1$, and the solid line is given by

$$-\frac{2}{\pi} \text{Im } F[u, \Sigma(u)a^2],$$

where

$$F = \frac{\text{sgn}[\text{Re } f]}{2\sqrt{f^2 - 1}},$$

$$f = u - \Sigma a^2/2,$$

$u = (\hbar\omega - \epsilon_0)/(\hbar^2/ma^2)$, and $\Sigma(u)$ is the solution to the CPA equation (3.64a) with the 1D Green's function.

In order to be consistent with the later development on wave diffusion, the expression for l is given in the next chapter [Eq. (4.101)]. In Figures 3.4, 3.5, and 3.6 the density of states for the disordered system [which can be obtained from the imaginary part of the Green's function as evaluated from (3.64b)] and its corresponding mean free path are plotted for 1D, 2D, and 3D, respectively, with the randomness parameter $\sigma_w a^2 = 0.5$. It is interesting to observe that in the discrete lattice case, the CPA equation has solution for all frequencies inside the band. Comparison with Figures 2.1, 2.3, and 2.5 shows that the effect of randomness is to both broaden the band and round off the singularities in the density of states.

3.7 CPA—The Case of Classical Waves

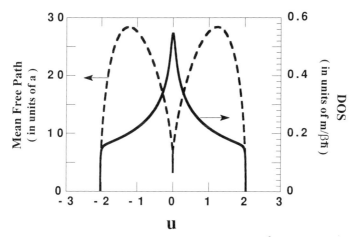

Figure 3.5 The density of states per site per unit frequency $\hbar/ma^2$ (solid line) and the mean free path l/a (dashed line) for a 2D square lattice with site randomness specified by $\sigma_w a^2 = 0.5$. β is taken to be 1. The density of states should be compared with Figure 2.3 for the perfect lattice case. Here the dashed line is given by Eq. (4.101) with $d = 2$ and the solid line is given by

$$-\frac{2}{\pi} \operatorname{Im} F[u, \Sigma(u)a^2],$$

where

$$F = \frac{1}{2\pi}\int_0^\pi d\theta \, \frac{\operatorname{sgn}[\operatorname{Re} f]}{\sqrt{f^2 - 4\cos^2\theta}},$$

$$f = u - \Sigma(u)a^2/2,$$

$u = (\hbar\omega - \epsilon_0)/(\hbar^2/ma^2)$, and $\Sigma(u)$ is the solution to the CPA equation (3.64a) with the 2D Green's function. The dip in the mean free path at the band center is due to the high density of scatterers at $u = 0$. The sharp decrease of the mean free path near the band edge, on the other hand, is attributable to the vanishing wave (group) velocity.

3.7 CPA—The Case of Classical Waves

For classical scalar waves, the scattering problem can be formulated from the discrete lattice point of view, where each scattering particle consists of infinitely many sites whose properties are correlated. The $\mathbf{t}_i$ for a single particle i is now given by

$$\mathbf{t}_i = \mathbf{v}_i(\mathbf{I}-\mathbf{G}_0\mathbf{v}_i)^{-1},$$

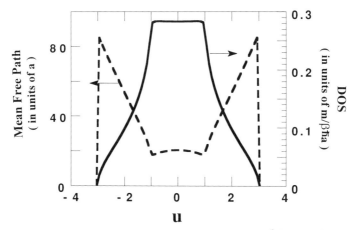

Figure 3.6 The density of states per site per unit frequency $\hbar/ma^2$ (solid line) and the mean free path l/a (dashed line) for a 3D simple cubic lattice with site randomness specified by $\sigma_w a^2 = 0.5$. β is taken to be 1. The density of states should be compared with Figure 2.5 for the perfect lattice case. Here the dashed line is given by Eq. (4.101) for $d = 3$, and the solid line is given by

$$-\frac{2}{\pi} \operatorname{Im} F[u, \Sigma(u)a^2],$$

where

$$F = \frac{1}{2\pi^2} \int_0^\pi d\theta_1 \int_0^\pi d\theta_2 \frac{\operatorname{sgn}[\operatorname{Re} f]}{\sqrt{f^2(\theta_1) - 4\cos^2\theta_2}},$$

$$f = u - \Sigma(u)a^2 + \cos\theta_1$$

$$u = (\hbar\omega - \epsilon_0)/(\hbar^2/ma^2),$$

and $\Sigma(u)$ is the solution to the CPA equation (3.64a) with the 3D Green's function. The small value of l/a at the band center is directly correlated with the high density of states around $u = 0$.

where

$$\mathbf{v}_i = \sum_{\mathbf{r}} (\delta r)^d \sigma_i(\mathbf{r}) |\mathbf{r}\rangle\langle\mathbf{r}|, \qquad \mathbf{r} \in \text{inside particle } i.$$

A straightforward generalization of the CPA condition, Eq. (3.64), is to treat a *single particle* in the same manner as a single site in the Anderson model. The resulting condition is the matrix equation

$$\Sigma = \frac{1}{2\sigma_w} \int_{-\sigma_w}^{\sigma_w} d\sigma\, \mathbf{v}_i(\sigma)\left[\mathbf{1} - (\mathbf{v}_i(\sigma) - \Sigma a^d)\tilde{\mathbf{G}}_c^+\right]^{-1}, \qquad (3.65)$$

3.7 CPA—The Case of Classical Waves

where we have used a function,

$$\tilde{G}_e^+(\omega, \mathbf{r}, \mathbf{r}') = H(\mathbf{r})G_e^+(\omega, \mathbf{r} - \mathbf{r}')H(\mathbf{r}'), \quad (3.66a)$$

in the equation. H is a particle function defined as

$$H(\mathbf{r}) = \begin{cases} 1 & \text{for } \mathbf{r} \text{ inside the particle} \\ 0 & \text{else} \end{cases} \quad (3.66b)$$

to restrict the action of the Green's function to be within a single particle.

From Eq. (3.65) it is difficult to see that Σ should be proportional to the identity operator $\mathbf{I}$, which is required if a single value of Σ were to satisfy the whole matrix equation. However, it turns out that there is indeed a low-frequency regime where the effective medium description holds. To demonstrate that, it is much simpler to look at the CPA condition in the k representation. Therefore we will not use Eq. (3.65) for our treatment of the CPA in the classical wave case. But the formal CPA condition, Eq. (3.65), is useful in demonstrating the Ward identity, shown in the next chapter.

Since the CPA condition is

$$\bar{\Sigma}^+ \cong \frac{N}{L^d} \langle \bar{\mathbf{T}}^+ \rangle_c \cong \sum_i \frac{N}{L^d} \langle \bar{\mathbf{t}}_i^+ \rangle_c = 0, \quad (3.67a)$$

and since the configurationally averaged T-matrix is a function of only one $\mathbf{k}$ as discussed before, it follows immediately that it is only the *forward scattering amplitudes that enters the CPA condition in the k representation* (because $\mathbf{k}' = \mathbf{k}$ implies forward scattering). Therefore,

$$\sum_i \frac{N}{L^d} \langle \bar{\mathbf{t}}_i^+ \rangle_c = \frac{N}{L^d} \sum_i \frac{1}{N} \sum_{\mathbf{\kappa}_e} \langle \mathbf{\kappa}_e | \bar{\mathbf{t}}_i^+ | \mathbf{\kappa}_e \rangle | \mathbf{\kappa}_e \rangle \langle \mathbf{\kappa}_e | = 0, \quad (3.67b)$$

where $\mathbf{\kappa}_e$ is a wave vector whose magnitude equals $\sqrt{\kappa_0^2 - \text{Re }\Sigma^+}$, which is the real part of κ_e^* when $\kappa_e l \gg 1$. By summing over the scatterer's index i and classifying the different scatterers into different species, one obtains the CPA condition as

$$\sum_m n_m \langle \mathbf{\kappa}_e | \bar{\mathbf{t}}_m^+ | \mathbf{\kappa}_e \rangle = 0, \quad (3.68)$$

where n_m means the *number density* of the mth species of scatterers. In connection with Eq. (3.68) it should be pointed out that, in general, the real part of $\langle \mathbf{\kappa}_e | \bar{\mathbf{t}}_m^+ | \mathbf{\kappa}_e \rangle$ is dominant at the long-wavelength limit, as seen by the solution to the single-scatterer problem. Therefore it is usually that part, rather than the imaginary part, which enters the CPA equation. Because the real part does not represent the total scattering cross section, as does the imaginary part, its sign can change from one species to the next, thus allowing the solution of Eq. (3.68) to exist.

The effective medium property of a continuous random medium is closely linked to its microstructure. Here the term microstructure refers to the volume fractions of the components plus the shapes and geometric arrangement of the inhomogeneities. While the effective medium generally exists in the low-frequency limit where the individual scatterers are beyond the wave's resolution, the effective medium parameters can nevertheless be affected by presence or absence of some particular type of *correlation* in a random composite that is represented by its microstructure. Such a correlation generally extends throughout the sample and can thus affect its macroscopic properties. Therefore an effective medium is uniquely determined only if the material properties of the components *and* their microstructure are both specified.

A useful approach to capturing the microstructural information of a random medium is through the concept of "structural unit(s)." This is illustrated in two commonly encountered microstructures, schematically depicted in Figures 3.7a and 3.7b. In Figure 3.7a, the two components of a

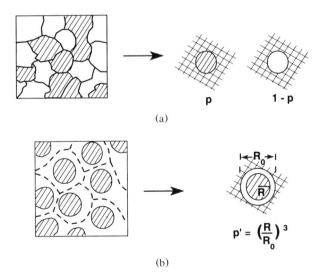

Figure 3.7 (a) The symmetric microstructure and its two structural units. On the right, the effective medium is indicated by the crosshatched region. (b) The dispersion microstructure. In this case the dashed lines indicate a schematic division of the composite into structural units. On the right, the effective medium is indicated by the crosshatched region.

3.7 CPA—The Case of Classical Waves

random medium are symmetrical in the statistical sense; i.e., an interchange of the two components results in the same type of medium with interchanged volume fractions. This is denoted the symmetric microstructure. The same is not true for the microstructure depicted in Figure 3.7b, where one component is always the dispersed phase in the matrix of the other. This is denoted the dispersion microstructure, like that of a colloid. For the symmetrical microstructure, there are two structural units, each approximated by a sphere (a circle in 2D, a line in 1D) of one material component. The volume fractions of the two units are p and $1 - p$. For the dispersion microstructure, the dashed lines in Figure 3.7b divide the medium into similar structural units, each consisting of a particle of dispersed phase coated by a layer of the matrix phase. By approximating the structural unit as a coated sphere, the coating layer thickness $h = R_0 - R$ may be determined by the local volume fraction of the dispersed components:

$$h = R_0 - R = \left((p')^{-1/d} - 1\right)R, \tag{3.69}$$

where R_0 is the radius of the total unit, R is the radius of the dispersed particle, and p' is its local volume fraction, which can fluctuate from one location to the next.

3.7.1 The Symmetric Microstructure

To calculate the forward scattering amplitudes of a two-component composite in the symmetric microstructure, each structural unit, characterized by $\kappa_m^2 = \epsilon_m \kappa_0^2$ and radius R_m ($m = 1, 2$), is individually embedded in an effective medium with $\kappa_e^2 = \bar{\epsilon}\kappa_0^2$. Since the calculation of the scattering amplitude has already been done in the last section, the answer may be directly quoted here:

$$\langle \kappa_e | t_m^+ | \kappa_e \rangle = \begin{cases} \dfrac{4\pi i}{\kappa_e} \sum\limits_{n=0}^{\infty} \overline{D}_n^{(m)}(2n + 1) & \text{3D}, \\ 4i \sum\limits_{n=-\infty}^{\infty} \overline{D}_n^{(m)} & \text{2D}, \\ 2i\kappa_e(\bar{\tau}^{(m)} - 1) & \text{1D}, \end{cases} \tag{3.70}$$

where θ has been set to 0 for the 3D and 2D cases, and $\overline{D}_n^{(m)}$ and $\bar{\tau}^{(m)}$ are given by Eqs. (3.36), (3.42), and (3.50), respectively, for 3D, 2D, and 1D, in which κ_0 is replaced by κ_e, and κ is replaced by κ_m. Combining Eqs. (3.68) and (3.70) gives the CPA equation for the symmetric microstructure.

In the $\kappa_e R_m \to 0$ limit, the CPA equation may be written down explicitly to the leading order in $(\kappa_e R_m)^d$:

$$C_d \sum_{m=1}^{2} n_m (\kappa_e R_m)^d \kappa_e^{2-d} \left[1 - \left(\frac{\kappa_m}{\kappa_e} \right)^2 \right] = 0, \quad (3.71)$$

where $C_d = 2$, π, and $4\pi/3$ for 1D, 2D, and 3D, respectively. Since $C_d n_m R_m^d = p_m$, the volume fraction of component m, we get

$$\bar{\epsilon} = p\epsilon_1 + (1-p)\epsilon_2 \quad (3.72)$$

in all spatial dimensions. Alternatively, one can express the CPA equation in terms of the wave speeds $\bar{v}$, v_1, and v_2:

$$\frac{1}{\bar{v}^2} = \frac{p}{v_1^2} + \frac{1-p}{v_2^2}. \quad (3.73a)$$

An important application of the effective medium equation is the prediction of sound speed in immiscible fluid–fluid mixtures, when the inhomogeneities are small compared to the wavelength. In this case, the sound speed in a uniform medium is given by $v^2 = B/\rho$, where B is the bulk modulus of the fluid and ρ here denotes its mass density. The problem of acoustic wave scattering from a spherical inclusion of one fluid in the matrix of another involves boundary conditions which are slightly different from what have been used so far. Namely, if ϕ = pressure P, then instead of $\partial \phi / \partial r$ being continuous across a spherical interface, it should be the continuity of $\rho^{-1} \partial \phi / \partial r$. In the solution to Problem 3.8, this slight change is shown to result in the CPA equation

$$\frac{1}{\bar{B}} = \frac{p}{B_1} + \frac{1-p}{B_2}, \quad (3.73b)$$

i.e., v^2 in Eq. (3.73a) is replaced by B. Equation (3.73b) is known as Wood's formula in the literature. A particularly interesting application of Eq. (3.73b) is to a bubbly liquid. Since B for a gas is on the order of its pressure, at atmospheric pressure this would mean three to four orders of magnitude difference with the B of a liquid. Therefore, even the presence of 1% gas bubbles in a liquid would lower $\bar{B}$ by two orders of magnitude from that of the liquid. This effect is analogous to squeezing a sandwich made of two steel plates with a sheet of foam rubber in between. The effective $\bar{B}$ in that case is clearly dominated by that of the foam rubber. Since $\bar{v} = \sqrt{\bar{B}/\bar{\rho}}$, and $\bar{\rho}$ is hardly altered from that of the fluid by the

3.7 CPA—The Case of Classical Waves

presence of 1% gas, the acoustic wave speed in a bubbly liquid can thus be drastically reduced from the sound speed of the pure liquid. This effect has indeed been observed extensively. While not so dramatic, Eq. (3.73b) is also reasonably accurate for predicting the $\bar{B}$ of mixtures consisting of solid particles dispersed in a fluid.

For higher frequencies, the CPA equations (3.68) and (3.70) have to be solved numerically. Past experience has shown that a unique solution can generally be found up to the frequency where $\kappa_e R_m$ is on the order of 1 but cannot be found beyond that because once the wave is able to discern the structure of individual scatterers, the effective medium description breaks down. Therefore, for classical waves in continuous random media the effective medium exists only when the condition $\lambda \gg R$ is satisfied. It should be noted that sometimes for R very small, this condition can be satisfied at high frequencies. Therefore, instead of stating the effective medium condition as being the low-frequency limit, it is perhaps more accurate to state it as being the *long-wavelength limit*, where "long" is measured in units of R.

The mean free path may also be explicitly calculated in the long-wavelength limit. In an effective medium, the scattering vanishes to the first order, so the averaged Green's function may be written as

$$\langle G \rangle_c = \frac{1}{\kappa_e^2(\omega) - k^2 - \bar{\Sigma}(\omega)} \frac{N}{L^d}, \qquad (3.74)$$

where $\mathrm{Re}\,\bar{\Sigma}(\omega) \equiv NL^{-d}\langle \bar{T} \rangle_c(\omega) \equiv 0$ to the leading order. However, $\langle \bar{t} \rangle_c$ does not vanish to all orders. The mean free path information is contained in the next higher-order term of the $\kappa_e R_m$ expansion of $\bar{\Sigma}(\omega)$, which is exactly the leading order *imaginary* part of $\bar{\Sigma}(\omega)$, given by

$$-\mathrm{Im}\,\bar{\Sigma}(\omega) = \Delta \cong -\mathrm{Im}\left[\sum_{m=1}^{2} n_m \langle \kappa_e | \bar{t}_m^+ | \kappa_e \rangle \right]. \qquad (3.75a)$$

By defining l as

$$\frac{1}{l} = 2\,\mathrm{Im}\,\kappa^* = 2\,\mathrm{Im}\,\sqrt{\kappa_e^2 + i\Delta} \simeq \frac{\Delta}{\kappa_e} = \sum_m n_m \bar{O}_m, \qquad (3.75b)$$

where the last equality is obtained through the optical theorem, with $\bar{O}$ defined as the total scattering cross section evaluated in the effective medium. By gathering the results on the leading imaginary term of the scattering cross section given in the last section, one gets

$$l = \frac{A_d \bar{\epsilon}^2}{\kappa_e \left[p(\kappa_e R_1)^d (\bar{\epsilon} - \epsilon_1)^2 + (1-p)(\kappa_e R_2)^d (\bar{\epsilon} - \epsilon_2)^2 \right]}, \quad (3.76)$$

where $A_d = 3, 4/\pi, 1$ in 3D, 2D, 1D, respectively; $\kappa_e^2 = \bar{\epsilon} \kappa_0^2$; and $\bar{\epsilon}$ is given by Eq. (3.72). It is noted that $l \sim \kappa_e^{-(d+1)} \sim \omega^{-(d+1)}$, which is precisely the inverse frequency dependence of Rayleigh scattering. Therefore, $l \to \infty$ as $\omega \to 0$, which implies that the effective medium is accurately valid at the $\omega \to 0$ limit since not only $\langle \overline{T} \rangle_c = 0$, but $\overline{T} = 0$ as well, so that the coherent wave propagation holds throughout the sample. Also, from Eq. (3.76) the mean free path is seen to be dependent on the particle size. If $R_1 = R_2 = R$, then

$$l = \frac{A_d \bar{\epsilon}^2}{\kappa_e^{d+1} R^d p(1-p)(\epsilon_1 - \epsilon_2)^2}, \quad (3.77)$$

where Eq. (3.72) has been used. The R^{-d} dependence is reasonable because the strength of Rayleigh scattering is sensitive to the ratio between the wavelength and the particle size.

For an important class of classical waves—electromagnetic waves—the effective medium equation described above has to be modified for the following two reasons. First, because of the vector character of the transverse electromagnetic wave, its 2D and 3D scattering amplitudes in the $\kappa_e R \to 0$ limit are always proportional to $\cos \theta$, i.e., dipolar in character, because the physical quantity here is the electric field $\mathbf{E}$, which polarizes the scatterer in the scattering process. Since the projection of a vector on the radial coordinate $\mathbf{r}$ always involves an angle, $P_1(\cos \theta) = \cos \theta$ is the lowest-order, nonisotropic term that enters. Second, in the electromagnetic equation the electric field may be written as $\mathbf{E} = -\nabla \phi$ at the low-frequency limit. As a result, it follows the from the Maxwell equations that

$$\nabla \cdot \epsilon \mathbf{E} = -\nabla \cdot \epsilon \nabla \phi = 0$$

for system without free changes. Therefore, across an interface, it is the *normal derivative of ϕ times ϵ* that is continuous. This can be seen from the usual arguments leading to the electromagnetic boundary conditions.

To incorporate these changes, we can utilize the single-scatterer solutions worked out in the last section, modify them slightly to take into account the new boundary condition, and expand the $n = 1$ term ($m = \pm 1$ in 2D), instead of the $n = 0$ term, in the $\kappa_e R_m \to 0$ limit. This has been

2.2 Green's Functions for Waves in a Uniform Medium

Green's function is one of the most useful tools in physics because of the physical information it can provide about the system described by the differential equation. Simply put, Green's function in the present context is just the solution to the wave equation with a specific initial source and a boundary condition at infinity. One might ask why that is more useful than just any solution $\phi(t, \mathbf{r})$ to the homogeneous wave equation. The answer could be illustrated by the following analogy.

Suppose there are N mechanical players in a game whose rules, while well defined, are not known a priori to an outside observer. The game involves the exchange of some currency between the players (such as in Monopoly, except here the players have no free will), which we just call "money." The purpose of the observer is to figure out the game rules and predict what is going to happen. The observer can record the money each player has at every instant and how it varies with time. The recorded quantity, $M(t, i)$, where i is the index for the players, may be compared to the wave function $\phi(t, \mathbf{r})$, since both give successive snapshots of the state of affairs. However, an intelligent observer might want to ask the question: What is the interrelationship between the amounts of money of any two players, if any? This important information cannot be obtained by simply looking at each player individually. Instead, one has to look at the *correlation* between pairs of players as a function of time separation, i.e., the correlation between $M(t, A)$ and $M(t + \Delta t, B)$. If the observer has the power to do a controlled experiment on the players, this correlation may be obtained by (1) setting $M = 0$ for all players, (2) giving an amount M_0 to A at $t = 0$, and (3) watching how much of M_0 is transferred to B at a later time t. This is precisely the sort of information that a Green's function would provide. By making A and B the same player, the Green's function can also monitor the information contained in the wave function. Therefore it is more general. In the case of the wave equation, the precise knowledge of the Green's function is sufficient to figure out all the rules of the system; i.e., it is equivalent to an exact solution of the problem. In the absence of exact solutions, which are often difficult beyond simple or periodic systems, Green's function also provides an approach to the solution of the wave equation that is amenable to systematic perturbation analysis. Moreover, the meaning of these analyses can be interpreted physically. Therein lies the attractiveness of the Green's function approach.

The specific initial condition for the Green's function is that at $t = 0$, the system is excited by a pulse localized at $\mathbf{r} = \mathbf{r}'$; i.e., the right-hand sides of Eqs. (2.4) and (2.7) are replaced by $\delta(t)\delta(\mathbf{r} - \mathbf{r}')$, where $\mathbf{r}'$ denotes the source position. The reason for such a source is that it can excite the natural resonances (or eigenfunctions) of *all* temporal and spatial frequencies in the system (since the source contains all temporal and spatial frequencies), so that the subsequent development may contain information about them. A useful approach for analyzing the Green's function is through its frequency components. The left-hand side of the wave equation for the frequency component ω is given by Eq. (2.8), and the same frequency component on the right-hand side is simply $\delta(\mathbf{r} - \mathbf{r}')$ because $\delta(t)$ gives 1 as the amplitude for every frequency component. Equating both sides yields

$$(\nabla^2 + \kappa^2)G(\omega, \mathbf{r}, \mathbf{r}') = \delta(\mathbf{r} - \mathbf{r}'), \qquad (2.16)$$

where $G(\omega, \mathbf{r}, \mathbf{r}')$ denotes the Green's function in the frequency and spatial domain and $\mathbf{r}'$ is used to mark the source point. The source point position is in general an important parameter for the Green's function if the system is inhomogeneous, since the response to a point source can depend on its position. In addition to Eq. (2.16), a unique determination of $G(\omega, \mathbf{r}, \mathbf{r}')$ also requires the boundary condition that $G(\omega, \mathbf{r}, \mathbf{r}') \to 0$ as $|\mathbf{r} - \mathbf{r}'| \to \infty$, as well as the physical condition of causality; i.e., cause always precedes effect. In this chapter, only the Green's function for the uniform medium is considered. The effects induced by $\sigma(\mathbf{r})$ are treated in the subsequent chapters.

For the uniform medium case of $\kappa = \kappa_0$, the response of the system is independent of the source position, and the Green's function depends only on the relative separation $\mathbf{r} - \mathbf{r}'$ between the source and the detector. Therefore,

$$(\nabla^2 + \kappa_0^2)G_0(\omega, \mathbf{r} - \mathbf{r}') = \delta(\mathbf{r} - \mathbf{r}'). \qquad (2.17)$$

A simple solution approach to Eq. (2.17) is to first solve for the spatial frequency component $G_0(\omega, \mathbf{k})$ by writing

$$G_0(\omega, \mathbf{r} - \mathbf{r}') = \frac{L^d}{N} \int \frac{d\mathbf{k}}{(2\pi)^d} \exp[i\mathbf{k} \cdot (\mathbf{r} - \mathbf{r}')] G_0(\omega, \mathbf{k}) \qquad (2.18)$$

and

$$\delta(\mathbf{r} - \mathbf{r}') = \int \frac{d\mathbf{k}}{(2\pi)^d} \exp[i\mathbf{k} \cdot (\mathbf{r} - \mathbf{r}')], \qquad (2.19a)$$

where d denotes the spatial dimensionality, L^d is the sample volume $[(2\pi)^d L^{-d}$ may be regarded as the smallest volume unit $(\delta k)^d$ in the wave

2.2 Green's Functions for Waves in a Uniform Medium

vector space], and N is the number of states, or atoms, in the sample. $L^d/N = (\delta r)^d$ is the smallest volume unit in the system. The notation here is based on viewing the continuum as the limit of infinitesmal discretization, so that both the continuum case and the lattice case may be treated on the same footing. The presence of N in Eq. (2.18) is due to the fact that a plane wave has magnitude $\sqrt{N}$. The convention in this book is to associate a factor $1/N$ with every **k**-integration (summation) to compensate for that fact. To be consistent, $\delta(\mathbf{k} - \mathbf{k}')$ as defined by

$$\int d\mathbf{k}\, \delta(\mathbf{k} - \mathbf{k}') = 1$$

has the following Fourier representation:

$$\delta(\mathbf{k} - \mathbf{k}') = \frac{1}{(2\pi)^d} \int d\mathbf{r}\, \exp[i(\mathbf{k} - \mathbf{k}') \cdot \mathbf{r}]. \tag{2.19b}$$

In Eq. (2.19a)

$$\int \frac{d\mathbf{k}}{(2\pi)^d}$$

denotes integration over all **k** vectors. For $d = 1$, that means

$$\int_{-\infty}^{\infty} \frac{dk}{2\pi}.$$

For $d = 2$, that means

$$\int_{-\infty}^{\infty} \int_{-\infty}^{\infty} \frac{dk_x\, dk_y}{(2\pi)^2}$$

in Cartesian coordinates and

$$\int_0^{2\pi} d\theta \int_0^{\infty} \frac{k\, dk}{(2\pi)^2}$$

in circular (cylindrical) coordinates. For $d = 3$, that means

$$\int_{-\infty}^{\infty} \int_{-\infty}^{\infty} \int_{-\infty}^{\infty} \frac{dk_x\, dk_y\, dk_z}{(2\pi)^3}$$

in Cartesian coordinates and

$$\int_0^{2\pi} d\varphi \int_0^{\pi} d\theta \int_0^{\infty} \frac{\sin\theta \, k^2 \, dk}{(2\pi)^3}$$

in spherical coordinates. Substitution of Eqs. (2.18) and (2.19a) into Eq. (2.17) gives

$$\frac{L^d}{N} \int \frac{d\mathbf{k}}{(2\pi)^d} \exp[i\mathbf{k} \cdot (\mathbf{r} - \mathbf{r}')][\kappa_0^2(\omega) - k^2] G_0(\omega, \mathbf{k})$$

$$= \int \frac{d\mathbf{k}}{(2\pi)^d} \exp[i\mathbf{k} \cdot (\mathbf{r} - \mathbf{r}')]. \quad (2.20)$$

Equating the amplitudes of each **k** component on both sides yields

$$[\kappa_0^2(\omega) - k^2] G_0(\omega, \mathbf{k}) = \frac{N}{L^d}, \quad (2.21)$$

or

$$G_0(\omega, \mathbf{k}) = \frac{1}{\kappa_0^2(\omega) - k^2} \frac{N}{L^d}. \quad (2.22)$$

The Green's function $G_0(\omega, \mathbf{k})$ gives the response of the homogeneous medium to a source excitation that has frequency ω and spatial frequency **k**. It is noted that if the source **k** and ω coincide with the wave dispersion relation of the medium, i.e., $k^2 = \kappa_0^2(\omega)$, then G_0 diverges. In analogy with the response of a forced oscillator which has its own resonance frequencies, this divergent response is expected, since the dispersion relation essentially gives the condition of natural resonance in a medium. On the other hand, when the source frequency and **k** do not coincide with any natural mode of the system, Eq. (2.22) indicates that it can still excite many modes just by its proximity to the resonance condition, as in the case of a forced oscillator.

The divergence in $G_0(\omega, \mathbf{k})$ would cause a problem in the transformation back into the spatial domain. To fix this problem, it is generally the practice to add a small imaginary constant to the denominator, then to take the limit as the constant approaches zero. This simple procedure is illustrated in the solution to Problem 2.1 at the end of this chapter. The as a column vector and $\langle\phi|$ means the same wave function but expressed as a row vector. $\langle\phi|\phi'\rangle$ means the inner product (dot product between two vectors), and $|\phi'\rangle\langle\phi|$ means the outer product. In Dirac's notation,

$$\mathbf{D} = \sum_{\mathbf{l}} \frac{2\beta d}{a^2} |\mathbf{l}\rangle\langle\mathbf{l}| + \sum_{\mathbf{l},\mathbf{n}} \left(-\frac{\beta}{a^2}\right) |\mathbf{l}\rangle\langle\mathbf{l}+\mathbf{n}|, \qquad (2.47)$$

where the second sum is over all the nearest neighbors of $\mathbf{l}$, for all $\mathbf{l}$, and $|\mathbf{l}\rangle$ means a column vector which has all zero components except for the component $\mathbf{l}$, which is 1. The same holds for the row vector $\langle\mathbf{l}|$. The outer product $|\mathbf{l}\rangle\langle\mathbf{l}+\mathbf{n}|$ is a matrix whose elements are all zero except the one at the row corresponding to $\mathbf{l}$ and the column corresponding to $\mathbf{l}+\mathbf{n}$, which has the value 1. $\mathbf{D}$ is expressed in Eq. (2.47) as the sum of such matrices. Equation (2.41) can then be written as

$$\mathbf{D}|\phi\rangle = \kappa_0^2 |\phi\rangle. \qquad (2.48)$$

There are two ways to introduce randomness into the lattice problem—either via random variation in the atomic energy level or via random variation in β. If $\Delta\epsilon_\mathbf{l}$ denotes the random deviation (from ϵ_0) of the atomic energy level at site $\mathbf{l}$, then $\Delta\epsilon_\mathbf{l}$ randomizes only the diagonal matrix elements. For the phonon case, the same effect is obtained by varying the mass of the particle at each site by $\Delta m_\mathbf{l}$. When there is randomness, κ_0^2 should be replaced by $\kappa_0^2 - \sigma(\mathbf{l})$ in Eq. (2.41), where

$$\sigma(\mathbf{l}) = \begin{cases} \dfrac{2m\,\Delta\epsilon_\mathbf{l}}{\hbar^2} & \text{quantum} \\ \dfrac{\omega^2}{v_0^2}\dfrac{\Delta m_\mathbf{l}}{m} & \text{classical}. \end{cases} \qquad (2.49)$$

Since one is free to define ϵ_0 and m as the mean values of the respective quantities, $\Delta\epsilon_\mathbf{l}$ and $\Delta m_\mathbf{l}$ may be regarded as having zero mean:

$$\langle \sigma(\mathbf{l}) \rangle_c = 0, \qquad (2.50)$$

where $\langle\ \rangle_c$ means configurational averaging, which is equivalent to averaging over all sites in the present case. By defining a new matrix $\mathbf{M}$ such that the diagonal elements consist of only the deviations, i.e.,

$$\mathbf{M} = \sum_{\mathbf{l}} \sigma(\mathbf{l}) |\mathbf{l}\rangle\langle\mathbf{l}| + \sum_{\mathbf{l},\mathbf{n}} \left(-\frac{\beta}{a^2}\right) |\mathbf{l}\rangle\langle\mathbf{l}+\mathbf{n}|, \qquad (2.51)$$

2.3 Waves on a Discrete Lattice

where κ_0^2 is now renormalized with respect to $\epsilon_0 - (\hbar^2 d/m^* a^2)$:

$$\kappa_0^2 = \frac{2m(E - \epsilon_0)}{\hbar^2} + \frac{2\beta d}{a^2}, \tag{2.42}$$

with

$$\beta = \frac{m}{m^*}. \tag{2.43}$$

So far we have discussed only quantum waves on a lattice. However, Eq. (2.41) is equally applicable to waves in a classical spring-and-mass lattice. This can be seen by letting $\beta = 1$ and $\kappa_0^2 = \omega^2/v_0^2$, where $v_0^2 = Ka^2/m$ is the wave speed in the long-wavelength limit, K being the spring constant of the springs linking identical nearest-neighbor particles of mass m. Equation (2.41) then becomes

$$Ka^2 \Delta^{(2)} \phi(\omega, \mathbf{l}) = m\omega^2 \phi(\omega, \mathbf{l}). \tag{2.44}$$

The left-hand side of Eq. (2.44) is the negative of the force on the particle situated at $\mathbf{l}$, if ϕ denotes the displacement of the particle away from its equilibrium position. For example, in 1D it is

$$K\{[\phi(\omega, l+1) - \phi(\omega, l)] - [\phi(\omega, l) - \phi(\omega, l-1)]\},$$

which gives the force on the particle at $\mathbf{l}$ due to the net contraction/extension of the springs on its two sides. The right-hand side of Eq. (2.44) is the negative mass times acceleration of the particle (at frequency ω). Therefore this is precisely the dynamical equation for the vibrational modes of an elastic lattice. One thus concludes that Eq. (2.41) is the general form for both the quantum and the classical waves on a lattice.

An alternative way of expressing Eq. (2.41) is to use a vector $\boldsymbol{\phi}$ to express the wave function, where each component of the vector corresponds to the value of the wave function at a site. The operation $-\beta \Delta^{(2)}$ can then be expressed by a matrix $\mathbf{D}$ whose diagonal elements are

$$(\mathbf{D})_{\mathbf{l},\mathbf{l}} = \frac{2\beta d}{a^2} \tag{2.45}$$

and whose off-diagonal elements are

$$(\mathbf{D})_{\mathbf{l},\mathbf{l}+\mathbf{n}} = -\frac{\beta}{a^2} \tag{2.46}$$

and zero otherwise. A succinct way of writing the matrix is by using Dirac's bra and ket notation, where $|\phi\rangle = \boldsymbol{\phi}$ means the wave function expressed equivalent to a state with $|k_\alpha| \le \pi/a$. A related second difference is that $e(\mathbf{k})$ has a maximum, defined by $4d/a^2$. The eigenvalues of $-\Delta^{(2)}$ thus form a band, extending from 0 to $4d/a^2$. At the maximum, which occurs at $|k_\alpha| = \pi/a$, the dispersion relation gives a group velocity $de(\mathbf{k})/dk_\alpha = 0$ in the direction of k_α, which indicates a standing wave caused by Bragg scattering from the lattice structure. At frequencies beyond the maximum so that $e(\mathbf{k}) > 4d/a^2$, i.e., outside the band, it is clear that the only way for Eq. (2.37) to be satisfied is by switching k_α from real to imaginary so that $\cos(k_\alpha a)$ becomes $\cosh(k_\alpha a)$. However, a "state" which is associated with an imaginary k must decay exponentially toward infinity. These waves are called evanescent waves and are analogous to what happens in waveguides below the cutoff frequency. The third difference is that on a lattice, the "mass" of a quantum wave/particle is not necessarily the same as that in the continuum. Since in the long-wavelength limit the dispersion relation of the lattice wave has the same form as that in the continuum, the *effective mass* m^* on a lattice may be defined by the curvature of the kinetic energy expression around $\mathbf{k} = 0$:

$$\frac{1}{m^*} = \frac{1}{\hbar^2} \frac{d^2}{dk^2} (\text{kinetic energy})_{\mathbf{k}=0}. \tag{2.39}$$

The kinetic energy is dependent on the strength of coupling between the neighboring sites. If the interaction between neighboring sites is weak, the electron tends to be bound at a given site, and the transfer between the sites would occur slowly, thus implying a large m^*. The effective mass m^* may therefore be regarded as a parameter for measuring the lattice effect on the electron.

The total energy E for a quantum wave on a lattice is

$$E = \frac{\hbar^2}{2m^*} e(\mathbf{k}) + \epsilon_0 - \frac{\hbar^2}{2m^*}\left(\frac{2d}{a^2}\right), \tag{2.40}$$

where ϵ_0 is the energy of the atomic level, defined relative to potential $V = 0$. The fact that ϵ_0 always defines the center of the energy band accounts for the last term, which is half of the energy bandwidth. Equation (2.40) relates $\mathbf{k}$ and $E = \hbar\omega$ for the quantum Schrödinger wave on a lattice. Just as in the case of waves in continuum, the corresponding wave equation may be inferred from the dispersion relation. By noting that $e(\mathbf{k})$ is obtained as the eigenvalue of $-\Delta^{(2)}$, it is easy to see that the lattice wave equation is

$$(\beta \Delta^{(2)} + \kappa_0^2)\phi(\omega, \mathbf{l}) = 0, \tag{2.41}$$

2.3 Waves on a Discrete Lattice

tion. Since in continuum $\exp(i\mathbf{k} \cdot \mathbf{r})$ is the eigenfunction of the Laplacian operator ∇^2, the obvious choice in the present case is the Laplacian defined on the lattice, i.e., a discrete Laplacian $\Delta^{(2)}$. In 1D, where $l = \cdots, -2, -1, 0, 1, 2, \cdots$ and the lattice constant is denoted by a, the operation of the discrete Laplacian is defined by

$$\Delta^{(2)}\phi(\omega, l) = \frac{1}{a^2}[\phi(\omega, l+1) + \phi(\omega, l-1) - 2\phi(\omega, l)]. \quad (2.34)$$

Equation (2.34) expresses the operation of taking the discrete second derivative of ϕ. For a simple square lattice in 2D and a simple cubic lattice in 3D, the operation of $\Delta^{(2)}$ is a direct extension from 1D:

$$\Delta^{(2)}\phi(\omega, \mathbf{l}) = \frac{1}{a^2}\left[-2d\phi(\omega, \mathbf{l}) + \sum_{\mathbf{n}} \phi(\omega, \mathbf{l}+\mathbf{n})\right], \quad (2.35)$$

where $\mathbf{n}$ denotes the vectors pointing from the lattice site $\mathbf{l}$ to its nearest neighbors. The sum therefore has four terms in 2D and six terms in 3D. By substituting $\exp(-i\omega t + i\mathbf{k} \cdot \mathbf{l})$ for $\phi(\omega, \mathbf{l})$ and similar forms for $\phi(\omega, \mathbf{l}+\mathbf{n})$ into Eqn (2.35), we get

$$\Delta^{(2)}\phi(\omega, \mathbf{l}) = -e(\mathbf{k})\phi(\omega, \mathbf{l}), \quad (2.36)$$

where $e(\mathbf{k})$ is given by

$$e(\mathbf{k}) = \frac{2}{a^2}\sum_{\alpha=1}^{d}(1 - \cos k_\alpha a). \quad (2.37)$$

Here $k_{1,2,3}$ means $k_{x,y,z}$. When $k_\alpha a \to 0$, i.e., in the long-wavelength limit, $1 - \cos k_\alpha a \cong k_\alpha^2 a^2/2$, and

$$e(\mathbf{k}) \cong \sum_{\alpha=1}^{d} k_\alpha^2 = k^2, \quad (2.38)$$

which corresponds precisely to the negative of the eigenvalue for the Laplacian. Since $\hbar^2 k^2/2m$ is the kinetic energy of a quantum wave/particle, $\hbar^2 e(\mathbf{k})/2m^*$ may thus be identified as the similar quantity on a lattice. Here m^* is the "effective mass," to be defined below.

Equation (2.37) indicates significant differences between lattice waves and waves in continuum. In the first place, $e(\mathbf{k})$ is periodic in $\mathbf{k}$. This is due to the fact that lattice has a minimum length scale, a, which means that there cannot be a wave with wavelength less than $2a$, or a $\mathbf{k}$ with $|k_\alpha| \geq \pi/a$. When $|k_\alpha|$ is formally extended beyond π/a, it becomes quadratic in ω for classical waves, the Green's functions for the two types of waves are expected to differ in the $(t, \mathbf{r})$ domain (Economou, 1983). However, they share the same causal characteristics: In the time domain $G_0^+(t, |\mathbf{r} - \mathbf{r}'|) = 0$ for $t < 0$ and $G_0^-(t, |\mathbf{r} - \mathbf{r}'|) = 0$ for $t > 0$. This fact is easy to see in the case of 3D classical waves, where the Fourier transform of the numerator of (2.31) gives $\delta(|\mathbf{r} - \mathbf{r}'|/v_0 \mp t)$. Thus, G_0^+ is usually denoted the retarded (with respect to $t = 0$) Green's function, whereas G_0^- is denoted the advanced Green's function.

2.3 Waves on a Discrete Lattice

In the case of electrons in solids, the interaction of an electron with the solid lattice is known to play a dominant role in the electron's resulting behavior. It is therefore desirable that the lattice effect be taken into account first, before examining any effect due to disorder. Basic to the consideration of the lattice effect is the lattice potential. If the potential V at each site of an ordered, perfect lattice is given by a Coulomb or some form of attractive potential, then the solution of the Schrödinger's equation for each site, considered in isolation from other sites, would result in many atomic energy levels. On adding to this problem the interaction between the sites, i.e., the effect of the lattice structure, the combined state of affair can become very complicated indeed. However, in many electronic problems of physical interest only one atomic state is important, e.g., the valence electronic state. If the energy separations between the valence state and other atomic states are large compared to the interaction energy between the sites, then the problem can be simplified by solution in two stages. The first stage considers only the atomic-level problem for each site, in isolation from other sites. The next stage considers the effect of lattice structure on each atomic level, without worrying about the intermixing of the atomic levels. This approach is the basis of the "tight-binding approximation" in which the lattice-structure effect is treated as a perturbation on the single-site atomic energy levels. In what follows, the description of the quantum Schrödinger wave is based on this tight binding approach, where the one relevant atomic level is assumed to be identified, and our attention is therefore focused on the effect of the lattice structure plus the randomness.

The question before us is: If $\phi(t, \mathbf{l})$ stands for the wave amplitude of the relevant atomic level at lattice site $\mathbf{l}$ and time t, then what is the effect of the lattice on its frequency component $\phi(\omega, \mathbf{l})$? To answer this question, let us start with the wave form in a perfect lattice, $\exp(-i\omega t + i\mathbf{k} \cdot \mathbf{l})$, and ask what general spatial operator would yield $\exp(i\mathbf{k} \cdot \mathbf{l})$ as its eigenfunc-

2.2 Green's Functions for Waves in a Uniform Medium

(note the interchange of $\mathbf{r}$ and $\mathbf{r}'$), and G_0^+, G_0^- are complex conjugates in both the real space and the momentum space. These and other facts about the Green's function are made clear in the solution to Problem 2.3.

The Green's function in the $(\omega, \mathbf{k})$ domain is noted to have the same expression in different spatial dimensionalities. However, $G_0^\pm(\omega, \mathbf{r} - \mathbf{r}')$ is different in different spatial dimensions. This can be seen by the Fourier transformation of $G_0^\pm(\omega, \mathbf{k})$. In 3D, we have

$$G_0^\pm(\omega, \mathbf{r} - \mathbf{r}') = \lim_{\eta \to 0} \frac{1}{4\pi^2} \frac{L^d}{N}$$

$$\times \int_0^\infty \frac{k^2 \, dk}{\kappa_0^2 - k^2 \pm i\eta} \int_0^\pi d(\cos\theta) \exp(ik|\mathbf{r} - \mathbf{r}'|\cos\theta) \frac{N}{L^d}$$

$$= \lim_{\eta \to 0} \frac{1}{4\pi^2 |\mathbf{r} - \mathbf{r}'|} \frac{1}{i} \int_{-\infty}^{\infty} \frac{k \exp(ik|\mathbf{r} - \mathbf{r}'|)}{\kappa_0^2 - k^2 \pm i\eta} \, dk$$

$$= -\frac{\exp[\pm i(\kappa_0 \pm i\eta)|\mathbf{r} - \mathbf{r}'|]}{4\pi|\mathbf{r} - \mathbf{r}'|} \quad (2.31)$$

through contour integration in the upper complex k plane (so that the integrand vanishes in the limit of $k \to +i\infty$). $G_0^\pm(\omega, \mathbf{r} - \mathbf{r}')$ is seen to correspond to outgoing ($+$) and incoming ($-$) spherical waves. The sign of $i\eta$ determines whether the wave is outgoing or incoming because G_0 must vanish as $|\mathbf{r} - \mathbf{r}'| \to \infty$, so that $\kappa_0 + i\eta$ must be outgoing whereas $\kappa_0 - i\eta$ must be incoming. In 2D, similar contour integration and comparison with the integral representation of the Hankel function yield $G_0^\pm$ as outgoing and incoming cylindrical waves:

$$G_0^\pm(\omega, \mathbf{r} - \mathbf{r}') = \mp \frac{i}{4} H_0^{(1,2)}(\kappa_0|\mathbf{r} - \mathbf{r}'|), \quad (2.32)$$

where $H_0^{(1,2)}$ denotes the zeroth-order Hankel function of the first (outgoing wave) kind and the second (incoming wave) kind. In 1D, $G_0^\pm$ is given by

$$G_0^\pm(\omega, \mathbf{r} - \mathbf{r}') = \mp \frac{i}{2\kappa_0} \exp(\pm i\kappa_0|\mathbf{r} - \mathbf{r}'|). \quad (2.33)$$

It is easy to check that the imaginary part of $G_0^\pm(\omega, \mathbf{r} = \mathbf{r}')$, when it is substituted into Eq. (2.29), gives the correct density of states. A further Fourier transform with respect to frequency ω would yield Green's function in the $(t, \mathbf{r})$ domain. Since κ_0^2 is linear in ω for quantum waves and for the quantum case, and

$$\rho_0(\omega) = \begin{cases} \dfrac{\omega^2}{2\pi^2 v_0^3} & \text{3D} \\ \dfrac{\omega}{2\pi v_0^2} & \text{2D} \\ \dfrac{1}{\pi v_0} & \text{1D} \end{cases} \qquad (2.27)$$

for the classical wave case. Here ω is defined to be positive.

In Eq. (2.24), the integral may be reexpressed as

$$\int \frac{d\mathbf{k}}{(2\pi)^d} \delta[\kappa_0^2(\omega) - k^2] = -\frac{1}{\pi} \frac{L^d}{N} \int \frac{d\mathbf{k}}{(2\pi)^d} \operatorname{Im} G_0^+(\omega, \mathbf{k})$$

$$= -\frac{1}{\pi} \operatorname{Im} G_0^+(\omega, \mathbf{r} = \mathbf{r}'), \qquad (2.28)$$

where Im means taking the imaginary part of G^+. The last equality in Eq. (2.28) follows because at $\mathbf{r} = \mathbf{r}'$, the phase factor $\exp[i\mathbf{k} \cdot (\mathbf{r} - \mathbf{r}')]$ in the Fourier transform of $G_0^+(\omega, \mathbf{k})$ becomes 1, so the real and imaginary parts of $G_0^+(\omega, \mathbf{r} = \mathbf{r}')$ are just the $\mathbf{k}$-integrals of the real and imaginary parts of $G_0^+(\omega, \mathbf{k})$. Substitution of Eq. (2.28) into Eq. (2.24) gives

$$\rho_0(\omega) = -\frac{d\kappa_0^2(\omega)}{d\omega} \frac{1}{\pi} \operatorname{Im} G_0^+(\omega, \mathbf{r} = \mathbf{r}'). \qquad (2.29)$$

The reason for expressing the density of states in terms of the imaginary part of the Green's function (in the ω and $\mathbf{r}$ representation) is the general validity of the formula not only for waves in a uniform medium but also for waves in random media. In solution to Problem 2.3, it is shown that

$$\rho(\omega, \mathbf{r}) = -\frac{d\kappa_0^2(\omega)}{d\omega} \frac{1}{\pi} \operatorname{Im} G^+(\omega, \mathbf{r} = \mathbf{r}') \qquad (2.30)$$

gives the "local density of states" at $\mathbf{r}$ in the general case. The fact that the density of states can always be obtained this way is due to the Green's function's general property that it diverges at the resonant modes of the system, and as a result its imaginary part offers a way of picking them out as seen from Eq. (2.23). Another general property of the Green's function is that $G^+(\omega, \mathbf{r}, \mathbf{r}')$ and $G^-(\omega, \mathbf{r}', \mathbf{r})$ are complex conjugates of each other

2.2 Green's Functions for Waves in a Uniform Medium

result is

$$\frac{L^d}{N} G_0^{\pm}(\omega, \mathbf{k}) = \lim_{\eta \to 0} \frac{1}{\kappa_0^2(\omega) - k^2 \pm i\eta}$$

$$= P \frac{1}{\kappa_0^2(\omega) - k^2} \mp i\pi\delta[\kappa_0^2(\omega) - k^2]. \quad (2.23)$$

Here the sign of the imaginary constant is important in determining the behavior of the Green's function in real space, as will be seen later. The symbol P means taking the principal value of $[\kappa_0^2(\omega) - k^2]^{-1}$ as that quantity diverges, which is defined in the solution to Problem 2.1. The imaginary part of the Green's function, on the other hand, is interesting because the delta function picks out exactly the modes of the system. If one wants to count the number of modes per unit frequency range and per unit volume L^d/N, i.e., to get the density of states $\rho_0(\omega)$, it can be achieved by integrating the delta function over all possible free-space eigenstates that are characterized by the $\mathbf{k}$ vectors, i.e.,

$$\rho_0(\omega) = \frac{d\kappa_0^2(\omega)}{d\omega} \int \frac{d\mathbf{k}}{(2\pi)^d} \delta[\kappa_0^2(\omega) - k^2]. \quad (2.24)$$

Here the factor

$$\frac{d\kappa_0^2(\omega)}{d\omega} = \begin{cases} \dfrac{2m}{\hbar} & \text{quantum} \\ \dfrac{2\omega}{v_0^2} & \text{classical} \end{cases} \quad (2.25)$$

accounts for the fact that the delta function counts the states in units of κ_0^2, so it is required for conversion to states per unit frequency. It is shown in Problem 2.2 that $\rho_0(\omega)$ is given by

$$\rho_0(\omega) = \begin{cases} \dfrac{m^{3/2}}{\sqrt{2}\,\hbar^{3/2}\pi^2} \sqrt{\omega} & \text{3D} \\ \dfrac{m}{2\pi\hbar} & \text{2D} \\ \dfrac{1}{\pi}\sqrt{\dfrac{m}{2\hbar}} \dfrac{1}{\sqrt{\omega}} & \text{1D} \end{cases} \quad (2.26)$$

3.7 CPA—The Case of Classical Waves

worked out in the solution to Problem 3.9. For $d = 3$ and 2, the resulting CPA equation is

$$p\frac{\bar{\epsilon} - \epsilon_1}{(d-1)\bar{\epsilon} + \epsilon_1} + (1-p)\frac{\bar{\epsilon} - \epsilon_2}{(d-1)\bar{\epsilon} + \epsilon_2} = 0. \quad (3.78)$$

Equation (3.78) is known in the literature as Bruggeman's symmetric effective medium equation (Bruggeman, 1935). Also shown in the solution to Problem 3.9 is that in 1D, the modifications result in no change to the final answer, so the CPA equation remains the same as the $n = 0$ scalar case. Through similar considerations leading to Eq. (3.76), the electromagnetic mean free path in a symmetric-microstructure composite is obtained as

$$l_{em} = \begin{cases} \kappa_e^{-1}\left[p(\kappa_e R_1)^3\left(\frac{\bar{\epsilon}-\epsilon_1}{2\bar{\epsilon}+\epsilon_1}\right)^2 + (1-p)(\kappa_e R_2)^3\left(\frac{\bar{\epsilon}-\epsilon_2}{2\bar{\epsilon}+\epsilon_2}\right)^2\right]^{-1} & \text{3D} \\ (2\pi\kappa_e)^{-1}\left[p(\kappa_e R_1)^2\left(\frac{\bar{\epsilon}-\epsilon_1}{\bar{\epsilon}+\epsilon_1}\right)^2 + (1-p)(\kappa_e R_2)^2\left(\frac{\bar{\epsilon}-\epsilon_2}{\bar{\epsilon}+\epsilon_2}\right)^2\right]^{-1} & \text{2D} \end{cases}$$

$$(3.79)$$

where $\bar{\epsilon}$ is the solution to Eq. (3.78).

For finite $\kappa_e R$, the electromagnetic wave scattering can have channels other than the one presented, which pertains to electric dipole scattering. Next in the order of importance is the magnetic dipole scattering (Bohren and Huffman, 1983). When $\kappa_e R$ exceeds a certain value, the magnetic dipole scattering becomes dominant. However, the electric dipole scattering always dominates in the $\kappa_e R \to 0$ limit.

It should be noted that the Rayleigh frequency dependence and the particle size dependence of l_{em} are the same here as for the scalar wave scattering. However, the CPA equation (3.78) gives a very different effective medium than that for the scalar wave CPA equation (3.72). In the solution to Problem 3.10, it is shown that Eq. (3.78) predicts a *percolation threshold* when the material properties of the two components have infinite contrast, e.g., when $\epsilon_1 = 1$, $\epsilon_2 = 0$, the effective medium $\bar{\epsilon} = 0$ below the threshold $p_c = 1/3$. This property has a simple geometric interpretation. Suppose a random medium is composed of an electrically conducting component whose material property is represented by $\epsilon_1 = 1$ and an insulating component whose material property is represented by $\epsilon_2 = 0$. The electromagnetic boundary condition $\bar{\epsilon}(\partial \bar{\phi}/\partial n) = \epsilon_2(\partial \phi_2/\partial n)$, where

$\partial/\partial n$ means the derivative normal to the interface, implies $\partial\bar{\phi}/\partial n = 0$ at the interface with component 2; i.e., no electrical current can flow into component 2. The electrical current is therefore confined to component 1. In order for $\bar{\epsilon} \neq 0$, it is clear that the ϵ_1 component must be geometrically connected to form an infinite conducting network that spans the sample. Statistically, this proves to be possible only if p is greater than a threshold value p_c, called the percolation threshold. In contrast, the CPA Eq. (3.72) gives no percolation threshold behavior because for the scalar wave, even if the electromagnetic boundary condition were applied so that $\partial\bar{\phi}/\partial n$ vanishes at the interface with component 2, the physical quantity in this case is ϕ, which is still continuous across the interface. In contrast, $\mathbf{E} = -\nabla\phi$ is the physical quantity in the electromagnetic case. The lesson here is that the (long-wavelength) expansion of the $n = 0$ term gives an effective medium theory for ϕ, but the $n = 1$ term expansion gives an effective medium theory for $\nabla\phi$.

3.7.2 The Dispersion Microstructure

In the case of the dispersion microstructure, the same mathematical considerations are applied to a single type of structural unit, shown in Figure 3.7b. In a random medium, the local volume fraction of the dispersed component can fluctuate from one region to the next. Therefore the CPA equation (3.68) should be expressed as

$$\int_0^1 \mathscr{D}(p') \langle \boldsymbol{\kappa}_e | \bar{\mathbf{t}}^+ | \boldsymbol{\kappa}_e \rangle_{p'} \, dp' = 0, \qquad (3.80)$$

where $\mathscr{D}(p')$ gives the distribution of local volume fraction p', and the forward scattering amplitude (noted to depend on p') is given in the solution to Problem 3.11. In the long-wavelength limit, Eq. (3.80) becomes

$$\bar{\epsilon} = p\epsilon_1 + (1-p)\epsilon_2 \qquad (3.81)$$

for 1D, 2D, and 3D, where p is the globally averaged volume fraction of the dispersed component. This equation is noted to be identical to that of the symmetric microstructure case. However, the mean free path expression is different:

$$l = \frac{B_d \bar{\epsilon}}{(\kappa_e R_0)^d \kappa_e} \left\{ \int_0^1 \mathscr{D}(p')[\bar{\epsilon} - (1-p')\epsilon_2 - p'\epsilon_1]^2 \, dp' \right\}^{-1}. \qquad (3.82)$$

3.7 CPA—The Case of Classical Waves

Here $B_d = 3\epsilon_2$, $4\epsilon_2/\pi$, $\bar{\epsilon}$ for $d = 3, 2, 1$, respectively. In Eqs. (3.80) and (3.82), a suitable form of the distribution function may be taken to be

$$\mathcal{D}(p')\,dp' = \mathcal{D}[\eta(p')]\frac{d\eta}{dp'}\,dp' = \frac{dp'}{\sqrt{2\pi}s_0 p'(1-p')}\exp\bigl[-(\eta-\eta_0)^2/2s_0^2\bigr], \tag{3.83a}$$

where

$$\eta(p') = \ln\frac{p'}{1-p'}. \tag{3.83b}$$

The two parameters η_0 and s_0 may be determined from the average volume fraction p and the variance of the local compositional fluctuations. In applying Eqs. (3.81) and (3.82) to acoustic wave propagation in colloidal systems, ϵ should be replaced by the inverse of the bulk modulus as discussed before. Also, if the dispersed particles have a well-defined size, then $R_0^d \approx R^d/p$ in Eq. (3.82) so that one may use the size of the dispersants directly. However, if the size of the particles varies over a large range, then R_0^d means its average value, i.e., the dth moment of the distribution function for R_0.

The electromagnetic case yields a different CPA equation for the dispersion microgeometry than that for the symmetric microgeometry in 3D and 2D. The details are worked out in the solution to Problems 3.12. The answer is,

$$\int_0^1 \mathcal{D}(p')\mathcal{Q}(p')\,dp' = 0, \tag{3.84a}$$

where

$$\mathcal{Q}(p') = \frac{(\bar{\epsilon}-\epsilon_2)[\epsilon_1+(d-1)\epsilon_2]+p'[\bar{\epsilon}+(d-1)\epsilon_2](\epsilon_2-\epsilon_1)}{\bigl[\sqrt{\epsilon_2\bar{\epsilon}}+(d-1)\epsilon_2\bigr][\epsilon_1+(d-1)\epsilon_2]+(d-1)p'\bigl(\sqrt{\epsilon_2\bar{\epsilon}}-\epsilon_2\bigr)(\epsilon_1-\epsilon_2)}. \tag{3.84b}$$

The corresponding mean free path expression is given by

$$l_{em} = \frac{B'_d \epsilon_2/\bar{\epsilon}}{\kappa_e(\kappa_e R_0)^d}\left[\int_0^1 \mathcal{D}(p')\mathcal{Q}^2(p')\,dp'\right]^{-1}, \tag{3.85}$$

where the $B'_d = 1, 2/\pi$ for $d = 3, 2$, respectively. When $\mathcal{D}(p') = \delta(p' - p)$, Eq. (3.84) reduces to

$$\frac{\bar{\epsilon} - \epsilon_2}{\bar{\epsilon} + (d-1)\epsilon_2} - p\frac{\epsilon_1 - \epsilon_2}{\epsilon_1 + (d-1)\epsilon_2} = 0, \quad (3.86)$$

which is known as the Maxwell-Garnett theory in the literature (Maxwell-Garnett, 1904). It was first derived through the Clausius–Massotti relation. In the present derivation, much more emphasis is given to the role of the microstructure.

In 1D, both the CPA equation and the expression for l_{em} are shown to be identical to Eq. (3.81) and (3.82).

Comparison of Eqs. (3.86) and (3.78) shows differences in their predictions, the most significant of which is the absence of a percolation threshold for $\bar{\epsilon}$ in the dispersion microgeometry. This is understandable because the dispersed phase is always prevented from forming a connected network by the presence of the coating, and therefore at the infinite contrast limit $\bar{\epsilon}$ reflects the property of the matrix phase (the coating component) only.

3.8 Accuracy of the CPA

The meaning of accuracy depends on the reference standard for comparison. Since the $\mathbf{G}_e$ derived from the CPA is an approximation to $\langle \mathbf{G} \rangle_c$, its accuracy should be judged by comparison with $\langle \mathbf{G} \rangle_c$. In this context, the crucial approximation involved is Eq. (3.56), where $\langle \bar{\mathbf{T}} \rangle_c$ is expressed as the sum of individual $\langle \bar{\mathbf{t}}_i \rangle_c$'s. In the solution to Problem 3.13, it is shown that the first few terms of $\langle \bar{\mathbf{T}} \rangle_c$ may be written as

$$\langle \bar{\mathbf{T}} \rangle_c = \sum_i \langle \bar{\mathbf{t}}_i \rangle_c + \sum_i \sum_{j \neq i} \langle \bar{\mathbf{t}}_i \mathbf{G}_e \bar{\mathbf{t}}_j \rangle_c + \sum_i \sum_{j \neq i} \sum_{k \neq j} \langle \bar{\mathbf{t}}_i \mathbf{G}_e \bar{\mathbf{t}}_j \mathbf{G}_e \bar{\mathbf{t}}_k \rangle_c$$
$$+ \sum_i \sum_{j \neq i} \sum_{k \neq j} \sum_{l \neq k} \langle \bar{\mathbf{t}}_i \mathbf{G}_e \bar{\mathbf{t}}_j \mathbf{G}_e \bar{\mathbf{t}}_k \mathbf{G}_e \bar{\mathbf{t}}_l \rangle_c + \cdots . \quad (3.87)$$

For the second term, the fact that $j \neq i$ means averaging over every subset of configurations with a fixed $\bar{\mathbf{t}}_i$ would give zero under the CPA condition. The third term can have $i = k$, but there is always a single $\bar{\mathbf{t}}$ left over to be separately averaged, thus also giving zero. Therefore, the first nonzero correction to the CPA occurs in the fourth-order term ($i = k, j = l$). The CPA is thus highly accurate. Physically, the higher moments missed by the

CPA represent multiple scatterings from the same site(s), which could be important if there are local resonances or bound states.

Another inaccuracy of the effective medium description arises from the inability to capture fully the microstructure information. What have been presented in this chapter are only two cases whose solutions can be easily handled analytically in the long-wavelength limit. There are many varieties of correlations in physical samples which could be difficult to capture within a simple analytical framework. As one can see from the two examples given, the electronmagnetic CPA equations demonstrate very different predictions for the two microstructures. Therefore any inaccuracy in the microstructure input to the theory can result in serious error. However, this kind of inaccuracy is not intrinsic to the CPA approach and therefore should not be regarded as an indictment of the effective medium description.

3.9 Extension of the CPA to the Intermediate Frequency Regime

It has been emphasized that for classical waves in a continuum, the CPA approach breaks down when the wavelength becomes comparable to the scale of the scatterers. In this so-called intermediate-frequency regime, the self-energy is $\mathbf{k}$ dependent, and the CPA condition $\langle \overline{\mathbf{T}} \rangle_c = 0$ has no solution (or no unique solution) in terms of the effective medium parameter. However, there may still be partially coherent, local excitations of the medium even in the absence of a valid CPA solution. An alternative, but fruitful, approach to looking at what happens in the intermediate-frequency regime is to abandon the CPA requirement that $\langle \overline{\mathbf{T}} \rangle_c = 0$ and to monitor instead the peak(s) of the so-called spectral function, $-\mathrm{Im}\, G_e^+(\omega, \mathbf{k})$.

We may recall that in a uniform medium, $-\mathrm{Im}\, G_0^+(\omega, \mathbf{k})$ is a delta function that picks out the dispersion relation of the medium. In a random medium,

$$\langle G^+ \rangle_c(\omega, \mathbf{k}) = \frac{1}{(\omega^2/v_e^2) - k^2 - \Sigma_{v_e}^+} \frac{N}{L^d}, \qquad (3.88)$$

where the choice of a reference v_e can be arbitrary because the wave equations

$$\left[\nabla^2 + \frac{\omega^2}{v_e^2} + \omega^2 \left(\frac{1}{v^2(\mathbf{r})} - \frac{1}{v_e^2} \right) \right] \phi = 0 \qquad (3.89)$$

is invariant under the addition and subtraction of the same constant term. $\Sigma_{v_e}^+$ means the self-energy is calculated with a particular choice of v_e. When the CPA works, a choice of $v_e = \bar{v}$ can be found which makes $\operatorname{Re} \bar{\Sigma}^+ = 0$, so that

$$G_e^+(\omega, k) = \frac{1}{(\omega^2/\bar{v}^2) - k^2 - i \operatorname{Im} \bar{\Sigma}^+} \frac{N}{L^d}, \qquad (3.90)$$

where $\operatorname{Im} \bar{\Sigma}^+$ defines the mean free path. If we look at $-\operatorname{Im} G_e^+(\omega, \mathbf{k})$ in this case, it is clear that the peak of the spectral function coincides exactly with the dispersion relation of the effective medium, although it is no longer a delta function.

When $\operatorname{Re} \bar{\Sigma}^+$ can no longer be made equal to zero, one can still look for the peak(s) in the spectral function and identify the excitations of the system. However, due to the expected broad width of the peak(s), these modes should be called the *quasi-modes*. Moreover, the whole process of a mode turning into a quasi-mode as ω increases may now be viewed as a continuous process, thus avoiding a pitfall of the CPA, where the abrupt termination of a solution at a certain frequency may mislead one into concluding wrongly that an excitation mode has vanished abruptly.

To implement this simple extension of CPA to the intermediate frequency regime, let us utilize the freedom in choosing a dummy v_e as

$$v_e = \frac{\omega}{k}, \qquad (3.91)$$

so that

$$-\frac{L^d}{N} \operatorname{Im} \langle G^+ \rangle_c(\omega, \mathbf{k}) = \operatorname{Im} \frac{1}{\Sigma^+} = \frac{-\operatorname{Im} \Sigma^+}{(\operatorname{Re} \Sigma^+)^2 + (\operatorname{Im} \Sigma^+)^2}. \qquad (3.92)$$

Here, the Σ^+ is understood to be evaluated in an effective medium with the wave speed ω/k. Σ^+ is thus implicitly a function of both ω and k. By plotting a two-dimensional map of the spectral function calculated in the above manner, e.g., by indicating the height of the spectral function by color, the peak(s) may be attributed the significance of being the dispersion relation(s) of the modes or quasi-modes of the system. In this manner, the spectral function approach not only reproduces all the CPA results when the CPA is valid but also extends its validity beyond the CPA regime. This approach is thus denoted the generalized CPA (GCPA).

Since Σ^+ is difficult to calculate exactly, one can use the approximation of Eq. (3.67a). This approximation is accurate when $\Sigma^+ \cong 0$, i.e., near the peak(s) of the spectral function as given by (3.92). Therefore the accuracy is best where it matters most and is worst where one does not care anyway. The predictive power and accuracy of the GCPA have been demonstrated by its application to the quantitative explanation of experimental data on acoustic excitations in colloidal systems. It was found experimentally (Liu *et al.*, 1990) that when the wavelength becomes comparable to the size of the scatterers in a colloidal suspension, two acoustic modes can appear. The dispersion relations of the two modes were accurately predicted by the GCPA approach in the regime where the CPA fails. The readers are referred to the literature for more details (Jing *et al.* 1992).

Problems and Solutions

3.1 Show that $\langle G \rangle_c$ is a function of $\mathbf{r} - \mathbf{r}'$ in the real-space representation and of only one $\mathbf{k}$ in the wave vector representation.

The general Green's function $G(\omega, \mathbf{r}, \mathbf{r}')$ may be expressed in the wave vector representation as

$$G(\omega, \mathbf{r}, \mathbf{r}') = \frac{L^{2d}}{N^2 (2\pi)^{2d}} \int \int d\mathbf{k} \, d\mathbf{k}' \, G(\omega, \mathbf{k}, \mathbf{k}') \exp(i\mathbf{k} \cdot \mathbf{r}) \exp(-i\mathbf{k}' \cdot \mathbf{r}').$$

(P3.1)

By transforming $\mathbf{r}$ and $\mathbf{r}'$ to the center-of-mass coordinates $\mathbf{R}_c = (\mathbf{r} + \mathbf{r}')/2$ and $\Delta \mathbf{R} = (\mathbf{r} - \mathbf{r}')/2$, Eq. (P3.1) becomes

$$G(\omega, \mathbf{r}, \mathbf{r}') = \frac{L^{2d}}{N^2 (2\pi)^{2d}} \int \int d\mathbf{k} \, d\mathbf{k}' \, G(\omega, \mathbf{k}, \mathbf{k}')$$
$$\times \exp[i(\mathbf{k} - \mathbf{k}') \cdot \mathbf{R}_c] \exp[i(\mathbf{k} + \mathbf{k}') \cdot \Delta \mathbf{R}].$$ (P3.2)

The action of averaging over configurations of $\sigma(\mathbf{r})$ consists of two steps. The first is the replacement of $G(\omega, \mathbf{k}, \mathbf{k}')$ by its averaged value, $\langle G(\omega, \mathbf{k}, \mathbf{k}') \rangle_c$. The second is to fix $\Delta \mathbf{R}$ and let the source–detector pair be situated in all possible environments. Mathematically, this is easily carried out by the operation

$$\sum_{\mathbf{R}_c} G(\omega, \mathbf{r}, \mathbf{r}') = \frac{N}{L^d} \int G(\omega, \mathbf{r}, \mathbf{r}') \, d\mathbf{R}_c,$$ (P3.3)

The integration over $\mathbf{R}_c$ inside the integral of Eq. (P3.2) is seen to give

$$\frac{N}{L^d} \int \exp[i(\mathbf{k} - \mathbf{k}') \cdot \mathbf{R}_c] d\mathbf{R}_c = \frac{N}{L^d}(2\pi)^d \delta(\mathbf{k} - \mathbf{k}'). \quad (P3.4)$$

By substituting the result of Eq. (P3.4) into Eq. (P3.2), the integration over $\mathbf{k}'$ can be immediately performed, with the result

$$\langle G(\omega, \mathbf{r}, \mathbf{r}') \rangle_c = \langle G \rangle_c(\omega, \mathbf{r} - \mathbf{r}')$$

$$= \frac{L^d}{N(2\pi)^d} \int d\mathbf{k} \, \langle G \rangle_c(\omega, \mathbf{k}, \mathbf{k}) \exp[i\mathbf{k} \cdot (\mathbf{r} - \mathbf{r}')]. \quad (P3.5)$$

The fact that $\langle G \rangle_c$ depends on only one $\mathbf{k}$ (i.e., $\mathbf{k} = \mathbf{k}'$) is clear from (P3.5), which proves our assertion.

3.2. Show that the convolution integral in real space means simple multiplication in the wave vector space.

A convolution integral is defined by

$$F(\mathbf{x}) = \frac{1}{L^d} \int G(\mathbf{x} - \mathbf{y}) H(\mathbf{y}) \, d\mathbf{y}. \quad (P3.6)$$

In Fourier components, each function may be written as

$$F(\mathbf{x}) = \frac{L^d}{N} \int \frac{d\mathbf{k}}{(2\pi)^d} F(\mathbf{k}) \exp(i\mathbf{k} \cdot \mathbf{x}), \quad (P3.7)$$

$$G(\mathbf{x} - \mathbf{y}) = \frac{L^d}{N} \int \frac{d\mathbf{k}_1}{(2\pi)^d} G(\mathbf{k}_1) \exp[i\mathbf{k}_1 \cdot (\mathbf{x} - \mathbf{y})], \quad (P3.8)$$

$$H(\mathbf{y}) = \frac{L^d}{N} \int \frac{d\mathbf{k}_2}{(2\pi)^d} H(\mathbf{k}_2) \exp(i\mathbf{k}_2 \cdot \mathbf{y}). \quad (P3.9)$$

By substituting Eq. (P3.7)–(P3.9) into Eq. (P3.6) and performing the $\mathbf{y}$ integral, one gets the factor

$$\delta(\mathbf{k}_2 - \mathbf{k}_1)(2\pi)^d/L^d$$

as the result. Carrying out the $\mathbf{k}_2$ integration then gives

$$L^d \int \frac{d\mathbf{k}}{(2\pi)^d} F(\mathbf{k}) \exp(i\mathbf{k} \cdot \mathbf{x})$$

$$= L^d \int \frac{d\mathbf{k}_1}{(2\pi)^d} G(\mathbf{k}_1) H(\mathbf{k}_1) \exp(i\mathbf{k}_1 \cdot \mathbf{x}). \quad (P3.10)$$

Thus, the Fourier amplitudes have the relation
$$F(\mathbf{k}) = G(\mathbf{k})H(\mathbf{k}), \tag{P3.11}$$
i.e., simple multiplication. By interchanging $\mathbf{k}$ and $\mathbf{x}$, it is clear that the converse is also true: convolution in the $\mathbf{k}$ domain means simple multiplication in the spatial domain.

3.3. Derive an expression for the total scattering cross section of a single impurity in a lattice, and verify the optical theorem for that case.

By starting from the $\mathbf{t}^+$ operator for the single impurity scatterer, Eq. (3.19),
$$\mathbf{t}^+ = |\mathbf{l}_0\rangle \frac{\sigma_0 a^d}{1 - \sigma_0 a^d G_0^+(\omega, \mathbf{l} = \mathbf{l}')} \langle \mathbf{l}_0|,$$
it is easy to get
$$\langle \mathbf{k}'_0 | \mathbf{t}^+ | \mathbf{k}_0 \rangle = \langle \mathbf{k}'_0 | \mathbf{l}_0 \rangle \frac{\sigma_0 a^d}{1 - \sigma_0 a^d G_0^+(\omega, \mathbf{l} = \mathbf{l}')} \langle \mathbf{l}_0 | \mathbf{k}_0 \rangle,$$
$$= \exp[i(\mathbf{k}_0 - \mathbf{k}'_0) \cdot \mathbf{l}_0] \frac{\sigma_0 a^d}{1 - \sigma_0 a^d G_0^+(\omega, \mathbf{l} = \mathbf{l}')}, \tag{P3.12}$$
where $\mathbf{k}'_0$ and $\mathbf{k}_0$ are wave vectors which satisfy the relations $\kappa_0^2 = \beta e(\mathbf{k}_0)$ and $\kappa_0^2 = \beta e(\mathbf{k}'_0)$. The total scattering cross section O is given by
$$O = \int_\omega |f|^2 \, d\Omega = |f|^2 \int_\omega d\Omega, \tag{P3.13}$$
where $|f|^2$ stands for the scattering cross section, which is angle independent as can be seen from Eq. (P3.12), and the integral is over all solid angles but conditioned on the availability of states at the given frequency ω. The expression for f in the scalar wave case has been given in the text by Eqs. (3.38) (3D), (3.45) (2D), and (3.47)–(3.50) (1D). In the long-wavelength limit they should agree with the similar quantities in the lattice case. That means we can write

$$f(\mathbf{k}'_0, \mathbf{k}_0) = \begin{cases} -\dfrac{1}{4\pi} \langle \mathbf{k}'_0 | \mathbf{t}^+ | \mathbf{k}_0 \rangle & \text{3D} \\[2mm] -\dfrac{i}{4} \sqrt{\dfrac{2}{\pi \bar{k}_0}} \exp\left(-i\dfrac{\pi}{4}\right) \langle \mathbf{k}'_0 | \mathbf{t}^+ | \mathbf{k}_0 \rangle & \text{2D} \\[2mm] -\dfrac{i}{2\bar{k}_0} \langle \mathbf{k}'_0 | \mathbf{t}^+ | \mathbf{k}_0 \rangle & \text{1D} \end{cases} \tag{P3.14}$$

where $\bar{k}_0 = \kappa_0/\sqrt{\beta}$ in the long-wavelength limit. The general definition of $\bar{k}_0$ is given below. To do the angular integral at a constant frequency, it is noted that whereas in the continuum case the phase space is isotropic, in the lattice case this is no longer true since the Brillouin zone is not spherical. Therefore, the angular integral should be expressed in terms of an integral over **k** as

$$\int_\omega d\Omega = \begin{cases} \dfrac{2\beta}{\bar{k}_0} \int d\mathbf{k}\, \delta[\kappa_0^2 - \beta e(\mathbf{k})] & \text{3D} \\ 2\beta \int d\mathbf{k}\, \delta[\kappa_0^2 - \beta e(\mathbf{k})] & \text{2D} \\ 2\bar{k}_0 \beta \int d\mathbf{k}\, \delta[\kappa_0^2 - \beta e(\mathbf{k})] & \text{1D} \end{cases} \quad \text{(P3.15)}$$

The validity of these expressions may easily be verified in the case when $\kappa_0^2 \to 0$ so that the delta function is operative in the regime of $|\mathbf{k}| \to 0$. Then $\beta e(\mathbf{k}) = \beta k^2$, $\bar{k}_0 = \kappa_0/\sqrt{\beta}$, and by separating $d\mathbf{k}$ into $d\Omega \cdot k^{d-1} dk$, the integral is seen to yield 4π, 2π, 2 in 3D, 2D, 1D, respectively. The factor 2 is present because $e(\mathbf{k})$ is a function of k^2 (from $\cos ka$). By combining Eqs. (P3.12), (P3.13), (P3.14), and (P3.15) and recognizing

$$\operatorname{Im} G_0^+(\omega, \mathbf{l} = \mathbf{l}') = -\pi \int \frac{d\mathbf{k}}{(2\pi)^d} \delta[\kappa_0^2 - \beta e(\mathbf{k})],$$

one immediately gets for all spatial dimensionalities,

$$O = -\frac{(\sigma_0 a^d)^2 \operatorname{Im} G_0^+(\omega, \mathbf{l} = \mathbf{l}')}{\bar{k}_0 |1 - \sigma_0 a^d G_0^+(\omega, \mathbf{l} = \mathbf{l}')|^2}. \quad \text{(P3.16a)}$$

From Eq. (P3.12), it is easily seen that

$$O = -\frac{\operatorname{Im}\langle \mathbf{k}_0 | t^+ | \mathbf{k}_0 \rangle}{\bar{k}_0}, \quad \text{(P3.16b)}$$

which is precisely the optical theorem.

To define $\bar{k}_0$, we note that an alternative definition of the angular integral is

$$\int_\omega d\Omega = \frac{1}{\bar{k}_0^{d-1}} \int_\omega dS, \quad \text{(P3.17a)}$$

where dS represents a constant frequency surface integration. But

$$\int dS = \int dS \int dk_\perp |\nabla_k \beta e(\mathbf{k})| \delta[\kappa_0^2 - \beta e(\mathbf{k})], \quad \text{(P3.17b)}$$

where $k_\perp$ is the wave vector perpendicular to the constant-frequency surface. Since a given value of $\beta e(\mathbf{k})$ defines the constant-frequency surface and $\nabla_k \beta e(\mathbf{k})$ gives the direction perpendicular to it, $dk_\perp |\nabla_k \beta e(\mathbf{k})|$ is essentially $d[\beta e(\mathbf{k})]$, so the integration over values of $\beta e(\mathbf{k})$ in (P3.17b) is recognized to give 1 from the delta function. Since, $dS\,dk_\perp = d\mathbf{k}$, a comparison between Eq. (P3.17b) and (P3.15) thus gives

$$\bar{k}_0 = \frac{1}{2} \frac{\int d\mathbf{k} |\nabla_k \beta e(\mathbf{k})| \delta[\kappa_0^2 - \beta e(\mathbf{k})]}{\beta \int d\mathbf{k}\, \delta[\kappa_0^2 - \beta e(\mathbf{k})]}. \quad \text{(P3.18)}$$

The right-hand side of (P3.18) is clearly an average of $|\nabla_k \beta e(\mathbf{k})|$ over the states of constant frequency, which is the physical meaning of $\bar{k}_0$.

3.4. Solve the scalar-wave scattering problem of a plane wave incident on a sphere with radius R and wave speed $v \neq v_0$. Find the expression for the scattering amplitude in the limit of $\kappa_0 R \to 0$, and verify the optical theorem for each angular momentum channel.

For the scalar wave equation $(\nabla^2 + \kappa^2)\phi = 0$, where $\kappa^2 = \omega^2/v^2$ inside the sphere, the solution may be expressed in terms of an expansion:

$$\phi^<(\mathbf{r}) = \sum_{n=0}^{\infty} A_n j_n(\kappa r) P_n(\cos\theta), \quad r \leq R. \quad \text{(P3.19)}$$

Outside the sphere, $\phi^>(\mathbf{r})$ is a sum of the incident plane wave and the scattered wave, with $\kappa^2 = \kappa_0^2 = \omega^2/v_0^2$:

$$\phi^>(\mathbf{r}) = \exp(i\boldsymbol{\kappa}_0 \cdot \mathbf{r}) + \sum_{n=0}^{\infty} D_n h_n^{(1)}(\kappa_0 r) P_n(\cos\theta) i^n (2n+1), \quad r > R.$$

$$\text{(P3.20)}$$

Here j_n and $h_n^{(1)}$ are the spherical Bessel function and the spherical Hankel function of the first kind, respectively. If the scalar wave equation is valid everywhere in space, then the boundary conditions have to be the continuity of ϕ and $\partial\phi/\partial r$ across the interface. This is because the jump in the value of κ from ω/v inside the sphere to ω/v_0 outside the sphere implies the second (normal) derivative of ϕ must have the same jump so that the scalar wave equation holds across the interface. It follows that the first normal derivative and ϕ itself must both be continuous because they are, respectively, the first and second integrals of the jump. The two equations that result from the application of the boundary conditions are

$$A_n j_n(\kappa R) P_n(\cos\theta) = i^n(2n+1) P_n(\cos\theta)[j_n(\kappa_0 R) + D_n h_n(\kappa_0 R)], \quad (P3.21)$$

$$\frac{\kappa}{\kappa_0} A_n j'_n(\kappa R) P_n(\cos\theta) = i^n(2n+1) P_n(\cos\theta)[j'_n(\kappa_0 R) + D_n h'_n(\kappa_0 R)]. \quad (P3.22)$$

Here the prime indicates derivative with respect to the argument of the function, and the superscript (1) has been dropped from the spherical Hankel function. In obtaining the two equations, an expansion of the plane wave has also been used:

$$\exp(i\boldsymbol{\kappa}_0 \cdot \mathbf{r}) = \sum_{n=0}^{\infty} i^n(2n+1) P_n(\cos\theta) j_n(\kappa_0 r), \quad (P3.23)$$

and the orthogonality of the different angular momentum channels has been utilized. By dividing Eq. (P3.22) by (P3.21), one immediately gets

$$D_n = \frac{\kappa_0 j_n(\kappa R) j'_n(\kappa_0 R) - \kappa j'_n(\kappa R) j_n(\kappa_0 R)}{\kappa j'_n(\kappa R) h_n(\kappa_0 R) - \kappa_0 h'_n(\kappa_0 R) j_n(\kappa R)}. \quad (P3.24)$$

From Eq. (3.35), we can write

$$f(\boldsymbol{\kappa}'_0, \boldsymbol{\kappa}_0) = \sum_n f_n(\boldsymbol{\kappa}'_0, \boldsymbol{\kappa}_0),$$

where

$$f_n(\boldsymbol{\kappa}'_0, \boldsymbol{\kappa}_0) = -\frac{i}{\kappa_0} D_n P_n(\cos\theta)(2n+1) \quad (P3.26)$$

is the scattering amplitude for the nth angular momentum channel. It is easy to show that

$$2\pi \int_{-1}^{1} |f_n|^2 \, d\cos\theta = \frac{4\pi}{\kappa_0^2} |D_n|^2 (2n+1) = O_n. \quad (P3.27)$$

From Eq. (P3.24) and the definition

$$\langle \kappa'_0 | \mathbf{t}^+ | \kappa_0 \rangle_n = -4\pi f_n(\kappa'_0, \kappa_0), \tag{P3.28}$$

it is easy to see that, for $\kappa'_0 = \kappa_0$, $P_n(\cos\theta) = 1$ and

$$\frac{-\operatorname{Im}\langle \kappa_0 | \mathbf{t}^+ | \kappa_0 \rangle_n}{\kappa_0} = \frac{4\pi}{\kappa_0^2}(2n+1)[-\operatorname{Im} iD_n]. \tag{P3.29}$$

Since $h_n = j_n + in_n$, where n_n is the spherical Neumann function, and the numerator of D_n is noted to be real, it follows that

$$-\operatorname{Im} iD_n = -\operatorname{Re} D_n = |D_n|^2. \tag{P3.30}$$

Therefore,

$$O_n = \frac{-\operatorname{Im}\langle \kappa_0 | \mathbf{t}^+ | \kappa_0 \rangle_n}{\kappa_0}, \tag{P3.31}$$

i.e., the optical theorem is valid for every angular channel.

To derive an expression for the scattering amplitude in the limit of $\kappa_0 R \to 0$, it is necessary only to consider D_0, because physically, the scattering should be isotropic for a point scatterer. By expanding all functions to the leading orders of $\kappa_0 R$ and κR, we have

$$j_0(x) \cong 1 - \frac{x^2}{6},$$

$$j'_0(x) \cong -\frac{x}{3} + \frac{x^3}{30},$$

$$h_0(x) \cong -\frac{i}{x}\left(1 + ix - \frac{x^2}{2} - i\frac{x^3}{6}\right),$$

$$h'_0(x) \cong \left(\frac{1}{x} + i\frac{1}{x^2}\right)\left(1 + ix - \frac{x^2}{2} - i\frac{x^3}{6}\right). \tag{P3.32}$$

Collecting all the leading-order real and imaginary parts of the D_0 expansion gives

$$D_0 = -\frac{i}{3}(\kappa_0 R)^3\left[1 - \left(\frac{\kappa}{\kappa_0}\right)^2\right] - \frac{1}{9}(\kappa_0 R)^6\left[1 - \left(\frac{\kappa}{\kappa_0}\right)^2\right]^2. \tag{P3.33}$$

Therefore, the scattering amplitude to the leading order is given by

$$f(\boldsymbol{\kappa}'_0, \boldsymbol{\kappa}_0) = -\frac{1}{3} \frac{(\kappa_0 R)^3}{\kappa_0} \left[1 - \left(\frac{\kappa}{\kappa_0}\right)^2 \right]. \tag{P3.34}$$

3.5. Give the solution of the previous problem in two dimensions. In 2D, the expansions are

$$\phi^<(\mathbf{r}) = \sum_{m=-\infty}^{\infty} A_m J_m(\kappa r) \exp(im\theta), \quad r \leq R, \tag{P3.35}$$

$$\phi^>(\mathbf{r}) = \exp(i\boldsymbol{\kappa}_0 \cdot \mathbf{r}) + \sum_{m=-\infty}^{\infty} i^m D_m H_m^{(1)}(\kappa_0 r) \exp(im\theta)$$

$$= \sum_{m=-\infty}^{\infty} i^m \exp(im\theta) \left[J_m(\kappa_0 r) + D_m H_m^{(1)}(\kappa_0 r) \right], \quad r > R. \tag{P3.36}$$

Here J_m and $H_m^{(1)}$ are the Bessel function and the Hankel function of first kind, respectively, and we have utilized the 2D expansion formula for $\exp(i\boldsymbol{\kappa}_0 \cdot \mathbf{r})$. The definitions of θ, κ, and κ_0 are the same as in Problem 3.4. The boundary conditions at $r = R$, continuity of ϕ and $\partial\phi/\partial r$, may be combined to give the continuity of $(\partial\phi/\partial r)/\phi = \partial(\ln\phi)/\partial r$, which can be written as

$$\frac{\kappa J'_m(\kappa R)}{\kappa_0 J_m(\kappa R)} = \frac{J'_m(\kappa_0 R) + D_m H'_m(\kappa_0 R)}{J_m(\kappa_0 R) + D_m H_m(\kappa_0 R)} \tag{P3.37}$$

for each angular momentum channel. Here we have dropped the superscript (1) on the Hankel function. D_m is thus given by the expression

$$D_m = \frac{\kappa J'_m(\kappa R) J_m(\kappa_0 R) - \kappa_0 J_m(\kappa R) J'_m(\kappa_0 R)}{\kappa_0 H'_m(\kappa_0 R) J_m(\kappa R) - \kappa H_m(\kappa_0 R) J'_m(\kappa R)}. \tag{P3.38}$$

The scattering amplitude is given by

$$f(\boldsymbol{\kappa}'_0, \boldsymbol{\kappa}_0) = \sum_{m=-\infty}^{\infty} f_m(\boldsymbol{\kappa}'_0, \boldsymbol{\kappa}_0), \tag{P3.39}$$

$$f_m(\boldsymbol{\kappa}'_0, \boldsymbol{\kappa}_0) = \sqrt{\frac{2}{\pi\kappa_0}} D_m \exp(im\theta) \exp\left(-i\frac{\pi}{4}\right). \tag{P3.40}$$

Thus

$$O_m = \int_0^{2\pi} |f_m|^2 \, d\theta = \frac{4}{\kappa_0} |D_m|^2. \quad (P3.41)$$

From Eq. (P3.14) and the condition of $\theta = 0$ for forward scattering, we have

$$\langle \kappa_0 | t^+ | \kappa_0 \rangle_m = 4 i D_m, \quad (P3.42)$$

from which it immediately follows that

$$O_m = -\frac{\operatorname{Im}\langle \kappa_0 | t^+ | \kappa_0 \rangle_m}{\kappa_0}, \quad (P3.43)$$

because $-\operatorname{Im} i D_m = -\operatorname{Re} D_m = |D_m|^2$ due to the fact that $H_m^{(1)}(x) = J_m(x) + i N_m(x)$, where $N_m(x)$ is the Neumann function.

For the $\kappa_0 R \to 0$ limit, only the $m = 0$ term has to be expanded due to the consideration of isotropic scattering. Retaining the leading orders of real and imaginary parts means expanding J_0 and H_0 to the leading orders of their arguments:

$$J_0(x) = 1 - \frac{x^2}{4},$$

$$J_0'(x) = -\frac{x}{2},$$

$$H_0(x) = i\frac{2}{\pi} \ln x + 1,$$

$$H_0'(x) = i\frac{2}{\pi}\frac{1}{x} - \frac{x}{2}. \quad (P3.44)$$

Collecting terms together yields

$$D_0 = -i\frac{\pi}{4}(\kappa_0 R)^2 \left[1 - \left(\frac{\kappa}{\kappa_0}\right)^2\right] - \frac{\pi^2}{16}(\kappa_0 R)^4 \left[1 - \left(\frac{\kappa}{\kappa_0}\right)^2\right]^2. \quad (P3.45)$$

It follows that to the leading order the scattering amplitude is

$$f(\kappa_0', \kappa_0) = -\frac{i}{4}\sqrt{\frac{2\pi}{\kappa_0}}(\kappa_0 R)^2 \left[1 - \left(\frac{\kappa}{\kappa_0}\right)^2\right] \exp\left(-i\frac{\pi}{4}\right). \quad (P3.46)$$

3.6. Give the solution of Problem 3.4 in one dimension.

In 1D, the solution expansion can be written as a function of the coordinate x:

$$\phi^<(x) = A\exp(i\kappa x) + B\exp(-i\kappa x) \quad |x| \leq R \quad \text{(P3.47)}$$

$$\phi^>(x) = \exp(i\kappa_0 x) + \rho\exp(-i\kappa_0 x), \quad x < -R$$

$$= \tau\exp(i\kappa_0 x), \quad x > R \quad \text{(P3.48)}$$

Here ρ denotes the reflection coefficient and τ the transmission coefficient. The boundary conditions at $x = \pm R$, the continuity of ϕ and $d\phi/dx$, give

$$A\exp(-i\kappa R) + B\exp(i\kappa R) = \exp(-i\kappa_0 R) + \rho\exp(i\kappa_0 R) \quad \text{(P3.49)}$$

$$\frac{\kappa}{\kappa_0}[A\exp(-i\kappa R) - B\exp(i\kappa R)] = \exp(-i\kappa_0 R) - \rho\exp(i\kappa_0 R) \quad \text{(P3.50)}$$

$$A\exp(i\kappa R) + B\exp(-i\kappa R) = \tau\exp(i\kappa_0 R) \quad \text{(P3.51)}$$

$$\frac{\kappa}{\kappa_0}[A\exp(i\kappa R) - B\exp(-i\kappa R)] = \tau\exp(i\kappa_0 R) \quad \text{(P3.52)}$$

After some algebra, ρ and τ are solved to yield the expressions given by Eqs. (3.49) and (3.50). In 1D, ρ and τ together constitute the two scattering channels, and the optical theorem in that case has been explicitly demonstrated in the text.

In the $\kappa_0 R \to 0$ limit, the forward scattering amplitude, $\tau - 1$, can be written as

$$\tau - 1 = -i\kappa_0 R\left[1 - \left(\frac{\kappa}{\kappa_0}\right)^2\right] - (\kappa_0 R)^2\left[1 - \left(\frac{\kappa}{\kappa_0}\right)^2\right]^2. \quad \text{(P3.53a)}$$

The backward scattering amplitude is

$$\rho = i\kappa_0 R\left[1 - \left(\frac{\kappa}{\kappa_0}\right)^2\right] + (\kappa_0 R)^2\left[1 - \left(\frac{\kappa}{\kappa_0}\right)^2\right]^2. \quad \text{(P3.53b)}$$

Problems and Solutions

The total scattering cross section in the $\kappa_0 R \to 0$ limit is therefore

$$O = 2(\kappa_0 R)^2 \left[1 - \left(\frac{\kappa}{\kappa_0}\right)^2\right]^2. \qquad \text{(P3.54)}$$

3.7. Show the optical theorem to be a statement of flux conservation.

Since only elastic scattering is considered, flux conservation here is the same as energy conservation. The starting point of the consideration is the expression

$$\phi(\mathbf{r}) \xrightarrow[r \to \infty]{} \exp(i\boldsymbol{\kappa}_0 \cdot \mathbf{r}) + f(\theta) \frac{e^{i\kappa_0 r}}{r^{(d-1)/2}}, \qquad \text{(P3.55)}$$

where for simplicity the scattering is assumed to be uniform in the azimuthal direction. The plane wave $\exp(i\boldsymbol{\kappa}_0 \cdot \mathbf{r})$ may be written as

$$\exp(i\boldsymbol{\kappa}_0 \cdot \mathbf{r}) = \begin{cases} \displaystyle\sum_{n=0}^{\infty} i^n (2n+1) j_n(\kappa_0 r) P_n(\cos\theta) & \text{3D} \\ \displaystyle\sum_{m=-\infty}^{\infty} i^m \exp(im\theta) J_m(\kappa_0 r) & \text{2D}. \end{cases} \qquad \text{(P3.56)}$$

In the 1D case, because the optical theorem is already explicitly demonstrated to be equivalent to energy conservation, it does not need to be considered further. By using the asymptotic expression for $j_n(\kappa_0 r)$ and $J_m(\kappa_0 r)$ as $r \to \infty$,

$$j_n(\kappa_0 r) \xrightarrow[r \to \infty]{} \frac{1}{2}\left[(-i)^{n+1}\frac{\exp(i\kappa_0 r)}{\kappa_0 r} + i^{n+1}\frac{\exp(-i\kappa_0 r)}{\kappa_0 r}\right], \qquad \text{(P3.57)}$$

$$J_m(\kappa_0 r) \xrightarrow[r \to \infty]{} \frac{1}{\sqrt{2\pi}}\left[(-i)^m \frac{\exp\left(i\left(\kappa_0 r - \frac{\pi}{4}\right)\right)}{\sqrt{\kappa_0 r}} + i^m \frac{\exp\left(-i\left(\kappa_0 r - \frac{\pi}{4}\right)\right)}{\sqrt{\kappa_0 r}}\right],$$

$$\text{(P3.58)}$$

and substituting them in Eq. (P3.56), one obtains

$$\exp(i\boldsymbol{\kappa}_0 \cdot \mathbf{r}) \xrightarrow[r \to \infty]{} \frac{\exp(i\kappa_0 r)}{i\kappa_0 r} \sum_{n=0}^{\infty} \frac{2n+1}{2} P_n(\cos\theta)$$

$$- \frac{\exp(-i\kappa_0 r)}{i\kappa_0 r} \sum_{n=0}^{\infty} (-1)^n \frac{2n+1}{2} P_n(\cos\theta) \quad \text{3D},$$

$$\exp(i\boldsymbol{\kappa}_0 \cdot \mathbf{r}) \xrightarrow[r \to \infty]{} \sqrt{\frac{2\pi}{\kappa_0 r}} \exp\left[i\left(\kappa_0 r - \frac{\pi}{4}\right)\right] \sum_{m=-\infty}^{\infty} \frac{\exp(im\theta)}{2\pi}$$

$$+ \sqrt{\frac{2\pi}{\kappa_0 r}} \exp\left[-i\left(\kappa_0 r - \frac{\pi}{4}\right)\right] \sum_{m=-\infty}^{\infty} (-1)^m \frac{\exp(im\theta)}{2\pi} \quad \text{2D}.$$

(P3.59)

The angular summations in Eq. (P3.59) are angular delta functions. This can be shown by their action on some arbitrary angular function $F(\theta)$, which may be expanded as

$$F(\theta) = \begin{cases} \sum_{n=0}^{\infty} A_n P_n(\cos\theta) & \text{3D}, \\ \sum_{m=-\infty}^{\infty} A_m \exp(im\theta) & \text{2D}. \end{cases} \quad \text{(P3.60)}$$

Integrating $F(\theta)$ with one of the angular summations gives

$$\int_{-1}^{1} F(\theta) \sum_{n=0}^{\infty} (\pm 1)^n \frac{2n+1}{2} P_n(\cos\theta) \, d(\cos\theta)$$

$$= \sum_{n,m} A_m (\pm 1)^n \frac{2n+1}{2} \int_{-1}^{1} P_n(x) P_m(x) \, dx$$

$$= \sum_m A_m (\pm 1)^m = \sum_m A_m P_m(\pm 1)$$

$$= F(0) \quad \text{if } +1$$

$$= F(\pi) \quad \text{if } -1. \quad \text{(P3.61)}$$

That means

$$\sum_{n=0}^{\infty} (\pm 1)^n \frac{2n+1}{2} P_n(\cos\theta) = \delta(\cos\theta \mp 1). \quad \text{(P3.62)}$$

Similar reasoning gives

$$\sum_{m=-\infty}^{\infty} (\pm 1)^m \frac{\exp(im\theta)}{2\pi} = \begin{cases} \delta(\theta) & \text{if } +1 \\ \delta(\theta-\pi) & \text{if } -1. \end{cases} \quad \text{(P3.63)}$$

Therefore, in the limit of $r \to \infty$, $\phi(\mathbf{r})$ has the following form:

$$\phi(\mathbf{r}) \xrightarrow[r\to\infty]{} \frac{\exp(i\kappa_0 r)}{i\kappa_0 r} \delta(\cos\theta - 1)$$

$$+ f(\theta) \frac{\exp(i\kappa_0 r)}{r} - \frac{\exp(-i\kappa_0 r)}{i\kappa_0 r} \delta(\cos\theta + 1) \quad \text{3D},$$

$$\phi(\mathbf{r}) \xrightarrow[r\to\infty]{} \sqrt{\frac{2\pi}{\kappa_0 r}} \exp\left[i\left(\kappa_0 r - \frac{\pi}{4}\right)\right]\delta(\theta) + f(\theta)\frac{\exp(i\kappa_0 r)}{\sqrt{r}}$$

$$+ \sqrt{\frac{2\pi}{\kappa_0 r}} \exp\left[-i\left(\kappa_0 r - \frac{\pi}{4}\right)\right]\delta(\theta - \pi). \quad \text{2D}$$

$$\text{(P3.64)}$$

They have the simple interpretation that if one draws a large sphere (a circle in 2D) of radius R_0 centered at the scatterer, then the incoming beam is represented by the $\theta = \pi$, $\exp(-i\kappa_0 r)/r^{(d-1)/2}$ term in Eq. (P3.64). The other two terms represent the outgoing part of the wave. In steady state, they must balance. That means in 3D,

$$\int_{-1}^{1} \left| \frac{\exp(i\kappa_0 R_0)}{i\kappa_0 R_0} \delta(\cos\theta - 1) + f(\theta) \frac{\exp(i\kappa_0 R_0)}{R_0} \right|^2 R_0^2 \, d(\cos\theta)$$

$$= \int_{-1}^{1} \left| \frac{\exp(-i\kappa_0 R)}{i\kappa_0 R_0} \delta(\cos\theta + 1) \right|^2 R_0^2 \, d(\cos\theta). \quad \text{(P3.65)}$$

If $f(\theta) = 0$ (no scattering), this equation still has to hold. That means, after squaring, the two terms involving $f(\theta)$ must cancel each other:

$$0 = \int_{-1}^{1} |f(\theta)|^2 \, d(\cos\theta) - \frac{2}{\kappa_0} \operatorname{Im} f(0)$$

$$= \frac{O}{2\pi} + \frac{1}{2\pi\kappa_0} \operatorname{Im}\langle \boldsymbol{\kappa}_0 | \mathbf{t}^+ | \boldsymbol{\kappa}_0 \rangle. \tag{P3.66}$$

Equation (P3.66) is exactly the optical theorem. Similarly in 2D,

$$0 = \int_0^{2\pi} |f(\theta)|^2 \, d\theta + f(0) \exp\left(i\frac{\pi}{4}\right)\sqrt{\frac{2\pi}{\kappa_0}} + f^*(0) \exp\left(-i\frac{\pi}{4}\right)\sqrt{\frac{2\pi}{\kappa_0}}$$

$$= O + \frac{1}{\kappa_0} \operatorname{Im}\langle \boldsymbol{\kappa}_0 | \mathbf{t}^+ | \boldsymbol{\kappa}_0 \rangle, \tag{P3.67}$$

which is again the optical theorem. This proves our assertion.

3.8. Determine the CPA (effective medium) equation for acoustic wave propagation in immiscible fluid–fluid mixtures (emulsions), where the droplet size is small compared to the wavelength.

For acoustic waves in fluid, the relevant equations are

$$\rho \frac{\partial \mathbf{u}_t}{\partial t} = -\nabla P, \tag{P3.68}$$

$$\frac{\partial P}{\partial t} = -B \nabla \cdot \mathbf{u}_t. \tag{P3.69}$$

Here P denotes pressure, ρ the fluid mass density, $\mathbf{u}_t$ is the displacement velocity (where the subscript t here means the time derivative of the displacement vector $\mathbf{u}$), and B is the bulk modulus. The first equation is recognized to be Newton's equation of force density = (mass density) times (acceleration). The second equation is the time derivative of the equation of state for a fluid: $P = -B\nabla \cdot \mathbf{u}$, where $\mathbf{u}$ is the displacement, and $\nabla \cdot \mathbf{u}$ represents the fractional volume change at a point. To obtain a wave equation from the two equations, one first divides both sides of Eq. (P3.68) by ρ, and then takes the divergence of both sides. That gives

$$\frac{\partial \nabla \cdot \mathbf{u}_t}{\partial t} = -\nabla \cdot \frac{1}{\rho} \nabla P. \tag{P3.70}$$

Substituting $\nabla \cdot \mathbf{u}_t$ from Eq. (P3.69) and recognizing B to be time independent, one gets

$$\frac{\partial^2 P}{\partial t^2} + B\nabla \cdot \frac{1}{\rho} \nabla P = 0. \tag{P3.71}$$

This equation reduces to the scalar wave equation, with $\phi = P$, if ρ is a constant. In that case the speed of sound is identified as $v = \sqrt{B/\rho}$. For inhomogeneous fluid systems, however, ρ is not a constant. The scalar wave equation is valid in each homogeneous region, but one of the boundary conditions across interfaces must be modified. Instead of requiring the normal derivative of P to be continuous across the interface, it is ρ^{-1} times the normal derivative. This can easily be seen from Eq. (P3.71) by the fact that since P (and therefore $\partial^2 P/\partial t^2$) is continuous across the boundary, the term $B\nabla \cdot \rho^{-1}\nabla P$ must also be continuous. As a result, $\nabla \cdot \rho^{-1}\nabla P$ can, at worst, only have a jump across the interface to compensate for the jump in B. Due to the fact that the integral of a jump is continuous (with a discontinuity in the slopes), it follows that $\rho^{-1}\nabla P$ must be continuous across the interface. Physically, the continuity of $\rho^{-1}\nabla P$ implies displacement continuity across the interface.

With the modification of the boundary condition, the first step in the derivation of an effective medium equation is the consideration of the scattering solution for a single scatterer in the $\kappa_0 R \to 0$ limit. In 3D, that means a modification of Eq. (P3.22) with the replacement of the κ/κ_0 factor by $(\kappa/\kappa_0)(\rho_0/\rho)$. Tracing the effect of this change through Eq. (P3.24), we get a new expression for D_n:

$$D_n = \frac{\kappa_0 \rho j_n(\kappa R) j_n'(\kappa_0 R) - \kappa \rho_0 j_n'(\kappa R) j_n(\kappa_0 R)}{\kappa \rho_0 j_n'(\kappa R) h_n(\kappa_0 R) - \kappa_0 \rho h_n'(\kappa_0 R) j_n(\kappa R)}, \tag{P3.72}$$

where ρ_0 is the mass density of the uniform fluid medium. In the $\kappa_0 R \to 0$ limit, only the D_0 expansion has to be considered. By using the identities $\kappa = \omega/\sqrt{B/\rho}$ and $\kappa_0 = \omega/\sqrt{B_0/\rho_0}$, the leading-order terms of the expansion are

$$D_0 \simeq -\frac{i}{3}(\kappa_0 R)^3 \left(1 - \frac{B_0}{B}\right) - \frac{(\kappa_0 R)^6}{9}\left(1 - \frac{B_0}{B}\right)^2. \tag{P3.73}$$

Comparison with Eq. (3.37) tells us that in order to obtain the CPA equation in the $\kappa_e R_m \to 0$ limit ($m = 1, 2$ for the two components), it is necessary only to replace the factor $[1 - (\kappa_m/\kappa_e)^2]$ in Eq. (3.71) by $(1 - \bar{B}/B_m)$, which gives

$$\frac{1}{\bar{B}} = \frac{p}{B_1} + \frac{1-p}{B_2}. \quad \text{(P3.74)}$$

The average density $\bar{\rho}$ is simply given by

$$\bar{\rho} = p\rho_1 + (1-p)\rho_2. \quad \text{(P3.75)}$$

Therefore, $\kappa_e = \omega/\sqrt{\bar{B}/\bar{\rho}}$, and the mean-free path l is given by Eq. (3.76) with ϵ replaced by $1/B$, i.e.,

$$l = \frac{3}{\kappa_e \left[p(\kappa_e R_1)^3 \left(1 - \frac{\bar{B}}{B_1}\right)^2 + (1-p)(\kappa_e R_2)^3 \left(1 - \frac{\bar{B}}{B_2}\right)^2 \right]}. \quad \text{(P3.76)}$$

The CPA equations remains the same as (P3.74) for 2D and 1D, and the mean free path can be similarly obtained from Eq. (3.76). In the solution to Problem 3.11 it is shown that the CPA equation is also the same for the dispersion microgeometry. However, the mean free path expression is different. For application to colloids or mixtures with a well-defined size for the dispersed particles, the mean free path expression of the dispersion microgeometry should be more accurate.

The effective medium equation (P3.74) is noted to have a simple physical derivation in the static limit, where $\nabla P = 0$ [from Eq. (P3.68)], so that P is a constant throughout space. From the equation of state, the fractional volume change is given by

$$\nabla \cdot \mathbf{u} = -\frac{P}{B}.$$

The (volume) averaged fractional volume change is therefore given by

$$\langle \nabla \cdot \mathbf{u} \rangle_V = \frac{1}{L^d} \int \nabla \cdot \mathbf{u} \, d\mathbf{r} = -P \frac{1}{L^d} \int \frac{d\mathbf{r}}{B(\mathbf{r})}$$

$$= -P \left[\frac{p}{B_1} + \frac{1-p}{B_2} \right]. \quad \text{(P3.77)}$$

Expressing $\langle \nabla \cdot \mathbf{u} \rangle_V = -P/\bar{B}$ yields Eq. (P3.74) directly.

3.9. Derive the CPA (effective medium) equation for electromagnetic waves in a composite with the symmetric microstructure, where the scale of the inhomogeneities is small compared to the wavelength.

In the Lorentz gauge, the electric field **E** is expressed as

$$\mathbf{E} = -\nabla\phi - \frac{1}{c}\frac{\partial \mathbf{A}}{\partial t}, \tag{P3.78}$$

where ϕ and **A** are the scalar potential and the vector potential, respectively, and c is the speed of light. In free space, ϕ and every component of **A** each satisfies a scalar wave equation. In the quasi-static limit where the wavelength is large compared to the size of the scatterers, one may approximate **E** by $-\nabla\phi$ because $c^{-1}\partial \mathbf{A}/\partial t$ yields a factor $i\omega/c = i\kappa$, and if R is the length unit, then $c^{-1}\partial \mathbf{A}/\partial t$ is on the order of $\kappa R \to 0$.

The boundary condition for **E** is dictated by the well-known electromagnetic boundary condition of normal $\epsilon\nabla\phi$ being continuous across the interface, instead of $\nabla\phi$ by itself. Together with the fact that the leading-order scattering term in the $\kappa R \to 0$ limit must be the $n = 1$ term as discussed in the text, in 3D the changes in the solution to D_1, Eq. (P3.22), are contained in the substitution of κ/κ_0 by $(\kappa/\kappa_0)^3$, since $\epsilon = (\kappa/\kappa_0)^2$. (P3.24) is therefore modified to yield for $n = 1$,

$$D_1 = \frac{\kappa_0^3 j_1(\kappa R) j_1'(\kappa_0 R) - \kappa^3 j_1'(\kappa R) j_1(\kappa_0 R)}{\kappa^3 j_1'(\kappa R) h_1(\kappa_0 R) - \kappa_0^3 h_1'(\kappa_0 R) j_1(\kappa R)}. \tag{P3.79}$$

For $x \to 0$, $j_1(x) \cong (x/3) - (x^3/30)$, $h_1(x) \cong (-i/x^2) + (x/3)$. By substituting these small-parameter expansions into Eq. (P3.79), D_1 becomes

$$D_1 \cong \frac{i}{3}(\kappa_0 R)^3 \left(\frac{1-\epsilon}{2+\epsilon}\right) - \frac{(\kappa_0 R)^6}{9}\left(\frac{1-\epsilon}{2+\epsilon}\right)^2. \tag{P3.80}$$

The scattering amplitude is therefore [from Eq. (3.35)]

$$f(\boldsymbol{\kappa}_0', \boldsymbol{\kappa}_0) = \frac{(\kappa_0 R)^3}{\kappa_0}\left(\frac{1-\epsilon}{2+\epsilon}\right)\cos\theta + i\frac{(\kappa_0 R)^6}{3\kappa_0}\left(\frac{1-\epsilon}{2+\epsilon}\right)^2 \cos\theta. \tag{P3.81}$$

For the CPA equation, only the forward-scattering ($\theta = 0$) amplitude is important. So in Eq. (3.71), the factor $[1 - (\kappa_m/\kappa_e)^2]$ should be replaced by

$$\left[\frac{1-(\epsilon_m/\bar{\epsilon})}{2+(\epsilon_m/\bar{\epsilon})}\right] = \frac{\bar{\epsilon}+\epsilon_m}{2\bar{\epsilon}+\epsilon_m},$$

which directly yields Eq. (3.78) with $d = 3$ for the symmetric microstructure.

In the 2D case, we have

$$D_1 = \frac{\kappa^3 J_1'(\kappa R) J_1(\kappa_0 R) - \kappa_0^3 J_1(\kappa R) J_1'(\kappa_0 R)}{\kappa_0^3 H_1'(\kappa_0 R) J_1(\kappa R) - \kappa^3 H_1(\kappa_0 R) J_1'(\kappa R)}. \quad \text{(P3.82)}$$

By using the small-parameter expansions $J_1(x) \cong (x/2) - (x^3/16)$ and $H_1 \cong -(i2/\pi x) + (x/2)$, Eq. (P3.82) becomes

$$D_1 \simeq \frac{i\pi}{4}(\kappa_0 R)^2 \left(\frac{1-\epsilon}{1+\epsilon}\right) - \frac{\pi^2}{4}(\kappa_0 R)^4 \left(\frac{1-\epsilon}{1+\epsilon}\right)^2. \quad \text{(P3.83)}$$

In terms of the $\mathbf{t}^+$-matrix element, we have

$$\langle \kappa_0' | \mathbf{t}^+ | \kappa_0' \rangle = -2\pi(\kappa_0 R)^2 \left(\frac{1-\epsilon}{1+\epsilon}\right) \cos\theta - i 2\pi^2 (\kappa_0 R)^4 \left(\frac{1-\epsilon}{1+\epsilon}\right)^2 \cos\theta. \quad \text{(P3.84)}$$

For the CPA equation, only the forward-scattering amplitude is important, so the factor $[1 - (\kappa_m/\kappa_e)^2]$ in Eq. (3.71) should be replaced by

$$\left[\frac{1 - (\epsilon_m/\bar{\epsilon})}{1 + (\epsilon_m/\bar{\epsilon})}\right] = \frac{\bar{\epsilon} - \epsilon_m}{\bar{\epsilon} + \epsilon_m}. \quad \text{(P3.85)}$$

This directly yields Eq. (3.78) with $d = 2$ for the symmetric microstructure.

For the 1D case, if the wave propagation direction is along the x direction, then $\mathbf{E}$ is along either the y or the z direction. The continuity of the tangential component of $\mathbf{E}$ across the interface means the CPA equation can be simply derived in the static limit as follows.

Since $\mathbf{E}$ is constant in each uniform segment and is continuous across the interface, $\mathbf{E}$ must be a constant throughout the sample. Therefore, from the equation $\mathbf{D} = \epsilon \mathbf{E}$, we get

$$\langle \mathbf{D} \rangle_V = \frac{1}{L} \int \mathbf{D}(x) \, dx = \frac{1}{L} \int \epsilon(x) \, dx \cdot \mathbf{E} = [p\epsilon_1 + (1-p)\epsilon_2]\mathbf{E}. \quad \text{(P3.86)}$$

Equation (3.72) results by calling $\langle \mathbf{D} \rangle_V = \bar{\epsilon} \mathbf{E}$.

3.10. Solve the 3D and 2D electromagnetic CPA equations for the symmetric microstructure and show that in the limit of infinite contrast between the two components, there is a percolation threshold.

The electromagnetic CPA for the symmetric microstructure is

$$p\frac{\bar{\epsilon} - \epsilon_1}{(d-1)\bar{\epsilon} + \epsilon_1} + (1-p)\frac{\bar{\epsilon} - \epsilon_2}{(d-1)\bar{\epsilon} + \epsilon_2} = 0. \quad \text{(P3.87)}$$

This may be rewritten as a quadratic equation for $\bar{\epsilon}$:

$$(d-1)\bar{\epsilon}^2 + [(dp - (d-1))\epsilon_2 + (1-dp)\epsilon_1]\bar{\epsilon} - \epsilon_1\epsilon_2 = 0. \quad \text{(P3.88)}$$

The solution is

$$\bar{\epsilon} = \frac{-q + \sqrt{q^2 + 4(d-1)\epsilon_1\epsilon_2}}{2(d-1)}, \quad \text{(P3.89)}$$

$$q = [dp - (d-1)]\epsilon_2 + (1-dp)\epsilon_1. \quad \text{(P3.90)}$$

Only the $+$ sign solution is chosen because that is the branch yielding the correct limits of $\bar{\epsilon} = \epsilon_1$ at $p = 1$ and $\bar{\epsilon} = \epsilon_2$ at $p = 0$.

In the limit of infinite contrast, we can let $\epsilon_1 = 1$ and $\epsilon_2 = 0$. Then

$$q = (1 - dp),$$

and

$$\bar{\epsilon} = \begin{cases} \dfrac{-(1-dp) + |1-dp|}{2(d-1)} = 0, & p < \dfrac{1}{d} \\ \dfrac{d}{d-1}\left(p - \dfrac{1}{d}\right), & p > \dfrac{1}{d}. \end{cases} \quad \text{(P3.91)}$$

It is seen that in 3D, $p_c = 1/3$ (or $2/3$ if $\epsilon_2 = 1$, $\epsilon_1 = 0$), and in 2D, $p_c = 1/2$.

3.11. Derive the CPA equation(s) for the dispersion microgeometry in 3D, 2D, 1D and give the respective expressions for the mean free path in the long-wavelength limit.

The basic structural unit of the dispersion microgeometry is a coated sphere in 3D and a coated circle in 2D. The structural unit in 1D is taken to be two neighboring components 1 and 2 as shown in Figure P3.1.

To obtain the CPA equations in 3D and 2D, it is necessary first to solve the single-scatterer problem where the structural unit is embedded in a homogeneous effective medium as shown in Figure 3.7. The radius of the outer sphere (circle) is denoted by R_0 and that of the inner sphere (circle) by R. The material constant of the dispersed component, represented by

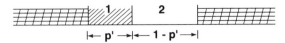

Figure P3.1 The structural unit for the 1D dispersion microgeometry. The effective medium is indicated by the crosshatched region. Intrinsically there is no difference between the two types of microstructure in 1D. However, the unit taken here should give a better description than the two-unit approach because the short-range correlation between the two components is explicitly taken into account. While the CPA equation remains the same, the expression for l is different.

the inner sphere (circle), is characterized by $\kappa_1 = \omega/v_1$, and that of the matrix component is characterized by $\kappa_2 = \omega/v_2$. R_0 and R are related by the local volume fraction p' of the dispersed component:

$$p' = \left(\frac{R}{R_0}\right)^d. \quad (P3.92)$$

The effective medium is characterized by $\kappa_e = \omega/v_e$. We define $\epsilon_1 = (\kappa_1/\kappa_0)^2$, $\epsilon_2 = (\kappa_2/\kappa_0)^2$, and $\bar{\epsilon} = (\kappa_e/\kappa_0)^2$ with respect to a homogeneous reference medium where $\kappa_0 = \omega/v_0$.

In 3D, the solutions in the three regions $r \leq R$, $R < r \leq R_0$, $r > R_0$ can be written as

$$\phi^{(1)} = \sum_{n=0}^{\infty} A_n j_n(\kappa_1 r) P_n(\cos\theta), \qquad r \leq R \quad (P3.93)$$

$$\phi^{(2)} = \sum_{n=0}^{\infty} [B_n j_n(\kappa_2 r) + C_n n_n(\kappa_2 r)] P_n(\cos\theta), \; R < r \leq R_0 \quad (P3.94)$$

$$\phi^{(3)} = \exp(i\boldsymbol{\kappa}_e \cdot \mathbf{r}) + \sum_{n=0}^{\infty} i^n(2n+1) D_n h_n^{(1)}(\kappa_e r) P_n(\cos\theta)$$

$$= \sum_{n=0}^{\infty} i^n(2n+1) P_n(\cos\theta)\left[j_n(\kappa_e r) + D_n h_n^{(1)}(\kappa_e r)\right] \quad r > R_0.$$

$$(P3.95)$$

Using the condition of $(\partial\phi/\partial r)/\phi$ being continuous across interfaces, two equations are obtained at the two interfaces $r = R$ and $r = R_0$:

$$\frac{\kappa_1}{\kappa_2} \frac{j'_n(\kappa_1 R)}{j_n(\kappa_1 R)} = \frac{B_n j'_n(\kappa_2 R) + C_n n'_n(\kappa_2 R)}{B_n j_n(\kappa_2 R) + C_n n_n(\kappa_2 R)}, \quad (P3.96)$$

$$\frac{\kappa_e}{\kappa_2} \frac{j'_n(\kappa_e R_0) + D_n h'_n(\kappa_e R_0)}{j_n(\kappa_e R_0) + D_n h_n(\kappa_e R_0)} = \frac{B_n j'_n(\kappa_2 R_0) + C_n n'_n(\kappa_2 R_0)}{B_n j_n(\kappa_2 R_0) + C_n n_n(\kappa_2 R_0)}, \quad (P3.97)$$

where the superscript (1) to the spherical Hankel function has been dropped. By defining the left-hand side of Eq. (P3.96) as χ_n, B_n can be solved in terms of C_n as

$$B_n = \mathscr{J}_n C_n, \tag{P3.98}$$

where

$$\mathscr{J}_n = \frac{-\chi_n n_n(\kappa_2 R) + n'_n(\kappa_2 R)}{\chi_n j_n(\kappa_2 R) - j'_n(\kappa_2 R)}. \tag{P3.99}$$

By substituting Eq. (P3.98) into Eq. (P3.97), D_n is solved to yield

$$D_n = \frac{-\kappa_e j'_n(\kappa_e R_0)/\kappa_2 + \mathscr{H}_n j_n(\kappa_e R_0)}{h'_n(\kappa_e R_0) - \mathscr{H}_n h_n(\kappa_e R_0)}, \tag{P3.100}$$

where

$$\mathscr{H}_n = \frac{\mathscr{J}_n j'_n(\kappa_2 R_0) + n'_n(\kappa_2 R_0)}{\mathscr{J}_n j_n(\kappa_2 R_0) + n_n(\kappa_2 R_0)}. \tag{P3.101}$$

Exactly the same expressions hold for 2D if all the spherical Bessel functions are replaced with their normal counterpart, i.e., $j_n \to J_n$, $n_n \to N_n$, and $h_n^{(1)} \to H_n^{(1)}$.

For 1D, a simple solution approach is the transfer matrix method. In Figure P3.1 there are four regions. By starting from the leftmost region and going toward the right side, the right-going and left-going waves in respective regions can be written as $\exp(i\kappa_e x) + \rho \exp(-i\kappa_e x)$, $A \exp(i\kappa_1 x) + B \exp(-i\kappa_1 x)$, $C \exp(i\kappa_2 x) + D \exp(-i\kappa_2 x)$, and $\tau \exp(i\kappa_e x)$. The coefficients between the neighboring regions are related to one another by the continuity of ϕ and $\partial\phi/\partial x$ across each interface. These conditions may be written as 2×2 matrices relating the wave amplitudes. Thus,

$$\begin{pmatrix} A \\ B \end{pmatrix} = \mathbf{M}_0 \begin{pmatrix} 1 \\ \rho \end{pmatrix}, \tag{P3.102}$$

$$\begin{pmatrix} C \\ D \end{pmatrix} = \mathbf{M}_1 \begin{pmatrix} A \\ B \end{pmatrix}, \tag{P3.103}$$

$$\begin{pmatrix} \tau \\ 0 \end{pmatrix} = \mathbf{M}_2 \begin{pmatrix} C \\ D \end{pmatrix}, \tag{P3.104}$$

where

$$\mathbf{M}_0 = \begin{bmatrix} \left(1+\dfrac{\kappa_e}{\kappa_1}\right)\exp[-2i(\kappa_e-\kappa_1)p'R_0] & \left(1-\dfrac{\kappa_e}{\kappa_1}\right)\exp[2i(\kappa_e+\kappa_1)p'R_0] \\ \left(1-\dfrac{\kappa_e}{\kappa_1}\right)\exp[-2i(\kappa_e+\kappa_1)p'R_0] & \left(1+\dfrac{\kappa_e}{\kappa_1}\right)\exp[2i(\kappa_e-\kappa_1)p'R_0] \end{bmatrix},$$

(P3.105)

$$\mathbf{M}_1 = \begin{bmatrix} \left(1+\dfrac{\kappa_1}{\kappa_2}\right) & \left(1-\dfrac{\kappa_1}{\kappa_2}\right) \\ \left(1-\dfrac{\kappa_1}{\kappa_2}\right) & \left(1+\dfrac{\kappa_1}{\kappa_2}\right) \end{bmatrix},$$

(P3.106)

and

$$\mathbf{M}_2 = \begin{bmatrix} \left(1+\dfrac{\kappa_2}{\kappa_e}\right) \\ \times\exp[2i(\kappa_2-\kappa_e)(1-p')R_0] & \left(1-\dfrac{\kappa_2}{\kappa_e}\right) \\ & \times\exp[-2i(\kappa_2+\kappa_e)(1-p')R_0] \\ \left(1-\dfrac{\kappa_2}{\kappa_e}\right) \\ \times\exp[2i(\kappa_2+\kappa_e)(1-p')R_0] & \left(1+\dfrac{\kappa_2}{\kappa_e}\right) \\ & \times\exp[-2i(\kappa_2-\kappa_e)(1-p')R_0] \end{bmatrix},$$

(P3.107)

By multiplying the matrices together, one gets

$$\begin{pmatrix} \tau \\ 0 \end{pmatrix} = \mathbf{M}_2\mathbf{M}_1\mathbf{M}_0\begin{pmatrix} 1 \\ \rho \end{pmatrix} = \mathbf{M}\begin{pmatrix} 1 \\ \rho \end{pmatrix}. \qquad (\text{P3.108})$$

Solution of (P3.108) gives τ and ρ in terms of the matrix elements of $\mathbf{M}$:

$$\rho = -\frac{(\mathbf{M})_{21}}{(\mathbf{M})_{22}}, \qquad (\text{P3.109})$$

$$\tau = \frac{\det(\mathbf{M})}{(\mathbf{M})_{22}}. \qquad (\text{P3.110})$$

Here det( ) means the determinant of the matrix.

From the solutions $D_n(p')$ in 3D and 2D, and $\tau(p') - 1$, in 1D, the forward scattering amplitudes can be obtained from Eq. (3.70). The consideration of different values of p' as constituting different structural units then yields the CPA condition as given by Eq. (3.80).

In the long-wavelength limit, expansion of the various functions in the $n = 0$ terms gives

$$\chi_0 \simeq -\frac{\kappa_1}{\kappa_2} \frac{\kappa_1 R}{d}, \tag{P3.111}$$

$$\mathscr{I}_0 \simeq \frac{A_d}{(\kappa_2 R)^d \left[1 - (\kappa_1/\kappa_2)^2\right]}, \tag{P3.112}$$

with

$$A_d = \begin{cases} 3 & \text{in 3D} \\ \dfrac{4}{\pi} & \text{in 2D,} \end{cases} \tag{P3.113}$$

$$\mathscr{H}_0 \cong -\frac{\kappa_2 R_0}{d}\left\{1 - p'\left[1 - \left(\frac{\kappa_1}{\kappa_2}\right)^2\right]\right\}, \tag{P3.114}$$

and

$$D_0 \cong -\frac{i}{A_d}\frac{\kappa_e}{\kappa_2}(\kappa_e R_0)^d \left\{1 - \left(\frac{\kappa_2}{\kappa_e}\right)^2 \right. $$
$$\left. + \left(\frac{\kappa_2}{\kappa_e}\right)^2 p'\left[1 - \left(\frac{\kappa_1}{\kappa_2}\right)^2\right]\right\}. \tag{P3.115}$$

In 1D, the expansion of the exponentials and the more tedious multiplication of matrices in the low frequency limit yield

$$\rho = \tau - 1 \cong -i(\kappa_e R_0)\left[1 - (1-p')\left(\frac{\kappa_2}{\kappa_e}\right)^2 - p'\left(\frac{\kappa_1}{\kappa_e}\right)^2\right]. \tag{P3.116}$$

By identifying $(\kappa_1/\kappa_e)^2 = \epsilon_1/\bar{\epsilon}$, $(\kappa_2/\kappa_e)^2 = \epsilon_2/\bar{\epsilon}$, and inserting the expressions of the forward scattering amplitudes into Eq. (3.80), the p'

integral may be performed to give

$$p = \int_0^1 \mathscr{D}(p') p' \, dp', \tag{P3.117}$$

and Eq. (3.81) results.

For the mean free path, it is noted that in the $\kappa_e R_0 \to 0$ limit, the total scattering cross section $O(p')$ may be expressed [from Eqs. (3.70) and (P3.14)] as

$$O(R_0, p') = \begin{cases} \dfrac{4\pi}{\kappa_e^2} |D_0|^2 & \text{3D} \\ \dfrac{4}{\kappa_e} |D_0|^2 & \text{2D} \\ |\rho|^2 + |\tau - 1|^2 & \text{1D.} \end{cases} \tag{P3.118}$$

By using the optical theorem and Eq. (3.75), l is given by

$$l = C_d R_0^d \left\{ \int_0^1 \mathscr{D}(p') O(R_0, p') \, dp' \right\}^{-1}, \tag{P3.119}$$

where $C_d = 4\pi/3$, π, 2, in 3D, 2D, 1D, respectively. Equation (3.82) results from the substitution of (P3.115) and (P3.116) into Eq. (P3.119).

3.12. Derive the 1D, 2D, and 3D electromagnetic CPA equations for the dispersion microgeometry in the long-wavelength limit, and give their respective mean free path expressions.

In the solution to Problem 3.9, details about the electromagnetic boundary condition and the requirement of the vector character of the electromagnetic wave scattering have been discussed. In short, instead of the $n = 0$ term, the lowest-order scattering in the long-wavelength limit is given by the $n = 1$ term. Also, the boundary condition change means κ_1/κ_2 in Eq. (P3.96) and κ_e/κ_2 in (P3.97) should be replaced by $(\kappa_1/\kappa_2)^3$ and $(\kappa_e/\kappa_2)^3$, respectively. In 1D the simple derivation of the CPA equation in the solution to Problem 3.9 shows it to be exact in the $\kappa_e R_0 \to 0$ limit. Therefore both the 1D electromagnetic CPA equation and its relevant expression for the mean free path are the same as their counterparts the scalar wave case. Only the 3D and 2D cases need to be detailed.

By expanding the various functions and making the necessary replacements consistently, it is found that

$$\chi_1 \cong \left(\frac{\kappa_1}{\kappa_2}\right)^3 \frac{1}{\kappa_1 R}, \tag{P3.120}$$

$$\mathscr{I}_1 \cong A_d \frac{1}{(\kappa_2 R)^d} \frac{(\kappa_1/\kappa_2)^2 + (d-1)}{(\kappa_1/\kappa_2)^2 - 1}, \tag{P3.121}$$

$$\mathscr{H}_1 \cong \frac{1}{\kappa_2 R_0} \frac{(\kappa_1/\kappa_2)^2 + (d-1) + (d-1)p'\left[(\kappa_1/\kappa_2)^2 - 1\right]}{(\kappa_1/\kappa_2)^2 + (d-1) - p'\left[(\kappa_1/\kappa_2)^2 - 1\right]}, \tag{P3.122}$$

and

$$D_1 \cong \frac{i}{A_d}\left(\frac{\kappa_e}{\kappa_2}\right)(\kappa_e R_0)^d$$

$$\times \frac{(\bar{\epsilon} - \epsilon_2)[(d-1)\epsilon_2 + \epsilon_1] + p'[(d-1)\epsilon_2 + \bar{\epsilon}](\epsilon_2 - \epsilon_1)}{\left[(d-1)\epsilon_2 + \sqrt{\epsilon_2 \bar{\epsilon}}\right][(d-1)\epsilon_2 + \epsilon_1] + (d-1)p'(\epsilon_1 - \epsilon_2)\left(\sqrt{\epsilon_2 \bar{\epsilon}} - \epsilon_2\right)}.$$

$$\tag{P3.123}$$

The assumption about local compositional fluctuations and the use of Eq. (3.80) directly result in the CPA equation as given by Eq. (3.84). As for the mean free path, it is noted that to the leading order,

$$f = \begin{cases} -\dfrac{3i}{\kappa_e} D_1 \cos\theta & \text{3D}, \\ 2\sqrt{\dfrac{2}{\pi \kappa_e}} \exp\left(-i\dfrac{\pi}{4}\right) D_1 \cos\theta & \text{2D}, \end{cases} \tag{P3.124}$$

and

$$O = \begin{cases} 2\pi \int_{-1}^{1} |f|^2 \, d\cos\theta & \text{3D}, \\ \int_{0}^{2\pi} |f|^2 \, d\theta & \text{2D}. \end{cases} \tag{P3.125}$$

It follows from Eq. (P3.119) that the mean free path expression has the form given by Eq. (3.85).

3.13. Express the **T** operator in terms of the $\mathbf{t}_i$'s for the individual scatterers.

From Eq. (3.8), it is only one step to see that $\mathbf{T} - \mathbf{VG}_0\mathbf{T} = \mathbf{V}$, so

$$\mathbf{T} = \mathbf{V}(\mathbf{I} + \mathbf{G}_0\mathbf{T}). \tag{P3.126}$$

Since

$$\mathbf{V} = \sum_i \mathbf{v}_i,$$

where $\mathbf{v}_i$'s are the scattering potential operators for individual scatterers, it follows that

$$\mathbf{T} = \sum_i \mathbf{R}_i, \tag{P3.127}$$

where

$$\mathbf{R}_i = \mathbf{v}_i(\mathbf{I} + \mathbf{G}_0\mathbf{T}) = \mathbf{v}_i\left(\mathbf{I} + \mathbf{G}_0 \sum_j \mathbf{R}_j\right) \tag{P3.128}$$

By subtracting $\mathbf{v}_i\mathbf{G}_0\mathbf{R}_i$ from both sides of this equation, it is seen that

$$(\mathbf{I} - \mathbf{v}_i\mathbf{G}_0)\mathbf{R}_i = \mathbf{v}_i\left(\mathbf{I} + \mathbf{G}_0 \sum_{j \neq i} \mathbf{R}_j\right), \tag{P3.129}$$

or

$$\begin{aligned}
\mathbf{R}_i &= (\mathbf{I} - \mathbf{v}_i\mathbf{G}_0)^{-1}\mathbf{v}_i\left(\mathbf{I} + \mathbf{G}_0 \sum_{j \neq i} \mathbf{R}_j\right), \\
&= \mathbf{t}_i\left(\mathbf{I} + \mathbf{G}_0 \sum_{j \neq i} \mathbf{R}_j\right) \\
&= \mathbf{t}_i + \mathbf{t}_i\mathbf{G}_0 \sum_{j \neq i} \mathbf{t}_j\left(\mathbf{I} + \mathbf{G}_0 \sum_{k \neq j} \mathbf{R}_k\right) \\
&= \mathbf{t}_i + \sum_{j \neq i} \mathbf{t}_i\mathbf{G}_0\mathbf{t}_j + \sum_{j \neq i}\sum_{k \neq j} \mathbf{t}_i\mathbf{G}_0\mathbf{t}_j\mathbf{G}_0\mathbf{t}_k + \cdots.
\end{aligned} \tag{P3.130}$$

The second line of the equation is noted to follow from Eq. (3.8). Substitution of Eq. (P3.130) into (P3.127) yields

$$\mathbf{T} = \sum_i \mathbf{t}_i + \sum_i \sum_{j \neq i} \mathbf{t}_i \mathbf{G}_0 \mathbf{t}_j + \sum_i \sum_{j \neq i} \sum_{k \neq j} \mathbf{t}_i \mathbf{G}_0 \mathbf{t}_j \mathbf{G}_0 \mathbf{t}_k + \cdots. \quad (P3.131)$$

Each succeeding index value is noted to be forced to differ from the one before it. However, there is no restriction on two indices not next to each other. Therefore, $i = k$ is allowed in the third term.

References

Arfken, G. (1970). "Mathematical Methods for Physicists," 2nd Ed. Academic Press, New York.
Bohren, C. F., and Huffman, D. R. (1983). "Absorption and Scattering of Light by Small Particles," p. 82. John Wiley & Sons, New York.
Bruggeman, D. A. G. (1935). *Ann. Phys.* (*Leipzig*) **24**, 636.
Economou, E. N. (1983). "Green's Functions in Quantum Physics," 2nd Ed. Springer-Verlag, Berlin.
Jackson, J. D. (1975). "Classical Electrodynamics," 2nd Ed. John Wiley & Sons, New York.
Jing, X., Sheng, P., and Zhou, M. (1992). *Phys. Rev. A* **46**, 6513.
Lax, M. (1951). *Rev. Mod. Phys.* **23**, 287.
Liu, J., Ye, L., Weitz, D. A., and Sheng, P. (1990). *Phys. Rev. Lett.* **65**, 2602.
Maxwell-Garnett, J. C., (1904). *Phil. Trans. Roy. Soc.* **203**, 385.

4

Diffusive Waves

4.1 Beyond the Effective Medium

In the last chapter it has been argued that a random medium can appear locally homogeneous to a propagating wave, with an effective material property that is a function of the material properties of the inhomogeneous components and their microstructure. This effective medium description can fail in two ways. The first is for the frequency to be high enough so that the CPA equation has no solution (or no unique solution), which occurs when the wave becomes capable of resolving the microstructure, and the self-energy thus acquires a **k** dependence. In this context, it is interesting to note that for the lattice model the CPA equation is always soluble (see Chapter 3) in the absence of site–site correlation; however, such is not the case for waves in continuum. The second is for the transport distance to be much larger than the mean free path l so that the description of wave properties is focused on a scale larger than that intended by the effective medium picture. This chapter is concerned with the wave transport behavior in the second scenario.

As a function of increasing observation scale, there is a smooth transition in the transport characteristics from the coherent (CPA) regime to the multiply scattered behavior, which is denoted the diffusive regime in anticipation of later developments. To better appreciate this transition, let us consider the situation where a plane wave has propagated the intermediate distance of two to three mean free paths into an inhomogeneous medium. With two to three scatterings, a certain fraction of the wave energy will have become incoherent, in the sense of departing from its original progpagation direction (but still at the same frequency, since all scatterings considered here are elastic). However, there is still a finite portion of the wave energy which remains coherent. This is clear from the calculation of the scattering amplitude $f(\theta)$: Each scattering takes away only a fraction of the wave energy from the forward direction. Therefore, as the wave encounters successive scatterings, the component that remains coherent decreases geometrically, i.e., decays exponentially with the number of scatterings. The possibility of a certain part of the incoherent wave energy being returned to the coherent state can be discounted, because to be coherent means propagation not only in the original direction but also with the same phase relation. The combined requirements make the return probability essentially nil. Therefore, the transition from the coherent, CPA-dominated regime to the diffusive regime is governed by a smooth exponential decay of the coherent component, and the growth and saturation of the incoherent component, as the number of scattering increases.

The approach of this book is to treat wave transport only in the two limits and to approximate the transition behavior as a superposition of the coherent and the diffusive components. The most serious error inherent in such an approximation is that the incoherent component may not be fully diffusive in the transition regime. It remains an open question as to how many scatterings are required before diffusive transport is firmly established. However, experimental evidence (see Section 4.7) seems to indicate well-established diffusive behavior after only a few scatterings. The caveat here is that such evidence applies only to *configurationally averaged intensity*, which is expected to differ from the single-configuration behavior.

4.2 Pulse Intensity Evolution in a Random Medium

To address quantitatively the question of what happens beyond the effective medium regime, let us approach the problem physically by following the *intensity* evolution of a pulse injected into a random medium at point $\mathbf{r}'$ and $t = 0$. The focus on intensity is intended to enable us to track the wave transport behavior after the phase coherence is destroyed by random scatterings. By definition, the pulse intensity is given by

$$P(t, \mathbf{r}, \mathbf{r}') = |G(t, \mathbf{r}, \mathbf{r}')|^2, \qquad (4.1)$$

where $|G|^2 = GG^*$. We are interested in the behavior of P for large t and $|\mathbf{r} - \mathbf{r}'|$. It is important to point out that although in that limit the wave energy is predominantly incoherent, the pulse front *is always coherent* because by definition it is the part of the wave train which reaches any given point the earliest in time. Since any scattering would slow down the wave by increasing its travel path, the pulse front thus consists of the

4.2 Pulse Intensity Evolution in a Random Medium

energy component that has suffered no scattering. Although this coherent component is small in amplitude, its importance lies in the fact that it is through pulse front propagation that the causal nature of the wave becomes manifest, and the propagation speed of the coherent wave can be measured by tracking the front. However, pulse-front tracking is *not* the approach developed below. Instead, we intend to track the dominant energy component of the wave, which has an entirely different transport behavior.

For various reasons that will become clear later, the information on intensity transport is easier to obtain in the Fourier-transformed domain. Since in the frequency domain

$$G(t,\mathbf{r},\mathbf{r}') = \frac{1}{2\pi} \int_{-\infty}^{\infty} G^+(\omega_1,\mathbf{r},\mathbf{r}') \exp(-i\omega_1 t) \, d\omega_1, \tag{4.2}$$

$$G^*(t,\mathbf{r},\mathbf{r}') = \frac{1}{2\pi} \int_{-\infty}^{\infty} G^-(\omega_2,\mathbf{r}',\mathbf{r}) \exp(i\omega_2 t) \, d\omega_2, \tag{4.3}$$

we have

$$|G(t,\mathbf{r},\mathbf{r}')|^2 = \frac{1}{(2\pi)^2} \int_{-\infty}^{\infty} \int_{-\infty}^{\infty} G^+(\omega_1,\mathbf{r},\mathbf{r}') G^-(\omega_2,\mathbf{r}',\mathbf{r})$$
$$\times \exp(-i\Delta\omega t) \, d\omega_1 \, d\omega_2. \tag{4.4}$$

Here $\Delta\omega = \omega_1 - \omega_2$ is the frequency difference; Eq. (4.3) follows from the solution to Problem 2.3 [Eqs. (P2.10) and (P2.12)], and G^+, G^- denote the retarded and the advanced Green's functions, respectively. In a uniform medium, they are seen to correspond to the outgoing (+) and the incoming (−) waves associated with a point source, as worked out in Chapter 2. The important point to note about Eq. (4.4) is that $\Delta\omega$ is the conjugate variable to the travel time t. This is reasonable because $\Delta\omega$ is the "beating frequency" relevant to the propagation of the wave packet *envelope*, which corresponds to the motion of the center of mass for a quantum particle (wave packet). Due to the nature of the conjugate variables, the $t \to \infty$ limit corresponds to the $\Delta\omega \to 0$ limit because for $\Delta\omega > 1/t$ the rapid oscillations of the factor $\exp(-i\Delta\omega t)$ in the integrand of (4.4) make the total integrated contribution negligible. The only significant contribution is thus from the region $\Delta\omega \leq 1/t$.

The spectral component of $P(t,\mathbf{r},\mathbf{r}')$, is given by

$$P(\bar{\omega},\mathbf{r},\mathbf{r}') = \int_{-\infty}^{\infty} P(t,\mathbf{r},\mathbf{r}') \exp(i\bar{\omega}t) \, dt. \tag{4.5}$$

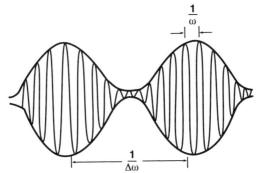

Figure 4.1 An amplitude-modulated wave. The faster variation inside corresponds to frequency ω. The frequency of the envelope corresponds to $\Delta\omega$.

The substitution of Eq. (4.4) into Eq. (4.5) gives

$$P(\bar{\omega},\mathbf{r},\mathbf{r}') = \frac{1}{2\pi} \int_{-\infty}^{\infty}\int_{-\infty}^{\infty} G^+(\omega_1,\mathbf{r},\mathbf{r}')G^-(\omega_2,\mathbf{r}',\mathbf{r})\delta(\bar{\omega}-\Delta\omega)\,d\omega_1\,d\omega_2,$$

$$= \frac{1}{2\pi}\int_{-\infty}^{\infty} G^+\left(\omega+\frac{\Delta\omega}{2},\mathbf{r},\mathbf{r}'\right)G^-\left(\omega-\frac{\Delta\omega}{2},\mathbf{r}',\mathbf{r}\right)d\omega, \quad (4.6)$$

where the integration variables have been changed from ω_1, ω_2 to $\omega=(\omega_1+\omega_2)/2$ and $\Delta\omega=\omega_1-\omega_2$. The spectral frequency $\bar{\omega}$ is seen to be identical to $\Delta\omega$. Since $\Delta\omega$ is related to the frequency of the wave packet envelope and ω to the wave frequency inside the wave packet as shown in Figure 4.1, sometimes $\Delta\omega$ is also called the external frequency, or the modulation frequency, and ω is called the internal frequency, or the carrier frequency.

For the spatial spectral component of P, it is simpler to examine the configurationally averaged quantity

$$\langle P(\Delta\omega,\mathbf{r},\mathbf{r}')\rangle_c = \frac{1}{2\pi}\int_{-\infty}^{\infty}\left\langle G^+\left(\omega+\frac{\Delta\omega}{2},\mathbf{r},\mathbf{r}'\right)\right.$$
$$\left.\times G^-\left(\omega-\frac{\Delta\omega}{2},\mathbf{r}',\mathbf{r}\right)\right\rangle_c d\omega. \quad (4.7)$$

4.2 Pulse Intensity Evolution in a Random Medium

In the solution to Problem 4.1, it is shown that

$$S(\Delta\omega, \mathbf{r} - \mathbf{r}'|\omega) = \left\langle G^+\left(\omega + \frac{\Delta\omega}{2}, \mathbf{r}, \mathbf{r}'\right) G^-\left(\omega - \frac{\Delta\omega}{2}, \mathbf{r}', \mathbf{r}\right) \right\rangle_c$$

$$= \frac{L^{3d}}{N^2(2\pi)^{3d}} \int\int\int \langle\langle \mathbf{k}_+|\mathbf{G}^+(\omega^+)|\mathbf{k}'_+\rangle\langle \mathbf{k}'_-|\mathbf{G}^-(\omega^-)|\mathbf{k}_-\rangle\rangle_c$$

$$\times \exp[i\,\Delta\mathbf{k}\cdot(\mathbf{r} - \mathbf{r}')]\,d\mathbf{k}\,d\mathbf{k}'\,d\Delta\mathbf{k},$$

$$= \frac{1}{N^3} \sum_{\mathbf{k}} \sum_{\mathbf{k}'} \sum_{\Delta\mathbf{k}} \langle\langle \mathbf{k}_+|\mathbf{G}^+|\mathbf{k}'_+\rangle\langle \mathbf{k}'_-|\mathbf{G}^-|\mathbf{k}_-\rangle\rangle_c$$

$$\times \langle \mathbf{r}|\Delta\mathbf{k}\rangle\langle\Delta\mathbf{k}|\mathbf{r}'\rangle, \qquad (4.8)$$

where all the **k** summations are understood to be within the first Brillouin zone in the lattice case, and

$$\mathbf{k}_+ = \mathbf{k} + \frac{\Delta\mathbf{k}}{2},$$

$$\mathbf{k}_- = \mathbf{k} - \frac{\Delta\mathbf{k}}{2},$$

$$\mathbf{k}'_+ = \mathbf{k}' + \frac{\Delta\mathbf{k}}{2},$$

$$\mathbf{k}'_- = \mathbf{k}' - \frac{\Delta\mathbf{k}}{2}. \qquad (4.9)$$

The abbreviations $\omega^+ = \omega + \Delta\omega/2$, $\omega^- = \omega - \Delta\omega/2$ have been used. Equation (4.8) tells us that the averaged quantity depends only on $\mathbf{r} - \mathbf{r}'$ and that the conjugate variable to the travel distance $|\mathbf{r} - \mathbf{r}'|$ is the difference in the momenta, $\Delta\mathbf{k}$, of the two Green's functions, which have a total of three distinct momenta after configurational averaging. Since $\Delta\mathbf{k}$ and $\mathbf{r} - \mathbf{r}'$ are conjugate variables, the $|\mathbf{r} - \mathbf{r}'| \to \infty$ limit therefore corresponds to the $|\Delta\mathbf{k}| \to 0$ limit. From Eq. (4.8) it is clear that the $\Delta\mathbf{k}$ Fourier component of $S(\Delta\omega, \mathbf{r} - \mathbf{r}'|\omega)$ is given by

$$S(\Delta\omega, \Delta\mathbf{k}|\omega) = \frac{1}{N^2} \sum_{\mathbf{k}} \sum_{\mathbf{k}'} \phi_{\mathbf{k},\mathbf{k}'}(\Delta\omega, \Delta\mathbf{k}|\omega), \qquad (4.10)$$

where

$$\phi_{\mathbf{k},\mathbf{k}'}(\Delta\omega, \Delta\mathbf{k}|\omega) = \langle\langle \mathbf{k}_+|\mathbf{G}^+(\omega^+)|\mathbf{k}'_+\rangle\langle \mathbf{k}'_-|\mathbf{G}^-(\omega^-)|\mathbf{k}_-\rangle\rangle_c. \quad (4.11)$$

The desired information is the behavior of $S(\Delta\omega, \Delta\mathbf{k}|\omega)$ in the limit of $\Delta\omega, |\Delta\mathbf{k}| \to 0$. It is important to be aware that in seeking the behavior of S, the approach to be followed below is analogous to that for obtaining the diffusive behavior from Brownian motion. Namely, instead of following the wave in a given medium for a long time and over many collisions, we look for *configurationally averaged* behavior and assume the two to be the same. The fact that this is true is not trivial; it is a reflection of the ergodicity of the system; i.e., the infinite-time average of the transported intensity in an (infinite) sample is equivalent to its spatial average over many configurations. Therefore, random scatterings will not appear explicitly in the subsequent development. Instead, the effect of random scatterings is captured by configurational averaging. The limitation of this approach is that the derived behavior is only in the *expected* sense mathematically. For any given configuration, the expected behavior can deviate from the actual behavior. Such a deviation may be especially large at short propagation times when the number of scatterings is small, and the effect of time averaging is therefore minimal. However, for the long-time, averaged property this approach should be accurate.

In anticipation of the result that pulse intensity evaluation becomes diffusive in the long-time, large-travel-distance limit, the desired form of $S(\Delta\omega, \Delta\mathbf{k}|\omega)$ may be obtained by assuming $\langle P(t, \mathbf{r}, \mathbf{r}')\rangle_c$ to be diffusive so that

$$(-i\Delta\omega - D(\omega)\nabla^2)S(\Delta\omega, \mathbf{r} - \mathbf{r}'|\omega) = \text{constant} \cdot \delta(\mathbf{r} - \mathbf{r}'), \quad (4.12)$$

where the factor $-i\Delta\omega$ is recognized to originate from the first-order time derivative $\partial/\partial t$ of the diffusion equation, $D(\omega)$ is the wave diffusion constant, and ∇^2 is understood to operate on $\mathbf{r}$. By writing $S(\Delta\omega, \mathbf{r} - \mathbf{r}'|\omega)$ and $\delta(\mathbf{r} - \mathbf{r}')$ as the sums of their spectral components, one immediately obtains

$$S(\Delta\omega, \Delta\mathbf{k}|\omega) \propto \frac{1}{-i\Delta\omega + D(\omega)|\Delta\mathbf{k}|^2}, \quad \Delta\omega, |\Delta\mathbf{k}| \to 0. \quad (4.13)$$

That is, $S(\Delta\omega, \Delta\mathbf{k}|\omega)$ has a *diffusive pole*. The derivation of this form and the calculation of $D(\omega)$ are detailed in the following.

4.3 The Bethe–Salpeter Equation and Its Solution by Moments

The evaluation of the configurational average of two Green's functions requires an equation that relates it to the effect of random scatterings. To that end, let us start with the operator formalism defined relative to the effect medium, where

$$\mathbf{G}^+ = \mathbf{G}_e^+ + \mathbf{G}_e^+ \bar{\mathbf{T}}^+ \mathbf{G}_e^+, \qquad (4.14a)$$

$$\mathbf{G}^- = \mathbf{G}_e^- + \mathbf{G}_e^- \bar{\mathbf{T}}^- \mathbf{G}_e^-, \qquad (4.14b)$$

and

$$\langle \bar{\mathbf{T}}^+ \rangle_c = \langle \bar{\mathbf{T}}^- \rangle_c = 0 \qquad (4.15)$$

to within the CPA accuracy. For intensity transport, what is required is the configurationally averaged outer product

$$\langle \mathbf{G}^+ \otimes \mathbf{G}^- \rangle_c = \mathbf{G}_e^+ \otimes \mathbf{G}_e^- + \langle \mathbf{G}_e^+ \bar{\mathbf{T}}^+ \mathbf{G}_e^+ \otimes \mathbf{G}_e^- \bar{\mathbf{T}}^- \mathbf{G}_e^- \rangle_c, \qquad (4.16)$$

where terms involving single $\bar{\mathbf{T}}^\pm$ are averaged to zero in accordance with the condition of Eq. (4.15). Here the concept of the outer product (of two vectors) has been discussed earlier in Chapter 2 in connection with the Dirac notation. For two vectors, the outer product yields a matrix. For the operators in Eq. (4.16), which are matrices, the outer product gives an object with four indices, i.e., a fourth-rank tensor. We denote each index as representing a "channel," which can be either in real space or in momentum space. Configurational averaging means that the number of independent channel variables is generally one less than the number of channels. This is shown in the solution to Problem 4.1.

The second term of Eq. (4.16) can be written as

$$\mathbf{G}_e^+ \langle \bar{\mathbf{T}}^+ \mathbf{G}_e^+ \otimes \mathbf{G}_e^- \bar{\mathbf{T}}^- \rangle_c \mathbf{G}_e^- = (\mathbf{G}_e^+ \otimes \mathbf{G}_e^-) : \langle \bar{\mathbf{T}}^+ \otimes \bar{\mathbf{T}}^- \rangle_c : (\mathbf{G}_e^+ \otimes \mathbf{G}_e^-),$$

$$= (\mathbf{G}_e^+ \otimes \mathbf{G}_e^-) : \Gamma : (\mathbf{G}_e^+ \otimes \mathbf{G}_e^-). \qquad (4.17)$$

Here an explanation of the notation is important for understanding the subsequent development. Γ is known as the "vertex" function. It has two input channels and two output channels. If Γ is expressed as $\langle \bar{\mathbf{T}}^+ \otimes \bar{\mathbf{T}}^- \rangle_c$, then the two input channels are on the outside left and right, whereas the two output channels are on the center left and the center right, facing each other across the outer product sign, i.e.,

$$\text{input} \to \bar{\mathbf{T}}^+ \overset{\text{outputs}}{\rightharpoonup} \otimes \overset{}{\leftharpoondown} \bar{\mathbf{T}}^- \leftarrow \text{input}.$$

In the convention used below, $(\bar{\mathbf{T}}^+ \otimes \bar{\mathbf{T}}^-):(\mathbf{G}_e^+ \otimes \mathbf{G}_e^-)$ means $\bar{\mathbf{T}}^+ \mathbf{G}_e^+ \otimes \mathbf{G}_e^- \bar{\mathbf{T}}^-$, whereas $(\mathbf{G}_e^+ \otimes \mathbf{G}_e^-):(\bar{\mathbf{T}}^+ \otimes \bar{\mathbf{T}}^-)$ means $\mathbf{G}_e^+ \bar{\mathbf{T}}^+ \otimes \bar{\mathbf{T}}^- \mathbf{G}_e^-$. One can see that graphically as

$$\text{input} \to \bar{\mathbf{T}}^+ \cdot \mathbf{G}_e^+ \overset{\text{outputs}}{\rightharpoonup} \otimes \overset{}{\leftharpoondown} \mathbf{G}_e^- \cdot \bar{\mathbf{T}}^- \leftarrow \text{input}$$

being equivalent to

$$+\text{channel input} \to \bar{\mathbf{T}}^+ \cdot \mathbf{G}_e^+ \to +\text{channel output}$$
$$\otimes \quad \otimes$$
$$-\text{channel input} \to \bar{\mathbf{T}}^- \cdot \mathbf{G}_e^- \to -\text{channel output}.$$

Here the inner product between two operators is denoted by a dot, which is usually omitted. Γ can be represented graphically by a box with four leads, two in and two out, as shown in Figure 4.2. The two top leads are the $+$ channels, indicated by the upper dots of the two colons, and the two bottom leads are the $-$ channels, indicated by the bottom dots of the two colons. In this alternative representation, the two channels to the left are the input ones, and the two to the right are the output ones. The purpose of the colons is to remind the readers that the channels indicated form inner products with the object(s) to its left or right in the manner just described. Here $+$ means (1) the frequency is $\omega + \Delta\omega/2$, (2) the wave vector is $\mathbf{k} + \Delta\mathbf{k}/2$, and (3) the Green's function is retarded. The $-$ has similarly implied meanings. The directions of the arrows, representing the Green's functions, are opposite for the $+$ and $-$ channels because $\mathbf{G}_e^+$

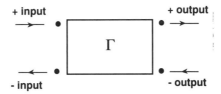

Figure 4.2 The schematic diagram for the vertex function Γ with its four channels, denoted by the two colons (solid circles).

4.3 The Bethe–Salpeter Equation and Its Solution by Moments

and $\mathbf{G}_e^-$ generally represent opposite directions of propagation. In Eq. (4.17), we have used the following general equivalence: $\mathbf{A}\bar{\mathbf{T}}^+\mathbf{B} \otimes \mathbf{B}'\bar{\mathbf{T}}^-\mathbf{A}' = (\mathbf{A} \otimes \mathbf{A}'){:}(\bar{\mathbf{T}}^+ \otimes \bar{\mathbf{T}}^-){:}(\mathbf{B} \otimes \mathbf{B}')$. At the risk of belaboring the point, this equivalence can also be demonstrated directly through the summation of indices. Let $(\mathbf{A})_{ij} \otimes (\mathbf{A}')_{kl} = (\mathbf{A} \otimes \mathbf{A}')_{iljk}$, where i, l are the input channels and j, k are the output channels. Then

$$(\mathbf{A}\bar{\mathbf{T}}^+\mathbf{B})_{ij} = \sum_{l,k} (\mathbf{A})_{il} (\bar{\mathbf{T}}^+)_{lk} (\mathbf{B})_{kj},$$

$$(\mathbf{B}'\bar{\mathbf{T}}^-\mathbf{A}')_{st} = \sum_{m,n} (\mathbf{B}')_{sm} (\bar{\mathbf{T}}^-)_{mn} (\mathbf{A}')_{nt},$$

and therefore

$$(\mathbf{A}\bar{\mathbf{T}}^+\mathbf{B} \otimes \mathbf{B}'\bar{\mathbf{T}}^-\mathbf{A}')_{itjs} = \sum_{l,k,m,n} (\mathbf{A})_{il} (\bar{\mathbf{T}}^+)_{lk} (\mathbf{B})_{kj} (\mathbf{B}')_{sm} (\bar{\mathbf{T}}^-)_{mn} (\mathbf{A}')_{nt}$$

$$= \sum_{l,n,k,m} (\mathbf{A} \otimes \mathbf{A}')_{itln} (\bar{\mathbf{T}}^+ \otimes \bar{\mathbf{T}}^-)_{lnkm} (\mathbf{B} \otimes \mathbf{B}')_{kmjs}$$

$$= [(\mathbf{A} \otimes \mathbf{A}'){:}(\bar{\mathbf{T}}^+ \otimes \bar{\mathbf{T}}^-){:}(\mathbf{B} \otimes \mathbf{B}')]_{itjs}.$$

Let us introduce a new vertex function $\mathbf{U}$ as

$$\Gamma = \mathbf{U} + \mathbf{U}{:}(\mathbf{G}_e^+ \otimes \mathbf{G}_e^-){:}\Gamma, \tag{4.18}$$

where the second term on the right-hand side means the $+$ and $-$ output channels of $\mathbf{U}$ form inner products with the respective $+$ and $-$ *input channels* of $\mathbf{G}_e^+ \otimes \mathbf{G}_e^-$, whose *output* channels form inner products with the respective $+$ and $-$ *input channels* of Γ. By substituting Eq. (4.18) into Eqs. (4.16) and (4.17), one gets inner products with the respective $+$ and $-$ *input channels* of Γ. By substituting Eq. (4.18) into Eqs. (4.16) and (4.17), one gets

$$\langle \mathbf{G}^+ \otimes \mathbf{G}^- \rangle_c = \mathbf{G}_e^+ \otimes \mathbf{G}_e^- + (\mathbf{G}_e^+ \otimes \mathbf{G}_e^-){:}\mathbf{U}{:}(\mathbf{G}_e^+ \otimes \mathbf{G}_e^-)$$

$$+ (\mathbf{G}_e^+ \otimes \mathbf{G}_e^-){:}\mathbf{U}{:}(\mathbf{G}_e^+ \otimes \mathbf{G}_e^-){:}\Gamma{:}(\mathbf{G}_e^+ \otimes \mathbf{G}_e^-)$$

$$= \mathbf{G}_e^+ \otimes \mathbf{G}_e^- + (\mathbf{G}_e^+ \otimes \mathbf{G}_e^-){:}\mathbf{U}{:}[\mathbf{G}_e^+ \otimes \mathbf{G}_e^-$$

$$+ (\mathbf{G}_e^+ \otimes \mathbf{G}_e^-){:}\Gamma{:}(\mathbf{G}_e^+ \otimes \mathbf{G}_e^-)].$$

Since the quantity in the square bracket is just $\langle \mathbf{G}^+ \otimes \mathbf{G}^- \rangle_c$, we have

$$\langle \mathbf{G}^+ \otimes \mathbf{G}^- \rangle_c = \mathbf{G}_e^+ \otimes \mathbf{G}_e^- + (\mathbf{G}_e^+ \otimes \mathbf{G}_e^-){:}\mathbf{U}{:}\langle \mathbf{G}^+ \otimes \mathbf{G}^- \rangle_c. \tag{4.19}$$

Equation (4.19), together with the definitions of $\mathbf{U}$ and Γ, is known as the Bethe–Salpeter equation. It is the analog of the Dyson equation for

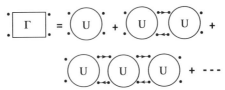

Figure 4.3 Diagrammatic representation of the reducible vertex function Γ in terms of repeated **U**'s, the irreducible vertex function. The channels are denoted by solid circles.

$\langle \mathbf{G}^+ \rangle_c$, and **U** is the two-Green's-function analog of self-energy. From Eq. (4.18), Γ may be expressed as a series of **U**:

$$\mathbf{\Gamma} = \mathbf{U} + \mathbf{U}{:}(\mathbf{G}_e^+ \otimes \mathbf{G}_e^-){:}\mathbf{U} + \mathbf{U}{:}(\mathbf{G}_e^+ \otimes \mathbf{G}_e^-){:}\mathbf{U}{:}(\mathbf{G}_e^+ \otimes \mathbf{G}_e^-){:}\mathbf{U} + \cdots . \quad (4.20)$$

This series is represented graphically in Figure 4.3. Since Γ can be represented as the sum of many terms, it is called the "reducible" vertex function. In contrast, **U** is called the "irreducible" vertex function because it cannot be further decomposed.

Leaving aside the calculation of **U** for the moment, let us explore the solution of Eq. (4.19) in the momentum representation, at the limit of small $\Delta\omega$ and $|\Delta\mathbf{k}|$. In k-space, $\langle \mathbf{G}^+ \otimes \mathbf{G}^- \rangle_c = \phi_{\mathbf{k},\mathbf{k}'}(\Delta\omega, \Delta\mathbf{k}|\omega)$, and $\mathbf{G}_e^+ \otimes \mathbf{G}_e^- = NG_e^+(\omega^+, \mathbf{k}^+)G_e^-(\omega^-, \mathbf{k}^-)\delta_{\mathbf{k},\mathbf{k}'}$. Therefore, Eq. (4.19) may be written as

$$\phi_{\mathbf{k},\mathbf{k}'} = (G_e^+ G_e^-)_{\mathbf{k}}\left[N\delta_{\mathbf{k},\mathbf{k}'} + \frac{1}{N}\sum_{\mathbf{k}_1} U_{\mathbf{k},\mathbf{k}_1}\phi_{\mathbf{k}_1,\mathbf{k}'} \right]. \quad (4.21)$$

Here **k** denotes the input channels ($\mathbf{k} + \Delta\mathbf{k}/2$ and $\mathbf{k} - \Delta\mathbf{k}/2$), and $\mathbf{k}'$ the output channels. The inner product between $(G_e^+ G_e^-)_{\mathbf{k}}$ and $U_{\mathbf{k},\mathbf{k}_1}$ has already been performed in Eq. (4.21) as shown by the identical **k** index, a result of summation over a delta function. From Eq. (4.21) it is clear that the $\mathbf{k}'$ variable may be summed to give an equation in terms of the quantity

$$\phi_{\mathbf{k}} = \frac{1}{N}\sum_{\mathbf{k}'} \phi_{\mathbf{k},\mathbf{k}'} . \quad (4.22)$$

The new equation for $\phi_{\mathbf{k}}(\Delta\omega, \Delta\mathbf{k}|\omega)$ is

$$\phi_{\mathbf{k}} = (G_e^+ G_e^-)_{\mathbf{k}}\left[1 + \frac{1}{N}\sum_{\mathbf{k}'} U_{\mathbf{k},\mathbf{k}'}\phi_{\mathbf{k}'} \right], \quad (4.23)$$

4.3 The Bethe–Salpeter Equation and Its Solution by Moments

where the last $\mathbf{k}_1$ index is relabeled as $\mathbf{k}'$. To proceed further, it is noted that since the product of two numbers a, b may be expressed as

$$ab = \frac{a - b}{b^{-1} - a^{-1}}, \tag{4.24}$$

it follows that

$$G_e^+ G_e^- = \frac{G_e^+ - G_e^-}{(G_e^-)^{-1} - (G_e^+)^{-1}}. \tag{4.25}$$

From the definition of $G_e^+(\omega, \mathbf{k})$ as an approximation to $\langle G^+ \rangle_c(\omega, \mathbf{K})$ with $\beta = 1$ [Eq. (3.51)],

$$\frac{N}{L^d}\left[(G_e^-)^{-1} - (G_e^+)^{-1}\right] = -\frac{d\kappa_0^2}{d\omega}\Delta\omega$$
$$+ \nabla_k e(\mathbf{k}) \cdot \Delta\mathbf{k} + \Sigma^+(k_+, \omega^+) - \Sigma^-(k_-, \omega^-), \tag{4.26}$$

where ∇_k means the gradient operator on the variables k_x, k_y, and k_z. For simplicity, the definition of $e(\mathbf{k})$ is generalized for the classical wave as $e(\mathbf{k}) = \mathbf{k} \cdot \mathbf{k}$, so that $\nabla_k e(\mathbf{k}) = 2\mathbf{k}$ in that case. On the other hand, in the limit of $|\Delta\mathbf{k}|, \Delta\omega \to 0$, $G_e^+ - G_e^-$ may be approximated as

$$\lim_{|\Delta\mathbf{k}|, \Delta\omega \to 0} (G_e^+ - G_e^-)$$

$$= (\Delta G_e)_\mathbf{k} = \frac{2i\,\mathrm{Im}\,\Sigma^+}{\left[\kappa_0^2 - e(\mathbf{k}) - \mathrm{Re}(\Sigma^+)\right]^2 + (\mathrm{Im}\,\Sigma^+)^2}\frac{N}{L^d}. \tag{4.27}$$

Equation (4.27) is obtained with the knowledge that the $\mathrm{Im}\,\Sigma^+$ and $\mathrm{Im}\,\Sigma^-$ differ by a sign. By defining $\kappa_e^2 = \kappa_0^2 - \mathrm{Re}(\Sigma^+)$ and *evaluating Σ^+ relative to the effective medium* κ_e, $(\Delta G_e)_\mathbf{k}$ may be approximated by a delta function when $|\mathrm{Im}\,\Sigma^+|$ is small (as shown in the solution to Problem 2.1), given by

$$(\Delta G_e)_\mathbf{k} = 2i\,\mathrm{Im}\,G_e^+(\omega, \mathbf{k})$$
$$\simeq -2\pi i\,\delta\left[\kappa_e^2 - e(\mathbf{k})\right]\frac{N}{L^d}. \tag{4.28}$$

Here the negative sign comes from the fact that $\mathrm{Im}\,\Sigma^+ < 0$, as required by the casual behavior of retarded Green's function. Note that although Σ^+ is evaluated relative to the effective medium, the overbar on Σ^+ has been dropped for simplicity. The last equality of Eq. (4.28) can be seen alternatively from the fact that in the limit of $|\Delta\mathbf{k}|, \Delta\omega \to 0$, G_e^+, G_e^- are complex conjugates of each other. The approximation expressed by Eq. (4.28) will be used extensively in the following calculations. The quantity $(\Delta G_e)_k$ is alternatively known as the spectral function because it peaks at the dispersion relation(s) of the effective medium excitation(s). Its use in extending the CPA to the intermediate-frequency regime has been presented in the last chapter.

With the conversion of $G_e^+ G_e^-$ to $(\Delta G_e)_k/[(G_e^-)^{-1} - (G_e^+)^{-1}]$ and the multiplication of both sides of Eq. (4.23) by $[(G_e^-)^{-1} - (G_e^+)^{-1}]$, a new form of the Bethe–Salpeter equation is obtained:

$$\frac{L^d}{N}\left(-\frac{d\kappa_0^2}{d\omega}\Delta\omega + \nabla_k e(\mathbf{k})\cdot\Delta\mathbf{k} + \Sigma^+ - \Sigma^-\right)\phi_\mathbf{k}$$
$$= (\Delta G_e)_\mathbf{k}\left[1 + \frac{1}{N}\sum_{\mathbf{k}'} U_{\mathbf{k},\mathbf{k}'}\phi_{\mathbf{k}'}\right]. \qquad (4.29)$$

Equation (4.29) is the dynamical transport equation for wave intensity in the limit of $\Delta\omega, |\Delta\mathbf{k}| \to 0$. It corresponds to the Boltzmann transport equation for classical particles. Just as in the classical case, one solution approach to Eq. (4.29) is by calculating the moments of $\phi_\mathbf{k}$. To do that, let us first sum both sides of Eq. (4.29) with respect to $\mathbf{k}$. This gives

$$-\frac{d\kappa_0^2}{d\omega}\Delta\omega S + 2\Delta\mathbf{k}\cdot\mathbf{J} + \frac{1}{N}\sum_\mathbf{k}(\Sigma^+ - \Sigma^-)\phi_\mathbf{k}$$
$$= \frac{2i}{L^d}\sum_\mathbf{k}\mathrm{Im}\, G_e^+(\omega,\mathbf{k}) + \frac{1}{L^d N}\sum_\mathbf{k}\sum_{\mathbf{k}'}(\Delta G_e)_\mathbf{k} U_{\mathbf{k},\mathbf{k}'}\phi_{\mathbf{k}'}, \quad (4.30)$$

where S and $\mathbf{J}$ are the moments of $\phi_\mathbf{k}$, defined as

$$S = \frac{1}{N}\sum_\mathbf{k}\phi_\mathbf{k},$$

which is exactly the energy density given by Eq. (4.10), and

$$\mathbf{J} = \frac{1}{2N}\sum_\mathbf{k}\nabla_k e(\mathbf{k})\phi_\mathbf{k}, \qquad (4.31)$$

4.3 The Bethe–Salpeter Equation and Its Solution by Moments

which may be regarded as the energy current density. In Eq. (4.30) the first term on the right-hand side can be immediately evaluated to give

$$\frac{2i}{L^d} \sum_{\mathbf{k}} \operatorname{Im} G_e^+(\omega, \mathbf{k}) = -2i \frac{d\omega}{d\kappa_e^2} \pi \rho_e(\omega) \frac{N}{L^d}. \quad (4.32)$$

Here $\rho_e(\omega)$ is the density of states of the effective medium. That leaves two terms in Eq. (4.30) that seem difficult to handle. However, by switching the summation labels $\mathbf{k}$ and $\mathbf{k}'$, which effects no real change, a powerful relationship, called the Ward identity, can be invoked which states that

$$\Sigma^+(\omega^+, k_+) - \Sigma^-(\omega^-, k_-) = \left(\Delta \Sigma\right)_{\mathbf{k}} = \frac{1}{N} \sum_{\mathbf{k}'} (\Delta G_e)_{\mathbf{k}'} U_{\mathbf{k}',\mathbf{k}} \frac{N}{L^d}. \quad (4.33)$$

The immediate result is

$$\frac{1}{N} \sum_{\mathbf{k}} \left(\Delta \Sigma\right)_{\mathbf{k}} \phi_{\mathbf{k}} = \frac{1}{L^d N} \sum_{\mathbf{k}'} \sum_{\mathbf{k}} (\Delta G_e)_{\mathbf{k}'} U_{\mathbf{k}',\mathbf{k}} \phi_{\mathbf{k}}, \quad (4.34)$$

so Eq. (4.30) becomes

$$-\frac{d\kappa_0^2}{d\omega} \Delta \omega S + 2 \Delta \mathbf{k} \cdot \mathbf{J} = -2i \frac{d\omega}{d\kappa_e^2} \pi \rho_e(\omega) \frac{N}{L^d}. \quad (4.35a)$$

The Ward identity is an important statement relating Σ, which is a quantity associated with the average of single Green's function to the vertex function U, which is associated with the average of two Green's functions. The physical basis of the identity is energy conservation, and the demonstration of its validity is deferred to a later section.

Equation (4.35a) has two unknowns, S and $\mathbf{J}$. A physical interpretation of the equation may be obtained by translating the multiplication by $-i \Delta \omega$ as the result of the $\partial/\partial t$ operation, and the dot product with $i \Delta \mathbf{k}$ as the result of the divergence operation ($\nabla \cdot$). These interpretations originate from the fact that $\Delta \omega$ and $\Delta \mathbf{k}$ are the conjugate variables to travel time and travel distance, respectively, as pointed out earlier. Then Eq. (4.35a) has the form of a continuity equation with a source term, i.e.,

$$\frac{1}{2}\left(\frac{d\kappa_0^2}{d\omega}\right)\frac{\partial S}{\partial t} + \nabla \cdot \mathbf{J} = \frac{d\omega}{d\kappa_e^2} \pi \rho_e(\omega) \frac{N}{L^d}. \quad (4.35b)$$

In light of this interpretation, a diffusive pole for S would be established if $\mathbf{J}$ can be expressed as the gradient of S, i.e., $\mathbf{J} \sim |\Delta \mathbf{k}| S$. In order to obtain an additional equation relating S and $\mathbf{J}$, one can take a different moment of Eq. (4.29) by multiplying both sides by $(\frac{1}{2}\nabla_k e(\mathbf{k}) \cdot \Delta \mathbf{k})$ and summing with respect to $\mathbf{k}$. However, since for $|\Delta \mathbf{k}| \to 0$ this would result in a higher-order equation than (4.35a), $G_e^+ - G_e^-$ has to be expanded consistently to the same order. That is, $(\Delta G_e)_\mathbf{k}$ in (4.29) has to be replaced by

$$G_e^+ - G_e^- = (\Delta G_e)_\mathbf{k} + \frac{\partial G_e^+}{\partial e(\mathbf{k})}\left[\nabla_k e(\mathbf{k}) \cdot \frac{\Delta \mathbf{k}}{2}\right]$$

$$- \frac{\partial G_e^-}{\partial e(\mathbf{k})}\left[\nabla_k e(\mathbf{k}) \cdot \left(\frac{-\Delta \mathbf{k}}{2}\right)\right]$$

$$= (\Delta G_e)_\mathbf{k} + \left[(G_e^+)_\mathbf{k}^2 + (G_e^-)_\mathbf{k}^2\right]\left[\frac{1}{2}\nabla_k e(\mathbf{k}) \cdot \Delta \mathbf{k}\right]\frac{L^d}{N}. \quad (4.36a)$$

The net result of the three operations [multiplication by $(\frac{1}{2}\nabla_k e(\mathbf{k}) \cdot \Delta \mathbf{k})$, expansion of $G_e^+ - G_e^-$ to higher order, and summation with respect to $\mathbf{k}$] is

$$-\frac{d\kappa_0^2}{d\omega}\Delta\omega\,\Delta\mathbf{k} \cdot \mathbf{J} + \frac{2}{N}\sum_\mathbf{k}[\tfrac{1}{2}\nabla_k e(\mathbf{k}) \cdot \Delta\mathbf{k}]^2 \phi_\mathbf{k}$$

$$+ \frac{1}{N}\sum_\mathbf{k} i(\operatorname{Im}\Sigma^+)[\nabla_k e(\mathbf{k}) \cdot \Delta\mathbf{k}]\phi_\mathbf{k}$$

$$= \frac{1}{2L^d}\sum_\mathbf{k}[\nabla_k e(\mathbf{k}) \cdot \Delta\mathbf{k}](\Delta G_e)_\mathbf{k}$$

$$+ \frac{1}{2N^2}\sum_\mathbf{k}\sum_{\mathbf{k}'}(\Delta G_e)_\mathbf{k}[\nabla_k e(\mathbf{k}) \cdot \Delta\mathbf{k}]\frac{N}{L^d}U_{\mathbf{k},\mathbf{k}'}\phi_{\mathbf{k}'}$$

$$+ \frac{1}{4N}\sum_\mathbf{k}[\nabla_k e(\mathbf{k}) \cdot \Delta\mathbf{k}]^2\left[(G_e^+)_\mathbf{k}^2 + (G_e^-)_\mathbf{k}^2\right]$$

$$+ \frac{L^d}{4N^3}\sum_\mathbf{k}\sum_{\mathbf{k}'}\left[(G_e^+)_\mathbf{k}^2 + (G_e^-)_\mathbf{k}^2\right][\nabla_k e(\mathbf{k}) \cdot \Delta\mathbf{k}]^2\frac{N}{L^d}U_{\mathbf{k},\mathbf{k}'}\phi_{\mathbf{k}'} \quad (4.36b)$$

The first term on the right-hand side of Eq. (4.36b) vanishes because $(\Delta G_e)_\mathbf{k}$ is an even function of $\mathbf{k}$ [$e(\mathbf{k})$ is an even function of $\mathbf{k}$ as can be

4.3 The Bethe–Salpeter Equation and Its Solution by Moments

seen from Eq. (2.37)], and $[\nabla_k e(\mathbf{k}) \cdot \Delta \mathbf{k}]$ is an odd function of $\mathbf{k}$ as can be seen from the solution to Problem 4.2, so the angular integration implicit in the $\mathbf{k}$ summation forces it to be zero. Let us label the third term on the right-hand size as $NR_0|\Delta \mathbf{k}|^2/L^d$. If Σ^+ is assumed to be $\mathbf{k}$ independent, as in the case of CPA, then the third term on the left-hand size may be written as

$$2i(\mathrm{Im}\,\Sigma^+)\Delta \mathbf{k} \cdot \mathbf{J}.$$

That leaves the second terms on the left-hand side and the second and the last terms on the right-hand side of Eq. (4.36b) to be treated, which we will label as $L2$, $R2$, and $R4$, respectively.

At this stage it is necessary to make a digression on the $\mathbf{k}$ dependence of $\phi_\mathbf{k}$. From Eq. (4.23), the dominant $\mathbf{k}$ dependence of $\phi_\mathbf{k}$ is seen to be contained in $(G_e^+ G_e^-)_\mathbf{k}$, provided the scattering is predominantly isotropic so that $U_{\mathbf{k},\mathbf{k}'}$ is a function of ω only, which is exactly true for point scatterers. Therefore, to get the $\mathbf{k}$ dependence to the order $|\Delta \mathbf{k}|$, it is necessary only to expand $(G_e^+ G_e^-)_\mathbf{k}$ to that order. Through similar operations as that used in Eqs. (4.36a), the result is

$$G_e^+ G_e^- = (G_e^+)_\mathbf{k}(G_e^-)_\mathbf{k}$$

$$+ \left[(G_e^-)_\mathbf{k}(G_e^+)_\mathbf{k}^2 - (G_e^+)_\mathbf{k}(G_e^-)_\mathbf{k}^2\right]\left[\frac{1}{2}\nabla_k e(\mathbf{k}) \cdot \Delta \mathbf{k}\right]\frac{L^d}{N}$$

$$= \frac{(\Delta G_e)_\mathbf{k}}{2i\,\mathrm{Im}\,\Sigma^+}\frac{N}{L^d} + \frac{(\Delta G_e)_\mathbf{k}^2}{2i\,\mathrm{Im}\,\Sigma^+}\left[\frac{1}{2}\nabla_k e(\mathbf{k}) \cdot \Delta \mathbf{k}\right], \quad (4.37a)$$

where the identity

$$\frac{L^d}{N}(G_e^+)_\mathbf{k}(G_e^-)_\mathbf{k} = \frac{(\Delta G_e)_\mathbf{k}}{2i\,\mathrm{Im}\,\Sigma^+} \quad (4.37b)$$

has been used. It follows that when $|\Delta \mathbf{k}| = 0$, the $\phi_\mathbf{k}$ may be written as the product of $(\Delta G_e)_\mathbf{k}$ with a function $\phi^{(0)}(\Delta \omega|\omega)$ which does not depend on $\mathbf{k}$ (consistent with the CPA assumption, here Σ^+ is taken to be only frequency dependent). For $|\Delta \mathbf{k}| \neq 0$ but small, the $\mathbf{k}$ dependence of the term linear in $\Delta \mathbf{k}$ is contained in $(\Delta G_e)_\mathbf{k}^2$, as seen from the second term of Eq. (4.37a). Therefore,

$$\phi_\mathbf{k} \cong (\Delta G_e)_\mathbf{k}\left[\phi^{(0)} + \tfrac{1}{2}(\nabla_k e(\mathbf{k}) \cdot \Delta \mathbf{k})(\Delta G_e)_\mathbf{k}\frac{L^d}{N}\phi^{(1)}\right], \quad (4.38a)$$

where $\phi^{(1)}$ is just another function of $\Delta\omega$ and ω. By summing both sides of (4.38a) with respect to $\mathbf{k}$, $\phi^{(0)}$ is identified as

$$\phi^{(0)} = \frac{i}{2\pi} \frac{d\kappa_e^2}{d\omega} \rho_e^{-1}(\omega) S, \qquad (4.38b)$$

where the second term involving $\nabla_k e(\mathbf{k}) \cdot \Delta\mathbf{k}$ vanishes for the same reason as stated earlier. By multiplying both sides of (4.38a) by $(1/2)(\nabla_k e(\mathbf{k}) \cdot \Delta\mathbf{k})$ and summing with respect to $\mathbf{k}$ again, the term involving $\phi^{(0)}$ now vanishes; what remains is

$$\Delta\mathbf{k} \cdot \mathbf{J} = -|\Delta\mathbf{k}|^2 K_0 \phi^{(1)}, \qquad (4.39a)$$

with

$$K_0 = -\frac{L^d}{4N^2} \sum_\mathbf{k} \left[\nabla_k e(\mathbf{k}) \cdot \Delta\hat{\mathbf{k}}\right]^2 (\Delta G_e)_\mathbf{k}^2, \qquad (4.39b)$$

where $\Delta\hat{\mathbf{k}}$ is the unit vector in the direction of $\Delta\mathbf{k}$. K_0 is noted to be a function of ω only. For the purpose of later calculations, it would be useful to give a weak-scattering approximation to Eq. (4.39b). In that limit the $(\Delta G_e)_\mathbf{k}^2$ in K_0 may be approximated by

$$(\Delta G_e)_\mathbf{k}^2 \cong -2(G_e^+)_\mathbf{k}(G_e^-)_\mathbf{k} = -i\frac{(\Delta G_e)_\mathbf{k}}{-\mathrm{Im}\,\Sigma^+}\frac{N}{L^d}$$

$$\cong \frac{-2\pi}{-\mathrm{Im}\,\Sigma^+} \delta[\kappa_e^2 - e(\mathbf{k})]\left(\frac{N}{L^d}\right)^2,$$

so that

$$K_0 = \frac{\pi}{2}\frac{1}{-\mathrm{Im}\,\Sigma^+}\sum_\mathbf{k}\left[\nabla_k e(\mathbf{k}) \cdot \Delta\hat{\mathbf{k}}\right]^2 \delta[\kappa_e^2 - e(\mathbf{k})]\frac{1}{L^d}. \qquad (4.39c)$$

The neglect of the $(G_e^+)_\mathbf{k}^2$ and $(G_e^-)_\mathbf{k}^2$ terms as being small compared to $(G_e^+)_\mathbf{k}(G_e^-)_\mathbf{k}$ is justified if their respective magnitudes were compared when summed with respect to $\mathbf{k}$. Whereas the summation of $(G_e^+)_\mathbf{k}^2 = -\partial(G_e^+)_\mathbf{k}/\partial\kappa_0^2$ gives something finite, the summation of $(G_e^+)_\mathbf{k}(G_e^-)_\mathbf{k} \propto (\Delta G_e^+)_\mathbf{k}/-\mathrm{Im}\,\Sigma^+$ gives $\sim \rho(\omega)/-\mathrm{Im}\,\Sigma^+$, which diverges as $-\mathrm{Im}\,\Sigma^+ \to 0$.

4.3 The Bethe–Salpeter Equation and Its Solution by Moments

The net result of the digression above is that $\phi_{\mathbf{k}}$ may be expanded as

$$\phi_{\mathbf{k}} \equiv (\Delta G_e)_{\mathbf{k}} \left[\frac{i}{2\pi} \frac{d\kappa_e^2}{d\omega} \rho_e^{-1}(\omega) S \right.$$
$$\left. - \frac{1}{2} |\Delta \mathbf{k}|^{-2} K_0^{-1} [\nabla_k e(\mathbf{k}) \cdot \Delta \mathbf{k}] (\Delta G_e)_{\mathbf{k}} \frac{L^d}{N} \Delta \mathbf{k} \cdot \mathbf{J} \right]. \quad (4.40)$$

With this form of $\phi_{\mathbf{k}}$, we can return to Eq. (4.36b) and continue the solution of the Bethe–Salpeter equation. The substitution of Eq. (4.40) into the second term on the left side of Eq. (4.36b) gives

$$L2 = \frac{1}{N} \sum_{\mathbf{k}} [\tfrac{1}{2} \nabla_k e(\mathbf{k}) \cdot \Delta \mathbf{k}]^2 2(\Delta G_e)_{\mathbf{k}} \frac{i}{2\pi} \frac{d\kappa_e^2}{d\omega} \rho_e^{-1}(\omega) S$$

as a result. $\mathbf{J}$ does not appear here because of the presence of the odd function $\nabla_k e(\mathbf{k}) \cdot (\Delta \mathbf{k})$. The substitution of (4.40) into the last term, on the right-hand side of (4.36b) gives

$$R4 = \frac{L^d}{N^3} \sum_{\mathbf{k}} \sum_{\mathbf{k}'} [\tfrac{1}{2} \nabla_k e(\mathbf{k}) \cdot \Delta \mathbf{k}]^2 \frac{N}{L^d} U_{\mathbf{k},\mathbf{k}'} (\Delta G_e^+)_{\mathbf{k}'}$$
$$\times \left[(G_e^+)_{\mathbf{k}}^2 + (G_e^-)_{\mathbf{k}}^2 \right] \frac{i}{2\pi} \frac{d\kappa_e^2}{d\omega} \rho_e^{-1}(\omega) S.$$

From the Ward identity, Eq. (4.33), the summation over $\mathbf{k}'$ directly gives $\Sigma^+ - \Sigma^-$. So a new form of the expression is

$$R4 = \frac{L^d}{N^2} \sum_{\mathbf{k}} [\tfrac{1}{2} \nabla_k e(\mathbf{k}) \cdot \Delta \mathbf{k}]^2 (2i \, \mathrm{Im}\, \Sigma^+)$$
$$\times \left[(G_e^+)_{\mathbf{k}}^2 + (G_e^-)_{\mathbf{k}}^2 \right] \frac{i}{2\pi} \frac{d\kappa_e^2}{d\omega} \rho_e^{-1}(\omega) S.$$

By noting that $(\Delta G_e)_{\mathbf{k}}/i \,\mathrm{Im}\, \Sigma^+ = 2(G_e^+)_{\mathbf{k}} (G_e^-)_{\mathbf{k}} L^d/N$, one can combine the two expressions $L2$ and $R4$ (with a minus sign), yielding

$$L2 - R4 = -\frac{L^d}{4N^2} \sum_{\mathbf{k}} [\nabla_k e(\mathbf{k}) \cdot \Delta \mathbf{k}]^2 (\Delta G_e)_{\mathbf{k}}^2 \left(\frac{-\mathrm{Im}\, \Sigma^+}{\pi \rho_e(\omega)} \frac{d\kappa_e^2}{d\omega} \right) S$$
$$= K_0 \left(\frac{-\mathrm{Im}\, \Sigma^+}{\pi \rho_e(\omega)} \frac{d\kappa_e^2}{d\omega} \right) |\Delta \mathbf{k}|^2 S.$$

On the other hand, the substitution of Eq. (4.40) into the second term on the right side of Eq. (4.36b) gives

$$R2 = -\frac{1}{4}\frac{1}{N^2}\sum_{\mathbf{k}}\sum_{\mathbf{k}'}(\Delta G_e)_{\mathbf{k}}(\Delta G_e)_{\mathbf{k}'}^2[\nabla_k e(\mathbf{k})\cdot \Delta \mathbf{k}]U_{\mathbf{k},\mathbf{k}'}$$
$$\times [\nabla_{k'} e(\mathbf{k}')\cdot \Delta \mathbf{k}]|\Delta \mathbf{k}|^{-2}K_0^{-1}\Delta \mathbf{k}\cdot \mathbf{J}$$
$$= -\frac{1}{4K_0}\left\{\frac{L^d}{N^3}\sum_{\mathbf{k}}\sum_{\mathbf{k}'}(\Delta G_e)_{\mathbf{k}}(\Delta G_e)_{\mathbf{k}'}^2[\nabla_k e(\mathbf{k})\cdot \Delta \hat{\mathbf{k}}]\right.$$
$$\left.\times \frac{N}{L^d}U_{\mathbf{k},\mathbf{k}'}[\nabla_{k'} e(\mathbf{k}')\cdot \Delta \hat{\mathbf{k}}]\right\}(\Delta \mathbf{k}\cdot \mathbf{J}). \qquad (4.41)$$

The $(\Delta G_e)_{\mathbf{k}}$'s in this equation ensure that the magnitudes of $\mathbf{k}'$ and $\mathbf{k}$ are both determined by the constant-energy condition, i.e., they are both on the "energy shell." It follows that the real degrees of freedom lie in the angular part, i.e., the angle θ between $\mathbf{k}$ and $\mathbf{k}'$, and the angle of $\mathbf{k}$ relative to the sample. If the sample is isotropic, only θ matters in the summation. It can be seen from Eq. (4.41) that if $U_{\mathbf{k},\mathbf{k}'}$ is independent of θ, then the double summations can be performed independently, each giving zero. Therefore, this term is nonzero only when $U_{\mathbf{k},\mathbf{k}'}$ is *anisotropic*. By defining

$$M_0 = i\frac{L^d}{N^3}\sum_{\mathbf{k}}\sum_{\mathbf{k}'}(\Delta G_e)_{\mathbf{k}}(\Delta G_e)_{\mathbf{k}'}^2[\nabla_k e(\mathbf{k})\cdot \Delta \hat{\mathbf{k}}]$$
$$\times \frac{N}{L^d}U_{\mathbf{k},\mathbf{k}'}[\nabla_{k'} e(\mathbf{k}')\cdot \Delta \hat{\mathbf{k}}], \qquad (4.42a)$$

the second term on the right-hand side of Eq. (4.36b) has the form

$$R2 = \frac{i}{4K_0}M_0\Delta \mathbf{k}\cdot \mathbf{J}.$$

In the weak-scattering limit, M_0 may be alternatively expressed as

$$M_0 = \frac{1}{N^2}\frac{1}{-\text{Im }\Sigma^+}\sum_{\mathbf{k}}\sum_{\mathbf{k}'}(\Delta G_e)_{\mathbf{k}}(\Delta G_e)_{\mathbf{k}'}[\nabla_k e(\mathbf{k})\cdot \Delta \hat{\mathbf{k}}]$$
$$\times \frac{N}{L^d}U_{\mathbf{k},\mathbf{k}'}[\nabla_{k'} e(\mathbf{k}')\cdot \Delta \hat{\mathbf{k}}], \qquad (4.42b)$$

4.3 The Bethe–Salpeter Equation and Its Solution by Moments

where $(\Delta G_e^+)_{\mathbf{k}}^2$ is again replaced by $-i(\Delta G_e)_{\mathbf{k},\mathbf{k}'} N/(-\operatorname{Im}\Sigma^+)L^d$ as in (4.39c). By combining terms, Eq. (4.36b) has the following form:

$$-\frac{d\kappa_0^2}{d\omega}\Delta\omega\,\Delta\mathbf{k}\cdot\mathbf{J} + \frac{1}{\pi\rho_e(\omega)}\frac{d\kappa_e^2}{d\omega}(-\operatorname{Im}\Sigma^+)|\Delta\mathbf{k}|^2 K_0 S + 2i\operatorname{Im}(\Sigma^+)\Delta\mathbf{k}\cdot\mathbf{J}$$
$$= \frac{i}{4K_0} M_0\,\Delta\mathbf{k}\cdot\mathbf{J} + \frac{N}{L^d}R_0|\Delta\mathbf{k}|^2, \tag{4.43}$$

or

$$\Delta\mathbf{k}\cdot\mathbf{J} = \frac{-i}{\pi\rho_e(\omega)}$$
$$\times\frac{K_0(d\kappa_e^2/d\omega)(-\operatorname{Im}\Sigma^+)|\Delta\mathbf{k}|^2 S - (N/L^d)\pi\rho_e(\omega)R_0|\Delta\mathbf{k}|^2}{(M_0/4K_0) - 2\operatorname{Im}(\Sigma^+)}, \tag{4.44}$$

where the term $-(d\kappa_0^2/d\omega)\Delta\omega\,\Delta\mathbf{k}\cdot\mathbf{J}$ has been dropped because it is small ($\Delta\omega \to 0$) compared to the other terms. The R_0 term is not proportional to S, therefore it may be viewed as an additional source term when (4.44) is combined with (4.35a). However, compared with the source term on the right side of (4.35a), the R_0 term is small because it is proportional to $|\Delta\mathbf{k}|^2$. Therefore, it can be dropped.

Because Eq. (4.44) states that $|\mathbf{J}| \sim |\Delta\mathbf{k}|S$ (without the R_0 term), its substitution into Eq. (4.35) immediately gives a diffusive pole for S, as anticipated:

$$S = \frac{2(d\omega/d\kappa_0^2)(d\omega/d\kappa_e^2)\rho_e(\omega)\pi}{-i\Delta\omega + D(\omega)|\Delta\mathbf{k}|^2}\frac{N}{L^d}, \tag{4.45}$$

with

$$D(\omega) = \frac{1}{\pi\rho_e(\omega)}\frac{8K_0^2(1 + d\operatorname{Re}(\Sigma^+)/d\kappa_0^2)(-\operatorname{Im}\Sigma^+)}{M_0 - 8K_0\operatorname{Im}(\Sigma^+)}, \tag{4.46}$$

where the definition $\kappa_e^2 = \kappa_0^2 - \operatorname{Re}(\Sigma^+)$ has been used. Earlier, Eq. (4.13) stated that if the intensity transport behavior were diffusive in the $\Delta\omega$, $|\Delta\mathbf{k}| \to 0$ limit, then S should have a diffusive pole. That assertion has now been substantiated. Since the derivation does not invoke the dispersion relation of the wave or whether it is on a lattice or in continuum, the diffusive behavior is thus a general property of multiply-scattered wave intensity transport in the limit of long propagation time and long travel distance. Also, since $U_{\mathbf{k},\mathbf{k}'}$ is not specified or approximated, the fact that S

has a diffusive pole is independent of the type of scattering. In fact, even if $U_{\mathbf{k},\mathbf{k}'} = 0$ so that $M_0 = 0$ (no vertex correction), $D(\omega)$ would still be finite. As $\text{Im}(\Sigma^+)$ gives the mean free path in the effective medium, the effect of M_0 may be viewed as providing a correction to the effective medium mean free path. This point will be made more explicit later in this chapter. The units of $D(\omega)$ can also be checked to be correct: Since $M_0 \sim [\text{length}]^{-d}$, $K_0, \sim [\text{length}]^{2-d}$, κ_0, $\Sigma \sim [\text{length}]^{-2}$, and $\rho_e \sim [\text{time}][\text{length}]^{-d}$, from Eq. (4.46) $D(\omega)$ has the units of $[\text{length}]^2/[\text{time}]$, independent of the spatial dimension. The evaluation of $D(\omega)$ is left to a later section, since it is necessary first to examine the vertex function and demonstrate the validity of the Ward identity. At this point it is instructive to reflect on the results obtained so far.

In electronic systems, the diffusion constant is known to be proportional to the electrical conductivity via the Einstein relation. Since electrical conductivity is associated with dissipation, it may seem odd that although in our calculations so far there has been no consideration of dissipative mechanisms (as evidenced by the fact that the basis of our models is the wave equation, in which energy conservation is an inherent attribute), the final derived expression is proportional to the electrical conductivity. The answer to this puzzle is contained in the subtle effect of configuration averaging. By smoothing out fluctuations through configurational averaging, irreversibility is implicitly introduced. As irreversibility is implied by entropy generation and dissipation, it is not altogether surprising that the diffusion constant and the electrical conductivity may be related to quantities calculated through the route of configurational averaging. The validity of this approach is in fact the basis of the linear response theory.

4.4 The Vertex Function

The reducible vertex function is defined by

$$\Gamma = \langle \bar{\mathbf{T}}^+ \otimes \bar{\mathbf{T}}^- \rangle_c, \qquad (4.47)$$

where the overbar indicates that the quantity is defined relative to a CPA effective medium (i.e., the embedding medium is the effective medium rather than the unperturbed uniform medium). In the solution to Problem 3.13, the $\bar{\mathbf{T}}$ operator is shown to be decomposable into operators involving individual scatterers:

$$\bar{\mathbf{T}} = \sum_i \bar{\mathbf{R}}_i, \qquad (4.48)$$

4.4 The Vertex Function

where i is the site index in the lattice model and is the (structural unit) particle index in the continuum case. Furthermore, it is also shown that

$$\bar{\mathbf{R}}_i = \bar{\mathbf{t}}_i \left(\mathbf{I} + \mathbf{G}_e \sum_{n \ne i} \bar{\mathbf{R}}_n \right) = \left(\mathbf{I} + \sum_{n \ne i} \bar{\mathbf{R}}_n \mathbf{G}_e \right) \bar{\mathbf{t}}_i . \quad (4.49)$$

Through iterative substitution of the right-hand side of Eq. (4.49) into the summation over $\bar{\mathbf{R}}_n$ (with a change of index labeling from i to n), an operator equation is obtained:

$$\bar{\mathbf{T}} = \sum_i \bar{\mathbf{t}}_i + \sum_i \sum_{j \ne i} \bar{\mathbf{t}}_i \bar{\mathbf{G}}_e \bar{\mathbf{t}}_j + \sum_i \sum_{j \ne i} \sum_{k \ne j} \bar{\mathbf{t}}_i \bar{\mathbf{G}}_e \bar{\mathbf{t}}_j \bar{\mathbf{G}}_e \bar{\mathbf{t}}_k + \cdots, \quad (4.50)$$

where each succeeding index can not coincide in value with the preceding index. So in the third term $j \ne i$ and $k \ne j$, but $k = i$ is allowed. In terms of individual scatterers, the structure of the vertex function is thus clear: It is the sum of configurational averages of infinite number of terms involving $\langle \bar{\mathbf{t}}_i^+ \bar{\mathbf{t}}_j^- \rangle_c$, $\langle \bar{\mathbf{t}}_i^+ \bar{\mathbf{G}}_e^+ \bar{\mathbf{t}}_j^+ \bar{\mathbf{t}}_k^- \rangle_c$, etc. However, certain terms can be eliminated due to the CPA condition of $\langle \bar{\mathbf{t}}_i^\pm \rangle_c = 0$. For example, terms involving an odd number of $\bar{\mathbf{t}}$'s must vanish because at least one $\bar{\mathbf{t}}$ can be independently averaged in the subsets of configurations where the other $\bar{\mathbf{t}}$'s are fixed. For terms involving an even number of $\mathbf{t}$'s, on the other hand, the indices must form pairs in order for them not to vanish. For example,

$$a_{123} = \langle \bar{\mathbf{t}}_1^+ \mathbf{G}_e^+ \bar{\mathbf{t}}_2^+ \mathbf{G}_e^+ \bar{\mathbf{t}}_3^+ \otimes \bar{\mathbf{t}}_3^- \mathbf{G}_e^- \bar{\mathbf{t}}_2^- \mathbf{G}_e^- \bar{\mathbf{t}}_1^- \rangle_c \quad (4.51)$$

would not vanish, because $\langle \bar{\mathbf{t}}_i^+ \bar{\mathbf{t}}_i^- \rangle_c \ne 0$. But if $\bar{\mathbf{t}}_1^-$ were replaced by $\bar{\mathbf{t}}_4^-$, the term would vanish because there is no corresponding $\bar{\mathbf{t}}_4^+$. However, even with this rule there can still be many possibilities for the three-scatterers case. This is because the *order* in which the operators multiply matters, so for example

$$b_{123} = \langle \bar{\mathbf{t}}_1^+ \mathbf{G}_e^+ \bar{\mathbf{t}}_2^+ \mathbf{G}_e^+ \bar{\mathbf{t}}_3^+ \otimes \bar{\mathbf{t}}_2^- \mathbf{G}_e^- \bar{\mathbf{t}}_3^- \mathbf{G}_e^- \bar{\mathbf{t}}_1^- \rangle_c \quad (4.52)$$

is different from a_{123}, even though the same three scatterers are involved. This may be visualized by diagrams as shown in Figure 4.4. Here a cross denotes a scatterer. The same scatterers related to two ($+$ and $-$) Green's functions are connected by a dashed "interaction" line. Together they represent $\langle \bar{\mathbf{t}}_i^+ \bar{\mathbf{t}}_i^- \rangle_c$. The Green's functions are represented by solid lines, with the $+$ channels on top and the $-$ channels on the bottom. Due to configurational averaging, the input and output channels associated with each scattering vertex must have the property given by Eq. (4.9), i.e., in the

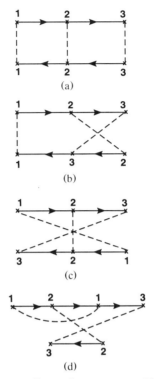

Figure 4.4 Four diagrams corresponding to three scatterers. The symbols are explained in the text.

k representation the total momentum of the input channels must equal to the sum of the output channels. In Figure 4.4a, which corresponds to a_{123}, the diagram is called reducible, as explained in relation to Figure 4.3. In Figure 4.4b, which corresponds to b_{123}, the diagram is also reducible. However, Figures 4.4c and 4.4d are irreducible, and they correspond, respectively, to

$$c_{123} = \langle \bar{\mathbf{t}}_1^+ \mathbf{G}_e^+ \bar{\mathbf{t}}_2^+ \mathbf{G}_e^+ \bar{\mathbf{t}}_3^+ \otimes \bar{\mathbf{t}}_1^- \mathbf{G}_e^- \bar{\mathbf{t}}_2^- \mathbf{G}_e^- \bar{\mathbf{t}}_3^- \rangle_c$$

and

$$d_{123} = \langle \bar{\mathbf{t}}_1^+ \mathbf{G}_e^+ \bar{\mathbf{t}}_2^+ \mathbf{G}_e^+ \bar{\mathbf{t}}_1^+ \mathbf{G}_e^+ \bar{\mathbf{t}}_3^+ \otimes \bar{\mathbf{t}}_2^- \mathbf{G}_e^- \bar{\mathbf{t}}_3^- \rangle_c.$$

4.4 The Vertex Function

With the understanding of the vertex function Γ, let us rewrite Eq. (4.47) by using Eq. (4.49) as

$$\Gamma = \left\langle \sum_i \bar{\mathbf{R}}_i^+ \otimes \sum_j \bar{\mathbf{R}}_j^- \right\rangle_c = \sum_i \sum_j \Gamma_{ij}, \quad (4.53a)$$

where

$$\begin{aligned}
\Gamma_{ij} &= \left\langle \bar{\mathbf{t}}_i^+ \left(\mathbf{I} + \mathbf{G}_e^+ \sum_{n \neq i} \bar{\mathbf{R}}_n^+ \right) \otimes \left(\mathbf{I} + \sum_{m \neq j} \bar{\mathbf{R}}_m^- \mathbf{G}_e^- \right) \bar{\mathbf{t}}_j^- \right\rangle_c \\
&= \delta_{i,j} \langle \bar{\mathbf{t}}_i^+ \otimes \bar{\mathbf{t}}_i^- \rangle_c + \left\langle \bar{\mathbf{t}}_i^+ \mathbf{G}_e^+ \left(\sum_{n \neq i} \bar{\mathbf{R}}_n^+ \otimes \sum_{m \neq j} \bar{\mathbf{R}}_m^- \right) \mathbf{G}_e^- \bar{\mathbf{t}}_j^- \right\rangle_c, \quad (4.53b)
\end{aligned}$$

and terms with an odd number of $\bar{\mathbf{t}}$ or $\bar{\mathbf{R}}$ vanish for the reason already given. It should be noted here that the subscripts ij on Γ are the scatter indices. They should not be confused with channel indices. The last term of Eq. (4.53b) may be approximated as

$$\begin{aligned}
&\left\langle \bar{\mathbf{t}}_i^+ \mathbf{G}_e^+ \left(\sum_{n \neq i} \bar{\mathbf{R}}_n^+ \otimes \sum_{m \neq j} \bar{\mathbf{R}}_m^- \right) \mathbf{G}_e^- \bar{\mathbf{t}}_j^- \right\rangle_c \\
&\cong \left\langle \bar{\mathbf{t}}_i^+ \mathbf{G}_e^+ \left\langle \sum_{n \neq i} \bar{\mathbf{R}}_n^+ \otimes \sum_{m \neq j} \bar{\mathbf{R}}_m^- \right\rangle_c \mathbf{G}_e^- \bar{\mathbf{t}}_j^- \right\rangle_c \\
&= \sum_{n \neq i} \sum_{m \neq j} \delta_{ij} \langle \bar{\mathbf{t}}_i^+ \otimes \bar{\mathbf{t}}_i^- \rangle_c : (\mathbf{G}_e^+ \otimes \mathbf{G}_e^-) : \Gamma_{mn}. \quad (4.54)
\end{aligned}$$

This approximation is in the same CPA spirit as discussed in the last chapter, because it reduces all calculations to one particle (or one site). As shall be seen below, the CPA approximation selects out a particular class of diagrams for the calculation of the vertex function.

Under the CPA approximation, Γ_{ij}'s satisfy the following equation:

$$\Gamma_{ij}^{(B)} = \delta_{i,j} \langle \bar{\mathbf{t}}_i^+ \otimes \bar{\mathbf{t}}_i^- \rangle_c + \delta_{i,j} \langle \bar{\mathbf{t}}_i^+ \otimes \bar{\mathbf{t}}_i^- \rangle_c : \sum_{n \neq i} \sum_{m \neq j} (\mathbf{G}_e^+ \otimes \mathbf{G}_e^-) : \Gamma_{nm}^{(B)}. \quad (4.55)$$

Here the superscript B is used to denote the quantities as evaluated under the CPA, which yields the classical Boltzmann diffusion constant, as will be seen later. The order of the terms has been rearranged in Eq. (4.55) for simplicity in manipulation. By adding the excluded term ($n \neq i, m \neq j$) to both sides of Eq. (4.55), the constraints in the summations are removed:

$$\left(\mathbf{I} + \langle \bar{\mathbf{t}}_i^+ \mathbf{G}_e^+ \otimes \mathbf{G}_e^- \bar{\mathbf{t}}_j^- \rangle_c \right) : \Gamma_{ij}^{(B)}$$
$$= \delta_{i,j} \langle \bar{\mathbf{t}}_i^+ \otimes \bar{\mathbf{t}}_i^- \rangle_c : \left[\mathbf{I} + \sum_n \sum_m (\mathbf{G}_e^+ \otimes \mathbf{G}_e^-) : \Gamma_{nm}^{(B)} \right], \quad (4.56)$$

where $\mathbf{I}$ is short for $\mathbf{I} \otimes \mathbf{I}$. The term $\langle \bar{\mathbf{t}}_i^+ \mathbf{G}_e^+ \otimes \mathbf{G}_e^- \bar{\mathbf{t}}_j^- \rangle_c$ in this equation is noted to have the input channels specified by $\bar{\mathbf{t}}_i^+$ and $\bar{\mathbf{t}}_j^-$, respectively, i.e., particle (site) i and j. The output channels are linked to $\Gamma_{ij}^{(B)}$, with i associated with the $+$ channel and j the $-$ channel. Since $\langle \bar{\mathbf{t}}_i^+ \otimes \bar{\mathbf{t}}_j^- \rangle_c \propto \delta_{i,j}$, however, all channels are reduced to the same particle (site).

The inverse of $\mathbf{A} \otimes \mathbf{B}$ is $\mathbf{A}^{-1} \otimes \mathbf{B}^{-1}$, since $(\mathbf{A} \otimes \mathbf{B}):(\mathbf{A}^{-1} \otimes \mathbf{B}^{-1}) = \mathbf{I} \otimes \mathbf{I}$. We wish to define

$$\frac{\mathbf{A}^+ \otimes \mathbf{B}^-}{\mathbf{C}^+ \otimes \mathbf{D}^-} \equiv (\mathbf{C}^+)^{-1} \mathbf{A}^+ \otimes \mathbf{B}^- (\mathbf{D}^-)^{-1}. \quad (4.57)$$

With this inverse notation, Eq. (4.56) may be written succinctly as

$$\Gamma_{ij}^{(B)} = \delta_{i,j} \mathbf{U}_i^{(B)} : \left[\mathbf{I} + \sum_n \sum_m (\mathbf{G}_e^+ \otimes \mathbf{G}_e^-) : \Gamma_{nm}^{(B)} \right], \quad (4.58)$$

where

$$\mathbf{U}_i^{(B)} = \frac{\langle \bar{\mathbf{t}}_i^+ \otimes \bar{\mathbf{t}}_i^- \rangle_c}{\mathbf{I} + \langle \bar{\mathbf{t}}_i^+ \mathbf{G}_e^+ \otimes \mathbf{G}_e^- \bar{\mathbf{t}}_i^- \rangle_c} = \mathbf{u}_i^+ \otimes \mathbf{u}_i^-, \quad (4.59)$$

and $\mathbf{u}_i^+, \mathbf{u}_i^-$ are associated with the $+$ and $-$ channels of particle (site) i.

Equation (4.58) can be expressed as

$$\Gamma_{ij}^{(B)} = \delta_{i,j} \mathbf{U}_i^{(B)} + \delta_{i,j} \sum_n \sum_m (\mathbf{u}_i^+ \mathbf{G}_e^+ \otimes \mathbf{G}_e^- \mathbf{u}_i^-) : \Gamma_{nm}^{(B)}. \quad (4.60)$$

Through iterative substitutions, Eq. (4.60) becomes

$$\Gamma_{ij}^{(B)} = \delta_{i,j} \left\{ \mathbf{u}_i^+ \otimes \mathbf{u}_i^- + \mathbf{u}_i^+ \left(\sum_n \mathbf{G}_e^+ \mathbf{u}_n^+ \otimes \mathbf{u}_n^- \mathbf{G}_e^- \right) \mathbf{u}_i^- \right.$$
$$\left. + \mathbf{u}_i^+ \left[\sum_n \mathbf{G}_e^+ \mathbf{u}_n^+ \left(\sum_m \mathbf{G}_e^+ \mathbf{u}_m^+ \otimes \mathbf{u}_m^- \mathbf{G}_e^- \right) \mathbf{u}_n^- \mathbf{G}_e^- \right] \mathbf{u}_i^- + \cdots \right\}, \quad (4.61)$$

4.4 The Vertex Function

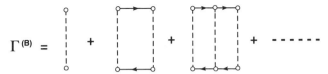

Figure 4.5 Diagrammatic representation of $\Gamma^{(B)}$. The symbols are explained in the text. Because the diagrams look like ladders, they are called the ladder diagrams.

where the δ function reduces every double sum to a single sum at each stage of the iteration. The structure of the CPA approximation is thus clear: It is composed of an infinite number of elementary diagrams like that shown in Figure 4.5. Since every one of these diagrams looks like a ladder lying on its side, they are called the "ladder diagrams." It should be noted that the diagrams here differ from those in Figure 4.4 because the vertex for each scattering is different. Here each $\mathbf{U}_i = \mathbf{u}_i^+ \otimes \mathbf{u}_i^-$ is represented by two open circles linked by a dashed line. It should be distinguished from the crosses shown in Figure 4.4. Also, whereas in Figure 4.4 one has to follow the rule that a succeeding particle (site) must differ from the preceding one, there is no such restriction here. The removal of this restriction is responsible for the denominator in the definition of $\mathbf{U}_i$. Without the restriction on the summation index means that the sum over i, j in Eq. (4.60) results in

$$\Gamma^{(B)} = \mathbf{U}^{(B)} + \mathbf{U}^{(B)}:(\mathbf{G}_e^+ \otimes \mathbf{G}_e^-):\Gamma^{(B)}. \tag{4.62}$$

Comparison with Eqs. (4.18) and (4.20) shows $\mathbf{U}^{(B)}$ to be the irreducible CPA vertex function. As the effect of $\mathbf{U}$ on the wave diffusion constant is embodied in the term M_0, Eq. (4.42), for $\mathbf{U} = \mathbf{U}^{(B)}$ the $M_0^{(B)}$ thus represents the total contribution of the ladder diagrams to the diffusion constant. The reducible vertex Γ, on the other hand, is useful in the evaluation of *scattered wave intensity*. From Eq. (3.11) it is seen that in the presence of scattering disorder,

$$\phi(\omega,\mathbf{r}) = \phi_0(\omega,\mathbf{r}) + \int G_e^+(\omega, \mathbf{r}-\mathbf{r}_1)\overline{T}^+(\omega,\mathbf{r}_1,\mathbf{r}_2)\phi_0(\omega,\mathbf{r}_2)\,d\mathbf{r}_1\,d\mathbf{r}_2.$$

If ϕ_0 is the incident wave and the scattered wave amplitude $\phi - \phi_0$ is denoted by Ψ_S, then the averaged scattering *intensity* may be expressed as

$$\langle |\Psi_S(\omega,\mathbf{r})|^2 \rangle_c = \int\int\int G_e^+(\mathbf{r}-\mathbf{r}_1)G_e^-(\mathbf{r}-\mathbf{r}_2)\Gamma(\mathbf{r}_1,\mathbf{r}_2,\mathbf{r}_3,\mathbf{r}_4)$$
$$\times \phi_0(\mathbf{r}_3)\phi_0^*(\mathbf{r}_4)\,d\mathbf{r}_1\,d\mathbf{r}_2\,d\mathbf{r}_3\,d\mathbf{r}_4,$$

where the ω labeling is suppressed on the right-hand side.

Let us now proceed to take a more detailed look at the irreducible vertex function $\mathbf{U}^{(B)}$. In the Anderson model, the vertex function is associated with a single site $\mathbf{l}$, and $\mathbf{U}_\mathbf{l}^{(B)}/a^d$ may be written as

$$\frac{1}{a^d}\mathbf{U}_\mathbf{l}^{(B)} = \frac{1}{a^d} U^{(B)}(\Delta\omega|\omega)(|\mathbf{l}\rangle_+\langle\mathbf{l}|_+ \otimes |\mathbf{l}\rangle_-\langle\mathbf{l}|_-). \tag{4.63}$$

At $\Delta\omega = 0$, Eq. (4.59) gives

$$\frac{1}{a^d} U^{(B)} = \frac{C_1}{1 + C_2}, \tag{4.64a}$$

with

$$C_1 = \frac{1}{a^d N} \sum_\mathbf{l} |\bar{t}_\mathbf{l}^+|^2$$

$$= \frac{a^d}{2\sigma_w} \int_{-\sigma_w}^{\sigma_w} d\sigma \left| \frac{\sigma - \Sigma^+}{1 - (\sigma - \Sigma^+)a^d G_0^+(\kappa_0^2 - \Sigma^+, \mathbf{l} = \mathbf{l}')} \right|^2, \tag{4.64b}$$

and

$$C_2 = \frac{1}{2\sigma_w} \int_{-\sigma_w}^{\sigma_w} d\sigma \left| \frac{(\sigma - \Sigma^+)a^d G_0^+(\kappa_0^2 - \Sigma^+, \mathbf{l} = \mathbf{l}')}{1 - (\sigma - \Sigma^+)a^d G_0^+(\kappa_0^2 - \Sigma^+, \mathbf{l} = \mathbf{l}')} \right|^2. \tag{4.64c}$$

Here Σ^+ is obtained from the solution of the CPA equation given in Chapter III. In real space, therefore,

$$\mathbf{U}^{(B)} = \sum_\mathbf{l} \mathbf{U}_\mathbf{l}^{(B)} = U^{(B)} \sum_\mathbf{l} |\mathbf{l}\rangle_+\langle\mathbf{l}|_+ \otimes |\mathbf{l}\rangle_-\langle\mathbf{l}|_-. \tag{4.65}$$

In the momentum representation, the input and output channels of $\mathbf{U}$ have the same structure as that of Γ because Eq. (4.18) shows

$$\Gamma_{ij} = \mathbf{U}_{ij} + \sum_n \sum_m (\mathbf{u}_i^+ \mathbf{G}_e^+ \otimes \mathbf{G}_e^- \mathbf{u}_j^-):\Gamma_{nm}, \tag{4.66}$$

where i, j are the particle (site) indices. Since $\mathbf{G}_e^\pm$ is diagonal in the $\mathbf{k}$ representation, the second term of Eq. (4.65) shows that the output momentum channels of $\mathbf{U}_{ij}$ must coincide with the input momentum channels of Γ_{nm}. That means the same momentum-channel structure as

4.4 The Vertex Function

that specified by Eq. (4.9). With this in mind, $U_{\mathbf{k},\mathbf{k}'}^{(B)}$ for the Anderson model may be written in the momentum-channel representation as

$$\frac{1}{a^d} U_{\mathbf{k},\mathbf{k}'}^{(B)}(|\mathbf{k}_+\rangle\langle \mathbf{k}'_+| \otimes |\mathbf{k}'_-\rangle\langle \mathbf{k}_-|)$$

$$= \frac{U^{(B)}}{a^d} \frac{1}{N} \sum_{\mathbf{l}} |\mathbf{k}_+\rangle\langle \mathbf{k}_+|\mathbf{l}\rangle\langle \mathbf{l}|\mathbf{k}'_+\rangle$$

$$\times \langle \mathbf{k}'_+| \otimes |\mathbf{k}'_-\rangle\langle \mathbf{k}'_-|\mathbf{l}\rangle\langle \mathbf{l}|\mathbf{k}_-\rangle\langle \mathbf{k}_-|$$

$$= \frac{U^{(B)}}{a^d} \frac{1}{N} \sum_{\mathbf{l}} \exp[i(\mathbf{k}'_+ - \mathbf{k}'_- - \mathbf{k}_+ + \mathbf{k}_-) \cdot \mathbf{l}](|\mathbf{k}_+\rangle$$

$$\times \langle \mathbf{k}'_+| \otimes |\mathbf{k}'_-\rangle\langle \mathbf{k}_-|). \qquad (4.67)$$

Equation (4.63) shows the sum of momenta in the exponent must be zero, so

$$U_{\mathbf{k},\mathbf{k}'}^{(B)} = U^{(B)}. \qquad (4.68)$$

The simplicity of this result is due to the fact that the $+$ and $-$ channels have the same site $\mathbf{l}$—they are correlated—so they may not be summed separately (otherwise it would yield the unphysical result of $\mathbf{k}' = \mathbf{k}$). Equation (4.68) shows $U^{(B)}$ to be independent of the scattering angle, i.e., $U^{(B)}$ is isotropic for the Anderson model.

For scalar waves in continuum, the situation is more complicated because in real space $\mathbf{u}_i^+$ and $\mathbf{u}_i^-$ are full matrices, since each scatterer may be regarded as consisting of infinitely many sites, so $\bar{\mathbf{t}}_i$ is no longer simple in real space as for the Anderson model. However, the calculation of $\bar{\mathbf{t}}_i$ is simpler in the momentum representation because it may be accomplished through the solution of boundary-value problems, as seen from the last chapter. Therefore, let us approach the evaluation of $\mathbf{U}^{(B)}$ directly in the momentum-channel representation. In order to do that, one should first fix the input and output channels as $\mathbf{k}_\pm$ and $\mathbf{k}'_\pm$, respectively, with $|\mathbf{k}_\pm| = |\mathbf{k}'_\pm| = \kappa_e$. Since the output channels of the term $\langle \bar{\mathbf{t}}_i^+ \mathbf{G}_e^+ \otimes \mathbf{G}_e^- \bar{\mathbf{t}}_i^-\rangle_c$ in the denominator of $\mathbf{U}^{(B)}$ [see Eq. (4.59)] are linked directly to the input channels of $\Gamma_{ij}^{(B)}$ [see Eq. (4.56)], they are thus fixed at $\mathbf{k}_\pm$. Moreover, the output channels of $\mathbf{G}_e^+$ and $\mathbf{G}_e^-$ in real space are *restricted to the same particle* due to the CPA condition. That means in momentum space they have to be convolved with the particle function H as described in the last chapter (Eq. (3.66)). The input channels of $\langle \bar{\mathbf{t}}_i^+ \mathbf{G}_e^+ \otimes \mathbf{G}_e^- \bar{\mathbf{t}}_i^-\rangle_c$ may be labeled as $(\mathbf{k}_1)_\pm$, which are summed over all possible values. For the $\langle \bar{\mathbf{t}}_i^+ \otimes \bar{\mathbf{t}}_i^-\rangle_c$ term in the numerator, its output channels are fixed at $\mathbf{k}'_\pm$, whereas its input channels are labeled as $(\mathbf{k}_2)_\pm$, which are summed over all possible values.

The numerator of $\mathbf{U}^{(B)} N/L^d$ with the specified output channels is given by

$$\frac{N}{L^d} \langle \bar{\mathbf{t}}_i^+ \otimes \bar{\mathbf{t}}_i^- \rangle_c = \frac{1}{N^2} \sum_{\mathbf{k}_{2+}} \sum_{\mathbf{k}_{2-}} \delta_{\mathbf{k}_{2+} - \mathbf{k}_{2-}, \Delta \mathbf{k}}$$

$$\times \left(\sum_m n_m \langle \mathbf{k}_{2+} | \bar{\mathbf{t}}_m^+ | \mathbf{k}'_+ \rangle \langle \mathbf{k}'_- | \bar{\mathbf{t}}_m^- | \mathbf{k}_{2-} \rangle \right)$$

$$\times (|\mathbf{k}_{2+}\rangle\langle \mathbf{k}'_+| \otimes |\mathbf{k}'_-\rangle\langle \mathbf{k}_{2-}|). \qquad (4.69)$$

Here the configuration averaging with the factor N/L^d yields the number density n_m for the mth scatterer species (obtained by fixing at a particular site and summing over all particles, with a probability $1/N$ that the center of a given particle would fall on that particular site), as well as a Kronecker delta relating $\mathbf{k}_{2+}$ and $\mathbf{k}_{2-}$. With fixed $\mathbf{k}'_\pm$, the $+$ channel operator may be regarded as a column vector and the $-$ channel operator a row vector.

The denominator of $\mathbf{U}^{(B)}$ may be expressed as

$$\mathbf{I} + \langle \bar{\mathbf{t}}_i^+ \mathbf{G}_e^+ \otimes \mathbf{G}_e^- \bar{\mathbf{t}}_i^- \rangle_c$$

$$= \frac{1}{N^2} \sum_{\mathbf{k}_{1+}} \sum_{\mathbf{k}_{1-}} \delta_{\mathbf{k}_{1+} - \mathbf{k}_{1-}, \Delta \mathbf{k}} \bigg\{ N \delta_{\mathbf{k}, \mathbf{k}_1}$$

$$+ \frac{1}{N} \sum_{\mathbf{k}''} \left(\sum_m n_m \langle \mathbf{k}_{1+} | \bar{\mathbf{t}}_m^+ | \mathbf{k}''_+ \rangle \langle \mathbf{k}''_- | \bar{\mathbf{t}}_m^- | \mathbf{k}_{1-} \rangle \right)$$

$$\times \frac{L^d}{N} G_e^+(\omega^+, \mathbf{k}''_+) G_e^-(\omega^-, \mathbf{k}''_-) H(\mathbf{k}'' - \mathbf{k}) \bigg\}$$

$$\times (|\mathbf{k}_{1+}\rangle\langle \mathbf{k}_+| \otimes |\mathbf{k}_-\rangle\langle \mathbf{k}_{1-}|). \qquad (4.70)$$

Here one factor of L^d/N (in front of G_e^+) has been generated by configurational averaging.

To evaluate $U^{(B)}_{\mathbf{k},\mathbf{k}'}$, it is simpler to invert the numerator and contract it with the denominator. By noting that the inverse of a column vector is a row vector and vice versa, and that $\langle \mathbf{k}_{2+} | \mathbf{k}_{1+} \rangle = N \delta_{\mathbf{k}_{2+}, \mathbf{k}_{1+}}$, for example,

4.4 The Vertex Function

the resulting expression is given by

$$\frac{L^d}{N}(U^{(B)}_{\mathbf{k},\mathbf{k}'})^{-1}(|\mathbf{k}'_+\rangle\langle\mathbf{k}_+|\otimes|\mathbf{k}_-\rangle\langle\mathbf{k}'_-|), \quad (4.71a)$$

where

$$\begin{aligned}\frac{L^d}{N}(U^{(B)}_{\mathbf{k},\mathbf{k}'})^{-1} &= \left(\sum_m n_m \langle\mathbf{k}_+|\bar{\mathbf{t}}^+_m|\mathbf{k}'_+\rangle\langle\mathbf{k}'_-|\bar{\mathbf{t}}^-_m|\mathbf{k}_-\rangle\right)^{-1}\\ &+ \frac{1}{N}\sum_{\mathbf{k}''}\frac{1}{N}\sum_{\mathbf{k}_1}\frac{\sum_m n_m\langle\mathbf{k}_{1+}|\bar{\mathbf{t}}^+_m|\mathbf{k}''_+\rangle\langle\mathbf{k}''_-|\bar{\mathbf{t}}^-_m|\mathbf{k}_{1-}\rangle}{\sum_M n_m\langle\mathbf{k}_{1+}|\bar{\mathbf{t}}^+_m|\mathbf{k}'_+\rangle\langle\mathbf{k}'_-|\bar{\mathbf{t}}^-_m|\mathbf{k}_{1-}\rangle}\\ &\times \frac{N}{L^d}\left[\frac{L^{2d}}{N^2}G^+_e(\omega^+,\mathbf{k}''_+)G^-_e(\omega^-,\mathbf{k}''_-)H(\mathbf{k}''-\mathbf{k})\right].\end{aligned} \quad (4.71b)$$

It should be noted that one of the Kronecker delta (e.g., $\delta_{\mathbf{k}_{1+}-\mathbf{k}_{1-},\Delta\mathbf{k}}$) is redundant because the inner product between $|\mathbf{k}_2\rangle$ and $|\mathbf{k}_1\rangle$ makes $\mathbf{k}_2 = \mathbf{k}_1$, so the two delta functions give the identical condition. One summation thus produces an extra factor of N, which cancels one N^{-1}. In the limit of $\Delta\omega, |\Delta\mathbf{k}| = 0$, the first term is

$$\left(\sum_m n_m |\langle\mathbf{k}|\bar{\mathbf{t}}^+_m|\mathbf{k}'\rangle|^2\right)^{-1}.$$

The second term may be estimated as follows. The summation over $\mathbf{k}_1$ consists of an integral over the magnitude of $\mathbf{k}_1$ and an angular integral over its directions. Since it is assumed that $|\mathbf{k}'| = \kappa_e$, and $|\mathbf{k}''|$ is also equal to κ_e as will be seen later, the magnitude dependence on $\mathbf{k}_1$ in the numerator and denominator of the integrand should be nearly identical. That means $(1/N)\Sigma_{\mathbf{k}_1}$ is just an angular average:

$$\frac{1}{S_d}\int d\Omega_{\hat{k}_1}\frac{\sum_m n_m|\langle\mathbf{k}_1|\bar{\mathbf{t}}^+_m|\mathbf{k}''\rangle|^2}{\sum_m n_m|\langle\mathbf{k}_1|\bar{\mathbf{t}}^+_m|\mathbf{k}'\rangle|^2} = \frac{1}{S_d}F(\hat{\mathbf{k}}''\cdot\hat{\mathbf{k}}'), \quad (4.72)$$

where $\hat{\mathbf{k}}''$ and $\hat{\mathbf{k}}'$ are the unit vectors in the directions of $\mathbf{k}''$ and $\mathbf{k}'$, $d\Omega_{\hat{k}_1}$ denotes angular integration over the direction of $\mathbf{k}_1$ (in 1D this would be just a summation over the two directions), and $S_d = 4\pi, 2\pi, 2$ in 3D, 2D,

1D, respectively. In the limit of $|\Delta \mathbf{k}|, \Delta \omega \to 0$, the product $L^{2d} G_e^+ G_e^-/N^2$ may be converted, through the use of Eqs. (4.25), (4.26), and (4.28), to the more usable form of

$$\frac{-2\pi i \, \delta\left[\kappa_e^2 - (k'')^2\right]}{2i \, \text{Im} \, \Sigma^+}.$$

Since $H(\mathbf{k}) = 1$ for $|\mathbf{k}|$ less than $\sim 1/R$, where R is the radius of the particle, the summation over $\mathbf{k}''$ may be performed with the assumption that $|\mathbf{k}'' - \mathbf{k}|$ is less than $1/R$. Together with the factors $(1/N)N/L^d$, the result is

$$\frac{1}{-\text{Im} \, \Sigma^+} \frac{1}{S_d} \hat{F} \frac{\pi \kappa_e^{d-2}}{2(2\pi)^d},$$

where both Σ^+ and $|\mathbf{k}''|$ in expression (4.72) are evaluated at $|\mathbf{k}''| = \kappa_e$, and

$$\hat{F} = \int d\Omega_{\hat{k}''} \, F(\hat{\mathbf{k}}'' \cdot \hat{\mathbf{k}}'). \tag{4.73}$$

Now (4.71a) can be inverted to give

$$\frac{N}{L^d} U_{\mathbf{k},\mathbf{k}'}^{(B)}(|\mathbf{k}_+\rangle\langle\mathbf{k}'_+| \otimes |\mathbf{k}'_-\rangle\langle\mathbf{k}_-|),$$

with

$$\frac{N}{L^d} U_{\mathbf{k},\mathbf{k}'}^{(B)} = \frac{\sum_m n_m \left|\langle \mathbf{k}|\bar{\mathbf{t}}_m^+|\mathbf{k}'\rangle\right|^2}{1 + \frac{\pi}{2} \frac{\kappa_e^{d-2}}{(2\pi)^d} \frac{1}{S_d} \hat{F} \frac{\sum_m n_m \left|\langle \mathbf{k}|\bar{\mathbf{t}}_m^+|\mathbf{k}'\rangle\right|^2}{-\text{Im} \, \Sigma^+}}. \tag{4.74}$$

From the optical theorem, $-\text{Im} \, \Sigma^+ \cong -\text{Im} \sum_m n_m \langle \mathbf{k}|\bar{\mathbf{t}}_m^+|\mathbf{k}\rangle = \kappa_e \sum_m n_m \bar{O}_m$, where

$$\bar{O}_m = \int d\Omega_{\hat{k}'} \left|\langle \mathbf{k}|\bar{\mathbf{t}}_m^+|\mathbf{k}'\rangle\right|^2 \cdot \frac{\kappa_e^{d-3}\pi}{2(2\pi)^d} \tag{4.75}$$

[see Eq. (P3.14)]. It follows that Eq. (4.74) may be reexpressed as

$$\frac{N}{L^d} U_{\mathbf{k},\mathbf{k}'}^{(B)} \cong \frac{\sum_m n_m \left|\langle \mathbf{k}|\bar{\mathbf{t}}_m^+|\mathbf{k}'\rangle\right|^2}{1 + Q(\mathbf{k},\mathbf{k}')} = \sum_m n_m \hat{O}_m(\mathbf{k},\mathbf{k}') \frac{2(2\pi)^d}{\pi \kappa_e^{d-3}}, \tag{4.76}$$

where $\hat{O}_m(\mathbf{k}, \mathbf{k}')$ is a renormalized scattering cross section, given by

$$\hat{O}_m(\mathbf{k}, \mathbf{k}') = \frac{\left|\langle \mathbf{k} | \bar{\mathbf{t}}_m^+ | \mathbf{k}' \rangle \right|^2}{1 + Q(\mathbf{k}, \mathbf{k}')} \frac{\kappa_e^{d-3} \pi}{2(2\pi)^d}, \qquad (4.77a)$$

and Q is a dimensionless number given by

$$Q(\mathbf{k}, \mathbf{k}') = \frac{1}{S_d} \hat{F} \frac{\sum_m n_m \left|\langle \mathbf{k} | \bar{\mathbf{t}}_m^+ | \mathbf{k}' \rangle\right|^2}{\sum_m n_m \int d\Omega_{\hat{k}} \left|\langle \mathbf{k}' | \bar{\mathbf{t}}_m^+ | \mathbf{k} \rangle\right|^2}. \qquad (4.77b)$$

In the limit of long wavelength where the scattering is *isotropic* and Rayleigh-like, then $\hat{F}/S_d \cong S_d$, and the denominator of Q is $\sim S_d \sum_m n_m \left|\langle \mathbf{k}' | \bar{\mathbf{t}}_m^+ | \mathbf{k} \rangle\right|^2$, so that

$$Q \cong 1. \qquad (4.78)$$

The meaning of the denominator in the vertex function may be traced to the removal of the restriction on *successive* scatterings from the same particle, which is a crucial step in the homogenization of the medium in the CPA sense. Therefore the denominator arises purely from the geometric effect that two different particles cannot overlap. In the generation of ladder diagrams, since it is implicitly assumed that the homogeneous medium outside the scattering particle already contains the same particle (successive scatterings from the same particle are allowed), the presence of the denominator is necessary to correct that.

4.5 The Ward Identity

The relationship between the irreducible vertex function and the self energy, known as the Ward identity (Ward, 1950; Mahan, 1981), is important to the diffusive solution of the Bethe–Salpeter equation. To demonstrate its validity, it has to be stressed that the Ward identity holds only when both the irreducible vertex function and the self-energy are calculated to the *same* level of approximation. In this section the Ward identity is derived within the CPA framework.

Since for classical waves in continuum each scatterer involves an infinite number of real-space sites, the CPA condition for Σ is in general an operator equation in real space. Therefore the following derivation is carried out in the operator notation for generality. Before starting, however, it is convenient to define $\tilde{\Sigma} = L^d \Sigma / N$, and note that the CPA condition on $\tilde{\Sigma}$ is given by either

$$\langle \bar{\mathbf{t}}_i \rangle_c = \left\langle (\mathbf{v}_i - \tilde{\Sigma}) \left[\mathbf{I} - \tilde{\mathbf{G}}_e (\mathbf{v}_i - \tilde{\Sigma}) \right]^{-1} \right\rangle_c = 0,$$

or equivalently

$$\langle \bar{\mathbf{t}}_i \rangle_c = \left\langle \left[\mathbf{I} - (\mathbf{v}_i - \tilde{\Sigma}) \tilde{\mathbf{G}}_e \right]^{-1} (\mathbf{v}_i - \tilde{\Sigma}) \right\rangle_c = 0.$$

Here $\tilde{\mathbf{G}}_e$ mean $\mathbf{H}\mathbf{G}_e\mathbf{H}$, where $\mathbf{H}$ is the particle function as defined in Chapter III. The purpose of using $\tilde{\mathbf{G}}_e$ is to restrict the action of $\mathbf{G}_e$ to sites inside the scattering particle. $\tilde{\mathbf{G}}_e$ is noted not to have an inverse in the general sense, because its matrix elements are zero outside the particle. However, $(\tilde{\mathbf{G}}_e)^{-1}$ may be defined on the sites inside the particle. In what follows this is taken to be the meaning of $\tilde{\mathbf{G}}_e$'s inverse. Also, $\tilde{\mathbf{G}}_e^{\pm} = \mathbf{H}\mathbf{G}_e^{\pm}\mathbf{H}$. For convenience, let us also define

$$\mathbf{L}^{\pm} = \left[(\tilde{\mathbf{G}}_e^{\pm})^{-1} - (\mathbf{v}_i^{\pm} - \tilde{\Sigma}^{\pm}) \right]^{-1}. \tag{4.79}$$

In this notation the CPA condition may be cast as

$$\langle \mathbf{L}^{\pm} \rangle_c = \tilde{\mathbf{G}}_e^{\pm}, \tag{4.80}$$

which is shown in the solution to Problem 4.3. In Eq. (4.79) the meaning of $\mathbf{v}_i^{\pm}$ is that the values of $\sigma_i^{\pm}$ are calculated at frequencies $\omega^{\pm} = \omega \pm \Delta\omega/2$. For the quantum case, σ_i is independent of the frequency, so $\mathbf{v}_i^+ = \mathbf{v}_i^-$ always. However, this is not so for classical waves, where $\sigma_i = (1 - \epsilon_i)\omega^2/v_0^2$ [see Eq. (2.15)]. This difference between the quantum case and the classical wave case has important consequences as demonstrated below.

The CPA irreducible vertex $\mathbf{U}^{(B)}$ is given by

$$\mathbf{U}^{(B)} = \frac{\langle \bar{\mathbf{t}}_i^+ \otimes \bar{\mathbf{t}}_i^- \rangle_c}{\mathbf{I} + \langle \bar{\mathbf{t}}_i^+ \tilde{\mathbf{G}}_e^+ \otimes \tilde{\mathbf{G}}_e^- \bar{\mathbf{t}}_i^- \rangle_c}. \tag{4.81}$$

This is essentially Eq. (4.59) except $\mathbf{G}_e^+$ is replaced with $\tilde{\mathbf{G}}_e^+$ to stress that the real-space input and output channels are within the same particle. The

4.5 The Ward Identity

numerator may be written as

$$
\begin{aligned}
\langle \bar{\mathbf{t}}_i^+ &\otimes \bar{\mathbf{t}}_i^- \rangle_c \\
&= \left\langle (\mathbf{v}_i^+ - \tilde{\Sigma}^+)\left[\mathbf{I} - \tilde{\mathbf{G}}_e^+(\mathbf{v}_i^+ - \tilde{\Sigma}^+)\right]^{-1} \otimes \left[\mathbf{I} - (\mathbf{v}_i^- - \tilde{\Sigma}^-)\tilde{\mathbf{G}}_e^-\right]^{-1}(\mathbf{v}_i^- - \tilde{\Sigma}^-) \right\rangle_c \\
&= \left\langle (\mathbf{v}_i^+ - \tilde{\Sigma}^+)\left[(\tilde{\mathbf{G}}_e^+)^{-1} - (\mathbf{v}_i^+ - \tilde{\Sigma}^+)\right]^{-1}(\tilde{\mathbf{G}}_e^+)^{-1} \right. \\
&\qquad \left. \otimes (\tilde{\mathbf{G}}_e^-)^{-1}\left[(\tilde{\mathbf{G}}_e^-)^{-1} - (\mathbf{v}_i^- - \tilde{\Sigma}^-)\right]^{-1}(\mathbf{v}_i^- - \tilde{\Sigma}^-) \right\rangle_c \\
&= \left\langle \left[-\mathbf{I} + (\tilde{\mathbf{G}}_e^+)^{-1}\mathbf{L}^+\right](\tilde{\mathbf{G}}_e^+)^{-1} \otimes (\tilde{\mathbf{G}}_e^-)^{-1}\left[-\mathbf{I} + \mathbf{L}^-(\tilde{\mathbf{G}}_e^-)^{-1}\right] \right\rangle_c \\
&= -(\tilde{\mathbf{G}}_e^+)^{-1} \otimes (\tilde{\mathbf{G}}_e^-)^{-1} \\
&\quad + (\tilde{\mathbf{G}}_e^+)^{-1}\langle \mathbf{L}^+(\tilde{\mathbf{G}}_e^+)^{-1} \otimes (\tilde{\mathbf{G}}_e^-)^{-1}\mathbf{L}^- \rangle_c (\tilde{\mathbf{G}}_e^-)^{-1}. \quad (4.82)
\end{aligned}
$$

The last line is obtained from the previous line by the application of Eq. (4.80) twice, and by noting that $(\tilde{\mathbf{G}}_e^\pm)^{-1}$ can be taken out of the configurational averaging brackets since it is already an averaged quantity. Now the term $\langle \bar{\mathbf{t}}_i^+ \tilde{\mathbf{G}}_e^+ \otimes \tilde{\mathbf{G}}_e^- \bar{\mathbf{t}}_i^- \rangle_c$ in Eq. (4.81) differs in treatment from $\langle \bar{\mathbf{t}}_i^+ \otimes \bar{\mathbf{t}}_i^- \rangle_c$ only slightly. From the last line of Eq. (4.82) it may be readily deduced that

$$\mathbf{I} + \langle \bar{\mathbf{t}}_i^+ \tilde{\mathbf{G}}_e^+ \otimes \tilde{\mathbf{G}}_e^- \bar{\mathbf{t}}_i^- \rangle_c = (\tilde{\mathbf{G}}_e^+)^{-1}\langle \mathbf{L}^+ \otimes \mathbf{L}^- \rangle_c (\tilde{\mathbf{G}}_e^-)^{-1}. \quad (4.83a)$$

The inverse of (4.83a) is

$$\left[(\tilde{\mathbf{G}}_e^+)^{-1}\langle \mathbf{L}^+ \otimes \mathbf{L}^- \rangle_c (\tilde{\mathbf{G}}_e^-)^{-1}\right]^{-1} = (\mathscr{L}^+)^{-1}\tilde{\mathbf{G}}_e^+ \otimes \tilde{\mathbf{G}}_e^-(\mathscr{L}^-)^{-1}, \quad (4.83b)$$

where $\mathscr{L}^\pm$ is introduced to mean

$$\langle \mathbf{L}^+ \otimes \mathbf{L}^- \rangle_c = \mathscr{L}^+ \otimes \mathscr{L}^-, \quad (4.84a)$$

$$\langle \mathbf{L}^+ \otimes \mathbf{L}^- \rangle_c^{-1} = (\mathscr{L}^+)^{-1} \otimes (\mathscr{L}^-)^{-1}. \quad (4.84b)$$

The operations of taking the average and taking the inverse are noted not to commute; i.e., their order can not be interchanged. Therefore $\langle \mathbf{L}^+ \otimes \mathbf{L}^- \rangle_c^{-1} \neq \langle (\mathbf{L}^+)^{-1} \otimes (\mathbf{L}^-)^{-1} \rangle_c$. From the definition of inverse multiplication, Eq. (4.57), an expression for $\mathbf{U}^{(B)}$ is obtained:

$$\begin{aligned}\mathbf{U}^{(B)} &= -(\mathscr{L}^+)^{-1}\tilde{\mathbf{G}}_e^+\left(\tilde{\mathbf{G}}_e^+\right)^{-1} \otimes \left(\tilde{\mathbf{G}}_e^-\right)^{-1}\tilde{\mathbf{G}}_e^-(\mathscr{L}^+)^{-1} \\ &\quad +(\mathscr{L}^+)^{-1}\tilde{\mathbf{G}}_e^+\left(\tilde{\mathbf{G}}_e^+\right)^{-1}\mathscr{L}^+\left(\tilde{\mathbf{G}}_e^+\right)^{-1} \\ &\quad \otimes \left(\tilde{\mathbf{G}}_e^-\right)^{-1}\mathscr{L}^-\left(\tilde{\mathbf{G}}_e^-\right)^{-1}\tilde{\mathbf{G}}_e^-(\mathscr{L}^-)^{-1} \\ &= -\langle \mathbf{L}^+ \otimes \mathbf{L}^-\rangle_c^{-1} + \left(\tilde{\mathbf{G}}_e^+\right)^{-1} \otimes \left(\tilde{\mathbf{G}}_e^-\right)^{-1}. \end{aligned} \quad (4.85)$$

To further simplify the term $\langle \mathbf{L}^+ \otimes \mathbf{L}^-\rangle_c^{-1}$ with the aim of getting rid of the $\mathbf{v}_i$'s, it is necessary first to examine the expression

$$\begin{aligned}\langle \mathbf{L}^+ - \mathbf{L}^-\rangle_c &= \langle \mathbf{L}^+ \otimes (\mathbf{L}^-)^{-1}\mathbf{L}^- - \mathbf{L}^+(\mathbf{L}^+)^{-1} \otimes \mathbf{L}^-\rangle_c \\ &= \left\langle \mathbf{L}^+ \otimes \left[\left(\tilde{\mathbf{G}}_e^-\right)^{-1} - (\mathbf{v}_i^- - \tilde{\Sigma}^-)\right]\mathbf{L}^-\right\rangle_c \\ &\quad - \left\langle \mathbf{L}^+\left[\left(\tilde{\mathbf{G}}_e^+\right)^{-1} - (\mathbf{v}_i^+ - \tilde{\Sigma}^+)\right] \otimes \mathbf{L}^-\right\rangle_c. \end{aligned} \quad (4.86)$$

In particular, let us pull out the two terms related to $\mathbf{v}_i^\pm$ and look at them in the form of fixed real-space input and output channels:

$$\begin{aligned}&\langle \mathbf{L}^+ \mathbf{v}_i^+ \otimes \mathbf{L}^-\rangle_c - \langle \mathbf{L}^+ \otimes \mathbf{v}_i^- \mathbf{L}^-\rangle_c \\ &= \langle (\mathbf{L}^+)_{r_1 r_2}(\mathbf{v}_i^+)_{r_2}(\mathbf{L}^-)_{r_3 r_4} - (\mathbf{L}^+)_{r_1 r_2}(\mathbf{v}_i^-)_{r_3}(\mathbf{L}^-)_{r_3 r_4}\rangle_c, \end{aligned} \quad (4.87)$$

where $\mathbf{r}_1, \mathbf{r}_2, \mathbf{r}_3, \mathbf{r}_4$ are all within the same particle, and the fact that $\mathbf{v}_i$ is diagonal in real space has been utilized. Here the two terms cancel if $\mathbf{v}_i$ is a constant inside the particle. But even if $\mathbf{v}_i$ is not constant, it is recalled from the definition of the pulse intensity, Eq. (4.7), that $\mathbf{r}_1 = \mathbf{r}_4$ and $\mathbf{r}_2 = \mathbf{r}_3$ because there are only two distinct spatial points, source position $\mathbf{r}'$ and receiver position $\mathbf{r}$, in the problem (see the solution to Problem 4.1). Therefore, the two terms *must cancel for the quantum case, and for the classical wave case they cancel only when* $\Delta\omega = 0$, which is assumed for the moment. This is a crucial step in the derivation of the Ward identity. With $\mathbf{v}_i^\pm$ eliminated from Eq. (4.86), the rest of the expression may be written as

$$\begin{aligned}\langle \mathbf{L}^+ \otimes \mathbf{I} - \mathbf{I} \otimes \mathbf{L}^-\rangle_c &= \tilde{\mathbf{G}}_e^+ \otimes \mathbf{I} - \mathbf{I} \otimes \tilde{\mathbf{G}}_e^- \\ &= \langle \mathbf{L}^+ \otimes \mathbf{L}^-\rangle_c : \left[\mathbf{I} \otimes \left(\tilde{\mathbf{G}}_e^-\right)^{-1} - \left(\tilde{\mathbf{G}}_e^+\right)^{-1} \otimes \mathbf{I} \right. \\ &\quad \left. + \mathbf{I} \otimes \tilde{\Sigma}^- - \tilde{\Sigma}^+ \otimes \mathbf{I}\right], \end{aligned} \quad (4.88)$$

4.5 The Ward Identity

where the first equality follows from the CPA relation, and the convention $(\mathbf{L}^+ \otimes \mathbf{L}^-):(\mathbf{I} \otimes (\tilde{\mathbf{G}}_e^-)^{-1}) = \mathbf{L}^+ \otimes (\tilde{\mathbf{G}}_e^-)^{-1} \mathbf{L}^-$ is followed; i.e., the quantity to the right always forms inner products with the output channels of the quantity to the left. From Eq. (4.88), one can use the inverse notation to get

$$-\langle \mathbf{L}^+ \otimes \mathbf{L}^- \rangle_c^{-1} = \left[\mathbf{I} \otimes (\tilde{\mathbf{G}}_e^-)^{-1} - (\tilde{\mathbf{G}}_e^+)^{-1} \otimes \mathbf{I} \right.$$
$$\left. + \mathbf{I} \otimes \tilde{\Sigma}^- - \tilde{\Sigma}^+ \otimes \mathbf{I} \right] : \left(\mathbf{I} \otimes \tilde{\mathbf{G}}_e^- - \tilde{\mathbf{G}}_e^+ \otimes \mathbf{I} \right)^{-1}. \quad (4.89)$$

This expression may be further simplified because

$$\mathbf{I} \otimes (\tilde{\mathbf{G}}_e^-)^{-1} - (\tilde{\mathbf{G}}_e^+)^{-1} \otimes \mathbf{I}$$
$$= (\tilde{\mathbf{G}}_e^+)^{-1} \tilde{\mathbf{G}}_e^+ \otimes \mathbf{I}(\tilde{\mathbf{G}}^-)^{-1} - (\tilde{\mathbf{G}}_e^+)^{-1} \mathbf{I} \otimes \tilde{\mathbf{G}}_e^- (\tilde{\mathbf{G}}_e^-)^{-1}$$
$$= -\left[(\tilde{\mathbf{G}}_e^+)^{-1} \otimes (\tilde{\mathbf{G}}_e^-)^{-1} \right] : \left(-\tilde{\mathbf{G}}_e^+ \otimes \mathbf{I} + \mathbf{I} \otimes \tilde{\mathbf{G}}_e^- \right). \quad (4.90)$$

Substitution of Eq. (4.90) into Eq. (4.89) gives

$$-\langle \mathbf{L}^+ \otimes \mathbf{L}^- \rangle_c^{-1} = -(\tilde{\mathbf{G}}_e^+)^{-1} \otimes (\tilde{\mathbf{G}}_e^-)^{-1}$$
$$+ (\mathbf{I} \otimes \tilde{\Sigma}^- - \tilde{\Sigma}^+ \otimes \mathbf{I}) : \left(\mathbf{I} \otimes \tilde{\mathbf{G}}_e^- - \tilde{\mathbf{G}}_e^+ \otimes \mathbf{I} \right)^{-1}. \quad (4.91)$$

Substitution of Eq. (4.91) into Eq. (4.85) yields

$$\mathbf{U}^{(B)} = \mathbf{u}^+ \otimes \mathbf{u}^- = (\mathbf{I} \otimes \tilde{\Sigma}^- - \tilde{\Sigma}^+ \otimes \mathbf{I}) : \left(\mathbf{I} \otimes \tilde{\mathbf{G}}_e^- - \tilde{\mathbf{G}}_e^+ \otimes \mathbf{I} \right)^{-1}, \quad (4.92)$$

or

$$\mathbf{u}^+ \otimes \tilde{\mathbf{G}}_e^- \mathbf{u}^- - \mathbf{u}^+ \tilde{\mathbf{G}}_e^+ \otimes \mathbf{u}^- = \mathbf{I} \otimes \tilde{\Sigma}^- - \tilde{\Sigma}^+ \otimes \mathbf{I}. \quad (4.93)$$

By remembering that $\tilde{\mathbf{G}}_e^\pm = \mathbf{H} \mathbf{G}_e^\pm \mathbf{H}$, this equation may be rewritten as

$$\mathbf{u}^+ \mathbf{G}_e^+ \mathbf{H} \otimes \mathbf{u}^- - \mathbf{u}^+ \otimes \mathbf{H} \mathbf{G}_e^+ \mathbf{u}^- = \tilde{\Sigma}^+ \otimes \mathbf{I} - \mathbf{I} \otimes \tilde{\Sigma}^-, \quad (4.94)$$

where $\mathbf{u}^+ \mathbf{H}$ and $\mathbf{H} \mathbf{u}^-$ are noted to be just $\mathbf{u}^+$ and $\mathbf{u}^-$ because they are already defined on the sites of the particle, so multiplying by the particle function $\mathbf{H}$ does not alter them in any way. The usual way of displaying the Ward identity is in the $\mathbf{k}$-representation, where $\tilde{\Sigma}^\pm$ are diagonal. For input and output channels of Eq. (4.94) characterized by $\mathbf{k}_\pm$, and with $\mathbf{u}^+ \otimes \mathbf{u}^-$ = $U^{(B)}_{\mathbf{k},\mathbf{k}'}(|\mathbf{k}_+\rangle\langle\mathbf{k}'_+| \otimes |\mathbf{k}'_-\rangle\langle\mathbf{k}_-|)$ the operator equation becomes

$$\left\{\frac{1}{N}\sum_{\mathbf{k}'} U^{(B)}_{\mathbf{k},\mathbf{k}'} G_e^+(\omega^+,\mathbf{k}'_+) H(\mathbf{k}'_+ - \mathbf{k}_+) - U^{(B)}_{\mathbf{k},\mathbf{k}'} G_e^-(\omega^-,\mathbf{k}'_-) H(\mathbf{k}'_- - \mathbf{k}_-)\right\}$$

$$(|\mathbf{k}_+\rangle\langle\mathbf{k}_+| \otimes |\mathbf{k}_-\rangle\langle\mathbf{k}_-|) = \left[(\tilde{\Sigma}^+(\omega^+,\mathbf{k}_+) - \tilde{\Sigma}^-(\omega^-,\mathbf{k}_-)\right]$$

$$\times (|\mathbf{k}_+\rangle\langle\mathbf{k}_+| \otimes |\mathbf{k}_-\rangle\langle\mathbf{k}_-|), \quad (4.95)$$

where the $\mathbf{k}$ dependence of $\tilde{\Sigma}$ is put in for accuracy, even though in the CPA it is $\mathbf{k}$ independent. By noting that $H(\mathbf{k}) = 1$ for $|\mathbf{k}| < \mathbf{k}_{max} \approx 1/R$, defining ΔG_e as $G_e^+ - G_e^-$, and returning $\tilde{\Sigma}^\pm$ to $\Sigma^\pm$, Eq. (4.95) becomes

$$\frac{1}{N}\sum_{\mathbf{k}'}' \frac{N}{L^d} U^{(B)}_{\mathbf{k},\mathbf{k}'}(\Delta G_e)_{\mathbf{k}'} = \Sigma^+(\omega^+,\mathbf{k}_+) - \Sigma^-(\omega^-,\mathbf{k}_-), \quad (4.96a)$$

which is the usual form of the Ward identity. The prime on the summation sign expresses the fact that the range of $\mathbf{k}'$ covers only $|\mathbf{k}' - \mathbf{k}| < k_{max}$. Since $(\Delta G_e)_\mathbf{k}$ is a peaked function, this restriction is essentially immaterial as long as the allowed $|\mathbf{k}'|$ regime overlaps with κ_e.

In the case of point scatterers such as encountered in the Anderson model and the long-wavelength limit of the CPA, the $U^{(B)}_{\mathbf{k},\mathbf{k}'}$ is $\mathbf{k}(\mathbf{k}')$ independent, and the summation over $\mathbf{k}'$ in Eq. (4.96a) produces ΔG_e for a single site, i.e., $\Delta G_e(\omega, \mathbf{l} = \mathbf{l}')$. The Ward identity of the Anderson model is therefore given by an algebraic equation

$$\frac{N}{L^d} U^{(B)} \Delta G_e(\omega, \mathbf{l} = \mathbf{l}') = \Sigma^+(\omega) - \Sigma^-(\omega). \quad (4.96b)$$

Since $\Delta G_e(\omega, \mathbf{l} = \mathbf{l}')$ is given by

$$\Delta G_e(\omega, \mathbf{l} = \mathbf{l}') = \frac{1}{N}\sum_\mathbf{k}(\Delta G_e)_\mathbf{k} = -2i\frac{d\omega}{d\kappa_e^2}\pi\rho_e(\omega),$$

it follows that

$$U^{(B)} = \frac{-\text{Im}\,\Sigma^+}{\pi\rho_e(\omega)} \frac{d\kappa_e^2}{d\omega} \frac{L^d}{N}. \quad (4.96c)$$

4.6 Modification of the Diffusion Constant 151

An important observation here is that the *independence* of $U^{(B)}$ from $\mathbf{k}, \mathbf{k}'$ makes the solution of S possible without using the small $\Delta\omega$, $|\Delta \mathbf{k}|$ expansion. This is clear from Eq. (4.23), where the summation of $\mathbf{k}$ in that case gives

$$S = \frac{(L^d/N^2)\Sigma_\mathbf{k}(G_e^+ G_e^-)_\mathbf{k}}{1 - (U^{(B)}/N)\Sigma_\mathbf{k}(G_e^+ G_e^-)_\mathbf{k}} \frac{N}{L^d}. \qquad (4.96\text{d})$$

It is straightforward to verify that by expanding $(G_e^+ G_e^-)_\mathbf{k}$ to order $\Delta\omega$ and $|\Delta \mathbf{k}|^2$, the diffusive form of S, Eq. (4.45), is recovered [with $D(\omega)$ replaced by $D^{(B)}(\omega)$]. However, terms higher-order in $\Delta\omega$ and $|\Delta \mathbf{k}|$ can also be obtained in this case.

In Eq. (4.96a), if $\Delta\omega$, $|\Delta \mathbf{k}| = 0$, then the right-hand side is $\sim 2i \operatorname{Im} \Sigma^+ \sim 2i \operatorname{Im} \langle T \rangle_c$. On the left-hand side, $(\Delta G_e)_{\mathbf{k}'}$ may be approximated as $-2\pi i$ times a delta function [see Eq. 4.28]. The summation over $\mathbf{k}'$ is therefore effectively an angular integration. If one looks only at the numerator of $U^{(B)}$, $\langle \bar{\mathbf{t}}_i^+ \otimes \bar{\mathbf{t}}_i^- \rangle_c$ [see Eq. (4.74)], the angular integration yields the total scattering cross section. The form of the Ward identity is therefore very close to the optical theorem. However, an important difference arises if the effective medium is allowed to be complex. In that case $\operatorname{Im}\langle \bar{\mathbf{t}}_i \rangle_c = 0$, as in the CPA for the Anderson model, and the optical theorem in the usual sense does not exist. Instead, energy conservation has to be expressed through the imaginary part of the self-energy, i.e., $-\operatorname{Im} \Sigma^+$, which is the quantity that appears in the Ward identity. Therefore the Ward identity may be regarded as a generalized optical theorem for a scattering medium.

4.6 Modification of the Diffusion Constant Due to Frequency-Dependent Scattering Potentials

The derivation of the CPA Ward identity clearly demonstrates the necessity for Σ and $\mathbf{U}$ to be evaluated to the same accuracy in order for the identity to hold. However, the condition of $\mathbf{v}_i$ being frequency independent is not something that can be satisfied by classical waves in random media, regardless of the accuracy of Σ and $\mathbf{U}$. Therefore, the Ward identity in its form of Eq. (4.33) is strictly valid for classical scalar waves only in the $\Delta\omega = 0$ limit ($\omega^+ = \omega^-$). When $\Delta\omega \neq 0$, $\mathbf{v}_i^+$ does not cancel $\mathbf{v}_i^-$ in Eq. (4.87), and in addition to $\Sigma^+ - \Sigma^-$ on the right-hand side of Eq. (4.96a), there is a term of the form

$$-\frac{1}{N^2} \sum_{\mathbf{k}'} \sum_{\mathbf{k}''} \frac{N}{L^d} \left\langle\!\!\left\langle \mathbf{k} \middle| \psi_{\mathbf{k}'}^+ \right\rangle\!\!\left\langle \psi_{\mathbf{k}'}^+ \middle| \frac{\partial \mathbf{v}_i}{\partial \omega} \middle| \psi_{\mathbf{k}''}^+ \right\rangle\!\!\left\langle \psi_{\mathbf{k}''}^+ \middle| \mathbf{k} \right\rangle\!\right\rangle_c \Delta\omega,$$

where $\psi_{\mathbf{k}'}^+$ is an eigenfunction with incident wave vector $\mathbf{k}'$ for a single configuration (the superscript $+$ denotes that the eigenfunction is calculated from the retarded Green's function $\mathbf{G}^+$). This term is evaluated at $|\Delta \mathbf{k}| = 0$. In the expression above, everything inside the outer $\langle \ \rangle_c$, except the state $|\mathbf{k}\rangle$, is subject to averaging. This can be done as follows. Since $\partial \mathbf{v}_i/\partial \omega$ is an operator that is *local* in character, one can fix $\mathbf{v}_i$ and average over configurations that change everything else. In the weak-scattering limit, the resulting averaged $\langle \mathbf{k} | \psi_{\mathbf{k}'}^+ \rangle$ may be approximated by the form $\exp(i\mathbf{k}' \cdot \mathbf{r}) + f(\theta)\exp(ik'r)/r^{(d-1)/2}$ away from $\mathbf{v}_i$, where $f(\theta)$ is the scattering amplitude for $\mathbf{v}_i$. This is reasonable because after configurational averaging, only the coherent part of $\psi_{\mathbf{k}'}^+$ is expected to remain, and the above approximation treats the surrounding (within a mean free path) of $\mathbf{v}_i$ as being homogeneous in the CPA sense. Since $\exp(i\mathbf{k}' \cdot \mathbf{r})$ has an intrinsic magnitude of N, $\langle\langle \mathbf{k} | \psi_{\mathbf{k}'}^+ \rangle\rangle_c \approx N\delta_{\mathbf{k},\mathbf{k}'}$, where the scattering term is assumed to be small compared to N. Using similar reasoning on $\langle\langle \psi_{\mathbf{k}''}^+ | \mathbf{k} \rangle\rangle$ and summing over $\mathbf{k}'$ and $\mathbf{k}''$ yield

$$-\frac{N}{L^d}\left\langle\left\langle \psi_{\mathbf{k}}^+ \left| \frac{\partial \mathbf{v}_i}{\partial \omega} \right| \psi_{\mathbf{k}}^+ \right\rangle\right\rangle_c \Delta\omega$$

as the extra term when $\Delta\omega \neq 0$. For classical waves, it is recalled that

$$\mathbf{v}_i = \frac{\omega^2}{v_0^2} \sum_{\mathbf{r}} [1 - \epsilon_i(\mathbf{r})](\delta r)^d \cdot |\mathbf{r}\rangle\langle\mathbf{r}|, \ \mathbf{r} \in \text{particle } i.$$

Using this form of $\mathbf{v}_i$ gives

$$-\frac{2\omega}{v_0^2} \sum_m n_m \sum_{\mathbf{r}} (\delta r)^d \langle \psi_{\mathbf{k}}^+ | \mathbf{r}\rangle\langle\mathbf{r}|[1 - \epsilon_m(\mathbf{r})]|\mathbf{r}\rangle\langle\mathbf{r} | \psi_{\mathbf{k}}^+ \rangle\Delta\omega$$
$$= (2\omega\Delta\omega/v_0^2)\delta \qquad (4.97\text{a})$$

where m is the index for the scatterer type, and

$$\delta = \int d\mathbf{r} \sum_m n_m |\psi_{\mathbf{k}}^{+(m)}(\mathbf{r})|^2 [\epsilon_m(\mathbf{r}) - 1]. \qquad (4.97\text{b})$$

Here the wave function may be obtained approximately by solving a scattering boundary value problem where a single scatterer of type m is embedded in an effective medium. This procedure has been demonstrated in the last chapter.

4.7 Evaluation of the Wave Diffusion Constant

Since the additional term is left over after the vertex function cancels with $\Sigma^+ - \Sigma^-$, it combines with the coefficient of S in Eq. (4.35) with a negative sign to give

$$-\frac{d\kappa_0^2}{d\omega}\left(1 + \frac{2\omega}{v_0^2}\frac{d\omega}{d\kappa_0^2}\delta\right)\Delta\omega S = -\frac{2\omega}{v_0^2}(1+\delta)\Delta\omega S = -\frac{2\omega}{v_0^2}\Delta\omega S'. \quad (4.98a)$$

Here $S' = (1 + \delta)S$, and we have specialized to classical waves (since this effect appears only for classical waves) so that $d\kappa_0^2/d\omega = 2\omega/v_0^2$. The modification of S thus leads directly to the modification of the wave diffusion constant (van Albada et al., 1991). This is easily seen if S is rewritten as $S'/(1 + \delta)$ in Eqs. (4.43) and (4.44); then for classical waves the diffusion constant expression becomes

$$D(\omega) = \frac{1}{\pi\rho_e(\omega)}\frac{8K_0^2(d\kappa_e^2/d\kappa_0^2)}{M_0 - 8K_0\,\mathrm{Im}(\Sigma^+)}\frac{-\mathrm{Im}\,\Sigma^+}{1+\delta}. \quad (4.98b)$$

Physically, δ can be large when the incident wave frequency coincides with an internal resonance of the scatterer. When that happens, $|\psi_\mathbf{k}^+|^2$ has large magnitude inside the scatterer, leading to a large δ. Another interpretation of resonant scattering is that the incident wave can get trapped inside the scatterer, bouncing around many times before emerging again. This picture tells us in an intuitive way that wave diffusion has to slow down when resonant scattering is present. However, Eq. (4.97b) would not be accurate in that case because the strong multiple scattering *between* the particles is not taken into account. These issues will be further addressed in Chapter VIII.

4.7 Evaluation of the Wave Diffusion Constant

The expressions for $D(\omega)$, given by Eqs. (4.46) and (4.98b), are to be evaluated for the Anderson model and for classical scalar waves, both to the accuracy of the CPA. Here the scattering is assumed to be weak, so the weak-scattering versions of the K_0 and M_0 expression, (4.39c) and (4.42b), will be used.

For the Anderson model, a great simplification is obtained in noting that $U_{\mathbf{k},\mathbf{k}'}^{(B)}$, given by Eq. (4.67), is isotropic. As a result, $M_0^{(B)}$ is zero; i.e., ladder diagrams sum to zero in the case of the Anderson model. That means

$$D^{(B)}(\omega) = \frac{K_0(d\kappa_e^2/d\kappa_0^2)}{\rho_e(\omega)\pi} = -\frac{d\kappa_e^2/d\kappa_0^2}{\rho_e(\omega)\pi} \frac{L^d}{4N^2} \sum_{\mathbf{k}} \left[\nabla_k e(\mathbf{k}) \cdot \Delta\hat{\mathbf{k}}\right]^2 (\Delta G)_{\mathbf{k}}^2.$$

(4.99a)

This expression for the diffusion constant is generally known as the Kubo–Greenwood formula (Kubo, 1956; Greenwood, 1958). By using the weak-scattering approximation for K_0, Eq. (4.39c), the expression becomes

$$D^{(B)}(\omega) = \frac{\frac{1}{2}\int_{\text{IBZ}} \frac{d\mathbf{k}}{(2\pi)^d} \left[\nabla_k e(\mathbf{k}) \cdot \Delta\hat{\mathbf{k}}\right]^2 \delta[\kappa_e^2 - e(\mathbf{k})](d\kappa_e^2/d\kappa_0^2)}{(-\text{Im}\,\Sigma^+)\int_{\text{IBZ}} \frac{d\mathbf{k}}{(2\pi)^d} \delta[\kappa_e^2 - e(\mathbf{k})](d\kappa_e^2/d\omega)}.$$

(4.99b)

The factor β has been taken to be 1 in this chapter. If $\beta \neq 1$, its effect is to modify κ_e^2 to give κ_e^2/β. In Eq. (4.99), $(d\kappa_e^2/d\kappa_0^2)/(d\kappa_e^2/d\omega)$ can be combined to give $d\omega/d\kappa_0^2 = \hbar/2m$, which is the unit of quantum diffusion constant. The rest of the expression represents an average of $|\nabla_k e(\mathbf{k}) \cdot \Delta\hat{\mathbf{k}}|^2$ on the energy shell (as required by the delta function), divided by $-\text{Im}\,\Sigma^+$. Since $e(\mathbf{k})$ represents the dispersion relation, $\nabla_k e(\mathbf{k})$ gives the group velocity. On the other hand, $-\text{Im}\,\Sigma^+$ gives the rate of scattering. Therefore the diffusion constant has the form

$$D^{(B)}(\omega) = \frac{1}{d} v_t l, \qquad (4.100)$$

where

$$l = \frac{\bar{\kappa}_e}{-\text{Im}\,\Sigma^+} \qquad (4.101)$$

is denoted the mean free path, and v_t is the transport velocity, given by

$$v_t = \frac{\hbar \bar{\kappa}_e}{m}, \qquad (4.102)$$

with

$$\bar{\kappa}_e^2 = \frac{d}{4a^2} \frac{\int_{\text{IBZ}} d\mathbf{k} \left[a\nabla_k e(\mathbf{k}) \cdot \Delta\hat{\mathbf{k}}\right]^2 \delta[\kappa_e^2 - e(\mathbf{k})]}{\int_{\text{IBZ}} d\mathbf{k}\, \delta[\kappa_e^2 - e(\mathbf{k})]}. \qquad (4.103)$$

From the definition of $\bar{\kappa}_e$, Eq. (4.103), it is clear the v_t is just the average group velocity on the constant frequency surface. The strong-scattering

4.7 Evaluation of the Wave Diffusion Constant

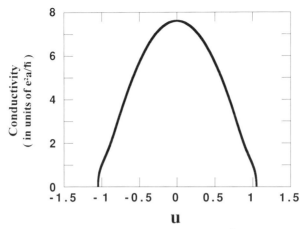

Figure 4.6 Conductivity of the 1D Anderson model for $\sigma_w a^2 = 0.5$, $\beta = 1$, plotted as a function of the nondimensionalized Fermi energy $u = (E_F - \epsilon_0)/(\beta \hbar^2/ma^2)$.

version of Eq. (4.101) is given in Chapter VIII [Eq. (8.21)]. In Eq. (4.103), the integral in the numerator is noted to be independent of the direction of $\Delta \hat{\mathbf{k}}$. This can be seen from the fact that $a\nabla_k e(\mathbf{k}) \cdot \Delta \hat{\mathbf{k}} = 2\Sigma_i(\sin k_i a)(\Delta \hat{\mathbf{k}})_i$. When squared, the cross terms integrate to zero because $\sin k_i a$ is an odd function of k_i, whereas $\delta[\kappa_c^2 - e(\mathbf{k})]$ is an even function. For the terms that remain, $4\Sigma_i(\sin^2 k_i a)(\Delta \hat{\mathbf{k}})_i^2$, each $\sin^2 k_i a$ term integrates to the same result independent of the direction i, so the $(\Delta \hat{\mathbf{k}})_i^2$ can be summed to give 1, the magnitude of $\Delta \hat{\mathbf{k}}$.

With the $\bar{\kappa}_c^2$ given by Eq. (4.103), the mean free path expression, Eq. (4.101), is what has been plotted in Figures 3.4, 3.5, and 3.6. For electrons, the diffusion constant is directly proportional to the electrical conductivity σ_E through the Einstein relation:

$$\sigma_E = \frac{e^2}{\hbar} \rho_e(\omega) D^{(B)}(\omega). \tag{4.104}$$

Here e is the electronic change and $e^2/\hbar$ has the units of conductance, given by $(4108 \text{ ohms})^{-1}$. The unit of conductivity is thus $(e^2/\hbar a^{d-2})$. In Figures 4.6, 4.7, and 4.8 the conductivity of the Anderson model is plotted as a function of the electron energy for $d = 1$, 2, and 3, respectively, with $a^2 \sigma_w = 0.5$. Since in actual materials only electrons at the Fermi surface can carry current, the electron energy may alternatively be regarded as the electron Fermi energy. The electronic spin degeneracy is not taken into account here.

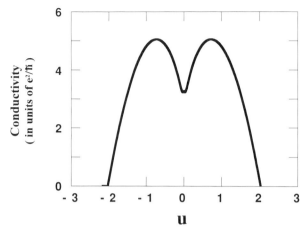

Figure 4.7 Conductivity of the 2D Anderson model for $\sigma_w a^2 = 0.5$, $\beta = 1$, plotted as a function of reduced Fermi energy $u = (E_F - \epsilon_0)/(\beta \hbar^2/ma^2)$. Here the dip in the band center is seen to reflect the effect of the small mean free path (see Figure 3.5).

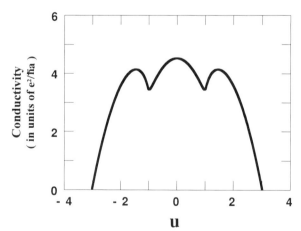

Figure 4.8 Conductivity of the 3D Anderson model for $\sigma_w a^2 = 0.5$, $\beta = 1$, plotted as a function of reduced Fermi energy $u = (E_F - \epsilon_0)/(\beta \hbar^2/ma^2)$. The structure in the vicinity of the band center is due to the interplay between l, v_l, and $\rho_e(\omega)$.

4.7 Evaluation of the Wave Diffusion Constant

For classical scalar waves in continuum, the function $e(\mathbf{k}) = k^2$, so the constants K_0 and $M_0^{(B)}$ may be explicitly evaluated. Since the weak-scattering versions of K_0 and M_0 are used, it is simpler to consider the combinations $(-\text{Im }\Sigma^+)K_0$ and $(-\text{Im }\Sigma^+)M_0$. With these combinations the diffusion constant expression may be rewritten as

$$D(\omega) = \frac{[\pi\rho_e(\omega)]^{-1}8[(-\text{Im }\Sigma^+)K_0]^2(d\kappa_e^2/d\kappa_0^2)(1+\delta)^{-1}}{(-\text{Im }\Sigma^+)M_0 + 8[(-\text{Im }\Sigma^+)K_0](-\text{Im }\Sigma^+)}. \quad (4.105)$$

Let us first calculate the denominator. For $[(-\text{Im }\Sigma^+)K_0]$, Eq. (4.39c) tells us that

$$(-\text{Im }\Sigma^+)K_0 = \frac{\pi}{2}\int\frac{d\mathbf{k}}{(2\pi)^d}\,4k^2\cos^2\theta_1\delta[\kappa_e^2 - k^2] = \frac{\pi}{(2\pi)^d}C_d\kappa_e^d, \quad (4.106a)$$

where θ_1 is the angle between $\mathbf{k}$ and $\Delta\mathbf{k}$, and $C_d = 4\pi/3$, π, and 2 in 3D, 2D, and 1D, respectively. For $M_0^{(B)}$, Eqs. (4.42b) may be written as

$$(-\text{Im }\Sigma^+)M_0^{(B)} = \int\frac{d\mathbf{k}}{(2\pi)^d}\int\frac{d\mathbf{k}'}{(2\pi)^d}(-4\pi^2)\delta[\kappa_e^2 - k^2]$$

$$\times \delta[\kappa_e^2 - (k')^2]4kk'\cos\theta_1\cos\theta_2 U_{\mathbf{k},\mathbf{k}'}^{(B)}\frac{N}{L^d}, \quad (4.106b)$$

where θ_2 is the angle between $\mathbf{k}'$ and $\Delta\hat{\mathbf{k}}$. The magnitude part of the integrals can be performed immediately with the help of the delta functions. Also, from Eq. (4.76) the vertex function is given by

$$U_{\mathbf{k},\mathbf{k}'}^{(B)}\frac{N}{L^d} = \sum_m n_m \hat{O}_m(\mathbf{k},\mathbf{k}')\frac{2(2\pi)^d}{\pi\kappa_e^{d-3}} = \sum_m n_m \hat{O}_{\kappa_e}^{(m)}(\theta)\frac{2(2\pi)^d}{\pi\kappa_e^{d-3}}, \quad (4.107)$$

where the magnitudes of $\mathbf{k}$ and $\mathbf{k}'$ are set at κ_e, and a slight change in the notation is meant to bring out explicitly the dependences on the scattering angle θ between $\mathbf{k}$ and $\mathbf{k}'$ as well as on the magnitude of the wave vector. The hat on O is a reminder that the quantity is evaluated with respect to the CPA effective medium and renormalized by $1 + Q$ (see Eq. (4.77)). Substitution of (4.107) into (4.106b) yields

$$(-\text{Im }\Sigma^+)M_0^{(B)} = -\frac{\kappa_e^{d+1}}{(2\pi)^{d-2}}\frac{2}{\pi}\int d\Omega_{\hat{k}}\int d\Omega_{\hat{k}'}$$

$$\times \cos\theta_1\cos\theta_2\sum_m n_m \hat{O}_{\kappa_e}^{(m)}(\theta), \quad (4.108)$$

where $d\Omega_{\hat{k}}$ and $d\Omega_{\hat{k}'}$ denote angular integrations. In the integrand, $\cos\theta_2$ can be expressed in terms of θ and θ_1 as

$$\cos\theta_2 = \begin{cases} \cos\theta\cos\theta_1 + \sin\theta\sin\theta_1\cos(\phi-\phi_1) & \text{3D} \\ \cos\theta\cos\theta_1 - \sin\theta\sin\theta_1 & \text{2D} \\ \pm 1. & \text{1D} \end{cases} \quad (4.109)$$

Substitution of Eq. (4.109) into (4.108) results in

$$(-\text{Im }\Sigma^+)M_0^{(B)} = -\frac{\kappa_e^{d+1}}{(2\pi)^{d-2}}$$

$$\times C_d \frac{2}{\pi} \begin{cases} 2\pi \int_{-1}^{1} d(\cos\theta)\cos\theta \sum_m n_m \hat{O}_{\kappa_e}^{(m)}(\theta) & \text{3D} \\ \int_0^{2\pi} d\theta \cos\theta \sum_m n_m \hat{O}_{\kappa_e}^{(m)}(\theta) & \text{2D} \\ \sum_m n_m \left[\hat{O}_{\kappa_e}^{(m)}(0) - \hat{O}_{\kappa_e}^{(m)}(\pi) \right] & \text{1D} \end{cases}$$

(4.110)

For the additional $8(-\text{Im }\Sigma^+)[(-\text{Im }\Sigma^+)K_0]$ term present in the denominator of the diffusion constant, we use the Ward identity in the $\Delta\omega = 0$ limit to evaluate $-\text{Im }\Sigma^+$:

$$-\text{Im }\Sigma^+ = \frac{i}{2}(\Sigma^+ - \Sigma^-) = \frac{i}{2} \int \frac{d\mathbf{k}}{(2\pi)^d} \sum_m n_m \hat{O}_{\kappa_e}^{(m)}(\theta)$$

$$\times \frac{2(2\pi)^d}{\pi\kappa_e^{d-3}} \left\{ -2\pi i \delta\left[\kappa_e^2 - k^2\right] \right\}$$

$$= \kappa_e \begin{cases} 2\pi \int_{-1}^{1} d(\cos\theta) \sum_m n_m \hat{O}_{\kappa_e}^{(m)}(\theta) & \text{3D} \\ \int_0^{2\pi} d\theta \sum_m n_m \hat{O}_{\kappa_e}^{(m)}(\theta) & \text{2D} \quad (4.111) \\ \sum_m n_m \left[\hat{O}_{\kappa_e}^{(m)}(0) - \hat{O}_{\kappa_e}^{(m)}(\pi) \right] & \text{1D.} \end{cases}$$

It is seen that the multiplication of $-\text{Im }\Sigma^+$ with $8(-\text{Im }\Sigma^+)K_0$ [Eq. (4.106a)] yields the exact prefactor (with a sign change) as that for

4.7 Evaluation of the Wave Diffusion Constant

$(-\operatorname{Im}\Sigma^+)M_0^{(B)}$, Eq. (4.110). Therefore the denominator of $D^{(B)}(\omega)$ (Eq. (4.105)), $(-\operatorname{Im}\Sigma^+)M_0^{(B)} + 8K_0(-\operatorname{Im}\Sigma^+)^2$, may be combined to give $(1 - \cos\theta)\sum_m n_m \hat{O}_{\kappa_e}^{(m)}(\theta)$ as the angular integral's integrand. What remains for the diffusion constant calculation is the numerator of Eq. (4.105), $8K_0^2(d\kappa_e^2/d\kappa_0^2)\pi^{-1}\rho_e^{-1}(\omega)(-\operatorname{Im}\Sigma^+)^2(1+\delta)^{-1}$, which is reduced straightforwardly to

$$\frac{8\pi}{(2\pi)^d}\frac{C_d}{d}\kappa_e^{d+2}\frac{v_0^2}{\omega}\frac{1}{1+\delta}.$$

By combining terms, a simple expression for $D^{(B)}(\omega)$ is obtained:

$$D^{(B)}(\omega) = \frac{1}{d}\frac{v_0^2}{\omega}\frac{\kappa_e}{1+\delta}l^* = \frac{1}{d}\frac{v_0^2}{v_e(1+\delta)}l^* = \frac{1}{d}v_t l^*, \quad (4.112)$$

where l^* is exactly the classical Boltzmann expression for the transport mean free path, given by

$$l^* = \left[\int d\Omega(1-\cos\theta)\sum_m n_m \hat{O}_{\kappa_e}^{(m)}(\theta)\right]^{-1}$$

$$= \begin{cases} \left[2\pi\int_{-1}^{1}d\cos\theta(1-\cos\theta)\sum_m n_m \hat{O}_{\kappa_e}^{(m)}(\pi)\right]^{-1} & \text{3D} \\ \left[\int_0^{2\pi}d\theta(1-\cos\theta)\sum_m n_m \hat{O}_{\kappa_e}^{(m)}(\theta)\right]^{-1} & \text{2D} \quad (4.113) \\ \left[2\sum_m n_m \hat{O}_{\kappa_e}^{(m)}(\theta)\right]^{-1} & \text{1D} \end{cases}$$

and

$$v_t = v_0^2 \frac{\kappa_e}{\omega(1+\delta)} = v_0\sqrt{1 - v_0^2\frac{\operatorname{Re}\Sigma^+}{\omega^2}}\frac{1}{1+\delta} \quad (4.114)$$

is defined as the transport velocity. The last equality follows from the definition of $\kappa_e = \sqrt{\kappa_0^2 - \operatorname{Re}\Sigma^+}$. Aside from the $(1+\delta)^{-1}$ factor, v_t is equal to the group velocity, given by $(d\kappa_e/d\omega)^{-1} = v_0^2\kappa_e/\omega = v_0^2/(\omega/\kappa_e)$, provided the CPA is valid and the frequency dependence of $\operatorname{Re}\Sigma^+$ is negligible compared to that of κ_0^2. The actual calculation and interpretation of v_t in a random medium are addressed in Chapter 8.

When the scattering is isotropic, the $\cos\theta$ term integrates to zero, and $l^* = l$, the usual mean free path. In that case a comparison with the former definition of l, given by Eq. (3.75), shows there to be a difference in the definition of the scattering cross sections. Here $\hat{O}_{\kappa_c}^{(m)}(\theta)$ is defined with a $1 + Q$ denominator [see Eq. (4.77a)], which is absent in the former definition because the geometric repulsion effect, related to the successive scattering from the same particle, has been ignored.

An interesting observation is that at low frequencies the scalar wave scattering is always isotropic; therefore the contribution of the ladder diagrams is zero, i.e., $M_0^{(B)} = 0$. On the other hand, for vector electromagnetic waves the long-wavelength scattering is always dipolar in its angular dependence, as noted in the last chapter. Therefore, in the present formalism $M_0^{(B)} \neq 0$ for electromagnetic waves even in the long-wavelength limit. The only catch here is that the formalism developed so far is for scalar waves. Therefore whether $M_0^{(B)}$ is truly nonzero for electromagnetic waves would have to be resolved by a fully vectorial calculation without any approximation that might destroy the polarization dependence of each scattering. Such a calculation is still lacking so far.

When the scattering is anisotropic, the transport mean free path l^* is always larger than l because the $(1 - \cos\theta)$ factor tends to emphasize the nonforward directions, and isotropic direction randomization would thus take more than one collision, hence $l^* > l$. Since nonisotropic scalar-wave scattering implies that the wave can already resolve some features of the scatterer, the assumption about the k independence of Σ, as well as the effective medium description, may both be called into question. When that happens, the diffusive behavior still persists at distances larger than l^*, but Σ has to be calculated with the GCPA approach described in Section 3.9. However, a more serious correction to the diffusion constant is the contributions to **U** beyond the ladder diagrams. It turns out that there are indeed such contributions due to the so-called maximally crossed diagrams, which are manifest physically as the coherent backscattering effect, to be described in the next chapter. But before leaving the subject of diffusive waves it is important to point out a novel application of diffusive waves for the optical measurement of dynamical properties in opaque materials.

4.8 Application: Diffusive Wave Spectroscopy

Most of the fluids we encounter daily are actually fluid suspensions. Coffee, milk, ink, mineral water, paint, etc. all contain a certain fraction of suspended solid particles which give the fluid some or most of its physical properties. In particular, when the concentration of the solid particles is high and their optical indices of refraction are sufficiently different from that of the solution liquid, the suspension would appear opaque due to the

4.8 Application: Diffusive Wave Spectroscopy

strong scattering by the particles, even if the solution liquid is transparent in its pure form. As a result, traditional optical measurement of suspensions is usually difficult except in the extremely dilute case.

In the past few years, a new optical measurement technique has emerged which is capable of determining the dynamic properties of the suspended particles, such as their diffusion constant. The technique relies on the fact that the suspended particles are generally not stationary but move stochastically as Brownian particles due to the thermal agitation by the solution molecules. As a result, the light which is scattered by the suspended particles must also exhibit some stochastic time dependence, i.e., fluctuations. The desired dynamic information about the suspension may be retrieved by an analysis of the time correlation function of the scattered light. The diffusive character of the light (Genack, 1990) enters precisely in such an analysis and is crucial for understanding the observed experimental results, hence the name of the technique—diffusive wave spectroscopy (DWS) (Pine et al., 1990a).

To give an understanding of the theoretical basis of DWS, let us start with what can be measured experimentally, which is $\langle I(\Delta t)I(0)\rangle_c/\langle I\rangle_c^2$. Here I is the intensity of the scattered light at a given point, Δt is the time separation between two intensity measurements, and $\langle\ \rangle_c$ denotes averaging, which is generally a time average experimentally but can also be interpreted as an average over different scattering configurations of the sample. At $\Delta t = 0$, this quantity becomes $\langle I^2\rangle_c/\langle I\rangle_c^2$, which is always greater than or equal to 1 and can be written as

$$\frac{\langle I^2\rangle_c}{\langle I^2\rangle_c^2} = 1 + \text{normalized variance of intensity fluctuations}$$

$$= 1 + \frac{\langle I^2\rangle_c - \langle I\rangle_c^2}{\langle I\rangle_c^2}. \tag{4.115}$$

In the solution to Problem 4.4, it is shown that for $\Delta t \neq 0$, a similar relation may be written as

$$\frac{\langle I(\Delta t)I(0)\rangle_c}{\langle I\rangle_c^2} = 1 + g_2(\Delta t), \tag{4.116}$$

where

$$g_2(\Delta t) = \left|\frac{\langle \phi(\Delta t)\phi^*(0)\rangle_c}{\langle |\phi|^2\rangle_c}\right|^2 = |g_1(\Delta t)|^2, \tag{4.117}$$

and ϕ is the wave amplitude (electric field amplitude of light). Equations (4.116) and (4.117) are known as the Siegert relation. This relation is widely used in optical measurements to relate the measured intensity to its theoretical interpretation, which generally begins with wave amplitudes. The crucial quantity here is therefore $\langle \phi(\Delta t)\phi^*(0)\rangle_c$. In order to arrive at a theoretical expression for $\langle \phi(\Delta t)\phi^*(0)\rangle_c$, the first question one may ask is: What is the difference between $\phi(\Delta t)$ and $\phi(0)$? Suppose the source of light is at a distance r away from the measuring point. In arriving at the measuring point the light would encounter many scatterings in a zigzag path. Let us for the moment focus on one class of light trajectories that consist of precisely n scatterings and label the field amplitude associated with that class of trajectories $\phi^{(n)}$. The difference between $\phi^{(n)}(0)$ and $\phi^{(n)}(\Delta t)$ is clearly caused by the movement of the intervening scatterers during the time period Δt. By assuming all scatterings to be elastic, this fact can be expressed mathematically as

$$\langle \phi^{(n)}(\Delta t)\phi^{(n)*}(0)\rangle_c = \langle |\phi^{(n)}(0)|^2\rangle_c \left\langle \exp\left\{-i\sum_{i=1}^n \mathbf{q}_i \cdot [\mathbf{r}_i(\Delta t) - \mathbf{r}_i(0)]\right\}\right\rangle_c. \tag{4.118}$$

Here $\mathbf{q}_i$ is the scattering wave vector and $\mathbf{r}_i$ the position of the scatterer for the ith scattering. If θ_i is the scattering angle, then

$$|\mathbf{q}_i| = 2\kappa_e \sin\frac{\theta_i}{2}. \tag{4.119}$$

By decomposing $g_1(\Delta t)$ as

$$g_1(\Delta t) = \sum_{n=1}^\infty g_1^{(n)}(\Delta t), \tag{4.120a}$$

from Eqs. (4.117) and (4.118) we have

$$g_1^{(n)}(\Delta t) = P(n)\left\langle \prod_{i=1}^n \exp\{-\mathbf{q}_i \cdot [\mathbf{r}_i(\Delta t) - \mathbf{r}_i(0)]\}\right\rangle_c, \tag{4.120b}$$

where $P(n) = \langle|\phi^{(n)}(0)|^2\rangle_c/\langle|\phi|^2\rangle_c$ is the probability that a trajectory of n scatterings ends at the measurement point. The averaging bracket $\langle\ \rangle_c$ in Eq. (4.120b) has the double function of averaging over $\mathbf{r}_i$ and over all possible directions of $\mathbf{q}_i$. Strictly speaking, for fixed incoming and outgoing wave vectors, $\mathbf{r}_i$ and $\mathbf{q}_i$ are correlated. However, for large n the correlation is weak. Here $\mathbf{r}_i$ and $\mathbf{q}_i$ will be treated as independent. This approximation

4.8 Application: Diffusive Wave Spectroscopy

allows the two averages to be done separately. To average over $\mathbf{r}_i$, the fact that the scatterers are undergoing Brownian (diffusive) motion means that the probability of $|\mathbf{r}_i(\Delta t) - \mathbf{r}_i(0)|$ is given by the Gaussian distribution:

$$\frac{1}{(4\pi D_p \Delta t)^{3/2}} \exp\left[-|\mathbf{r}(\Delta t) - \mathbf{r}_i(0)|^2 / 4D_p \Delta t\right], \quad (4.121)$$

where $d = 3$ is assumed since experiments so far have all been done on 3D systems, and D_p is the diffusion constant of the suspended particles. Averaging each term of (4.120b) with respect to the position of particle i means integrating the product of the Gaussian distribution with each term. That gives

$$g_1^{(n)}(\Delta t) = P(n)\left\langle \prod_{i=1}^n \exp\left(-D_p \Delta t\, q_i^2\right)\right\rangle_c$$

$$= P(n)\left\langle \prod_{i=1}^n \exp\left(-4D_p \kappa_e^2 \Delta t \sin^2 \frac{\theta_i}{2}\right)\right\rangle_c \quad (4.122)$$

through the well-known trick of completing the squares in the exponent so that the term left over is precisely that shown in Eq. (4.122). Since $(D_p \kappa_e^2)^{-1}$ has the units of time, it is denoted t_0. Physically, t_0 is the time scale for a suspended particle to diffuse a distance on the order of a wavelength. It also determines the decay rate of the autocorrelation function $g_1(\Delta t)$.

To the leading order of $\Delta t / t_0$, i.e., small Δt, the average over the scattering angle θ_i in Eq. (4.122) may be approximated by

$$\left\langle \prod_{i=1}^n \exp\left(-4D_p \kappa_e^2 \Delta t \sin^2 \frac{\theta_i}{2}\right)\right\rangle_c \approx \exp\left[-2n \frac{\Delta t}{t_0}\langle 1 - \cos\theta_i\rangle_c\right], \quad (4.123)$$

which is equivalent to retaining the first term in the cumulant expansion. Since the average of $1 - \cos\theta_i$ is precisely the factor that differentiates l^* from l, it follows from Eq. (4.113) that

$$\langle 1 - \cos\theta_i\rangle_c = \frac{l}{l^*}, \quad (4.124)$$

so that

$$g_1^{(n)}(\Delta t) \cong P(n)\exp\left[-\frac{2\Delta t}{t_0}\frac{l}{l^*}n\right] = P(s)\exp\left[-\frac{2\Delta t}{t_0}\frac{s}{l^*}\right], \quad (4.125)$$

where the discrete variable n is converted into the continuous path-length variable $s = nl$ for ease of calculation later on.

A crucial step in the evaluation of $g_1(\Delta t)$ involves the calculation of $P(s)$. Consistent with the fact that waves behave diffusively after many scattering, it is reasonable to approximate $P(s)$ by the solution to the diffusion equation, i.e.,

$$P(s) = \left(\frac{1}{4\pi D^{(B)}t}\right)^{3/2} \exp\left(-\frac{r^2}{4D^{(B)}t}\right)$$

$$= \left(\frac{3}{4\pi sl^*}\right)^{3/2} \exp\left(-\frac{3r^2}{4sl^*}\right). \qquad (4.126)$$

Here t is the photon travel time (which is very small because v_t is so large), and in the last equality $D^{(B)}t$ is expressed as $v_t t l^*/3 = s l^*/3$. One should be aware that although the expression (4.126) is very similar to Eq. (4.121), physically (4.121) is for the *diffusion of suspended particles*, whereas (4.126) is for the *diffusion of light*, which are two very different phenomena. The diffusion property of light is embodied in l^* plus the form of $P(s)$ used. From Eqs. (4.120a) and (4.125), we have

$$g_1(\Delta t) = \sum_{n=1}^{\infty} g_1^{(n)}(\Delta t) = \int_0^{\infty} ds\, P(s) \exp\left(-\frac{2\Delta t}{t_0}\frac{s}{l^*}\right)$$

$$= \int_0^{\infty} ds \left(\frac{3}{4\pi sl^*}\right)^{3/2} \exp\left[-\frac{3r^2}{4sl^*} - \frac{2\Delta t}{t_0}\frac{s}{l^*}\right]. \qquad (4.127)$$

The dominant behavior of $g_1(\Delta t)$ may be readily inferred from Eq. (4.127) because the expression in the exponent has a peak, located at the point $s = s_0$ where the derivative of the exponent vanishes, or

$$\frac{3r^2}{4l^*}\frac{1}{s_0^2} - \frac{2\Delta t}{t_0}\frac{1}{l^*} = 0,$$

from which one obtains

$$s_0 = r\sqrt{\frac{3t_0}{8\Delta t}}. \qquad (4.128)$$

Since the peak is in the exponent, the position of the peak dominates the behavior of $g_1(\Delta t)$. Therefore, substitution of s_0 back into the exponent

4.8 Application: Diffusive Wave Spectroscopy

gives

$$g_1(\Delta t) \sim \exp\left[-\frac{r}{l^*}\sqrt{\frac{6\Delta t}{t_0}}\right], \qquad (4.129)$$

which implies

$$g_2(\Delta t) = |g_1(\Delta t)|^2 \sim \exp\left[-\frac{2r}{l^*}\sqrt{\frac{6\Delta t}{t_0}}\right]. \qquad (4.130)$$

Equation (4.128) and the resulting form of $g_1(\Delta t)$ can be interpreted physically as follows. For Δt small, the dominant path length s_0 is large [as seen from (4.128)] because at short times the suspended particles move very little, and only through the long paths can these little changes add up to anything that can cause the decay of the autocorrelation function. One has to remember that light travels essentially instantaneously compared with the movement of the particles; therefore even a small Δt on the scale of t_0 means an almost infinite travel time for light. Through this interpretation it is easy to see that one can measure the displacements of the suspended particles that are much smaller than the light wavelength, although in this case the measurement represents an average over many particles rather than that of a single particle, and therefore does not represent a violation of the wave resolution limit.

The form of Eq. (4.129), the exponential decay with a square-root dependence on Δt, is noted to arise from the competition between the term that describes particle diffusion, $\exp[-(2\Delta t/t_0)s/l^*]$, which favors short paths because they dephase much less, and the term that describes light diffusion, $\exp[-3r^2/4sl^*]$, which favors long path because there are so many more of them. The final result represents a compromise between the two.

To compare theory with experimental measurement, the geometry of the experiment has to be taken into account. For this purpose it is much more convenient to retain the form of $g_1(\Delta t)$ as

$$g_1(\Delta t) = \int_0^\infty ds\, P(s) \exp\left(-\frac{2\Delta t}{t_0}\frac{s}{l^*}\right), \qquad (4.131)$$

where $P(s)$ has to be obtained from the solution to a boundary value problem of the diffusion equation. Equation (4.131) states that g_1 is the Laplace transform of $P(s)$. It turns out that the relevant mathematics is more accessible in the Laplace-transformed domain. Therefore, let us now solve for $g_1(\Delta t)$ directly with the geometry in which a coherent laser beam is uniformly incident on the front surface of a slab sample with thickness

L, extending from $z = 0$ to $z = L$, and measurement is done either at the front surface (reflected light) or the back surface (transmitted light). Before the calculation, however, it is necessary to clarify two points.

The first point is that the measured quantity is the *flux* of diffusive light. In the solution to Problem 4.5 it is shown that the flux of diffusing photons in the across the sample boundary $+z$ direction is given by

$$J_+ = \frac{Pv_t}{4} - \frac{D^{(B)}}{2}\left(\frac{\partial P}{\partial z}\right). \tag{4.132a}$$

Similarly, the flux in the $-z$ direction is given by

$$J_- = \frac{Pv_t}{4} + \frac{D^{(B)}}{2}\left(\frac{\partial P}{\partial z}\right). \tag{4.132b}$$

Here P is the *diffusing* photon density. It is a function of photon path length and detection location; i.e., P is a function of s and $\mathbf{r}$, and $D^{(B)} = v_t l^*/3$ is the light diffusion constant. The boundary condition at $z = L$ is that there should be no net flux of diffusing photons coming into the sample from the outside. The only J_- measured inside the sample has to be that reflected from the sample surface at $z = L$. If the reflection coefficient is ρ, then at $z = L$

$$J_- = \rho J_+,$$

which means

$$P + \left(\frac{2l^*}{3}\frac{1+\rho}{1-\rho}\right)\frac{\partial P}{\partial z} = 0. \tag{4.133}$$

If this expression is used to eliminate $\partial P/\partial z$ in J_+, then the result is

$$J_+ = \frac{Pv_t}{2(1+\rho)} \tag{4.134a}$$

at $z = L$. Similarly,

$$J_- = \frac{Pv_t}{2(1+\rho)} \tag{4.134b}$$

at $z = 0$. Therefore, what can be measured is proportional to P. This discussion also clarifies the boundary conditions at $z = 0, L$, which may be

4.8 Application: Diffusive Wave Spectroscopy

summarized as

$$P + C \frac{\partial P}{\partial z} = 0 \quad \text{at } z = L, \quad (4.135a)$$

$$P - C \frac{\partial P}{\partial z} = 0 \quad \text{at } z = 0, \quad (4.135b)$$

where

$$C = \frac{2l^*}{3} \frac{1+\rho}{1-\rho}. \quad (4.135c)$$

The second point is about the source of the diffusing light. Although experimentally the source is a steady laser beam, in the solution for $P(s)$, which essentially counts the number of path with a given path length s, we need a time-dependent solution of P where the source is a pulse of light. It is noted that the parameter s, the path length, is directly proportional to the photon travel time t as $s = v_t t$. On the scale of t_0, the photon travel time t is minuscule for a given s. We will denote the photon travel time by s/v_t so as to distinguish it from the particle diffusion time. Another aspect of the source is its location. Since the light entering the sample is coherent, it must take a few scatterings to make it diffusive. Therefore the diffusive light source should be located at some distance $z = z_0 \approx l^*$ inside the sample.

With the above boundary conditions and the initial condition, the equation for P is

$$D^{(B)} \frac{\partial^2}{\partial z^2} P(z,s) - v_t \frac{\partial P(z,s)}{\partial s} = \delta(z-z_0)\delta\left(\frac{s}{v_t}\right), \quad (4.136)$$

where the dependence on the x, y coordinates is absent because both the incident beam and the measurement points are assumed to cover whole surfaces, so the problem becomes mathematically one-dimensional. The solution of Eq. (4.136) consists of the addition of two parts, $P = P^{(1)} + P^{(2)}$. One part is that of the Green's function in an infinite 1D medium, given by the well-known solution

$$P^{(1)}(z,t) = \frac{1}{(4\pi D^{(B)} s/v_t)^{1/2}} \exp\left[-(z-z_0)^2 v_t/4D^{(B)} s\right]. \quad (4.137)$$

The other part of the solution, $P^{(2)}$, is that of the homogeneous diffusion equation [i.e., right-hand side of Eq. (4.136) is zero], with coefficients that may be adjusted to satisfy the boundary conditions at $z = 0, L$ given by Eqs. (4.135a, b). Since in the end it is the Laplace-transformed solution that is needed, the problem should be reformulated directly in the Laplace-transformed domain, where $P^{(2)}$ is easy to obtain. If

$$\bar{P}^{(1,2)}(z,q) = \int_0^\infty P^{(1,2)}(z,s)\exp(-qs)\,ds, \qquad (4.138a)$$

then

$$q = \frac{2\,\Delta t}{t_0 l^*} \qquad (4.138b)$$

by comparison with Eq. (4.131). It is straightforward to get

$$\bar{P}^{(1)} = \exp\left[-\frac{1}{l^*}\sqrt{\frac{6\,\Delta t}{t_0}}|z-z_0|\right]\bigg/\frac{2}{3}\sqrt{\frac{6\,\Delta t}{t_0}} \qquad (4.139)$$

by either performing the integral, Eq. (4.138a), or looking up a Laplace transform table. $\bar{P}^{(2)}$ is the solution of

$$D^{(B)}\frac{\partial^2}{\partial z^2}\bar{P}^{(2)} - v_t q \bar{P}^{(2)} = 0, \qquad (4.140)$$

or

$$\bar{P}^{(2)} = A\cosh\left(\sqrt{\frac{6\,\Delta t}{t_0}}\frac{z}{l^*}\right) + B\sinh\left(\sqrt{\frac{6\,\Delta t}{t_0}}\frac{z}{l^*}\right). \qquad (4.141)$$

The transformed boundary conditions for $\bar{P} = N_0(\bar{P}^{(1)} + \bar{P}^{(2)})$ are

$$\bar{P} + C\frac{\partial \bar{P}}{\partial z} = 0 \quad \text{at } z = L,$$

$$\bar{P} - C\frac{\partial \bar{P}}{\partial z} = 0 \quad \text{at } z = 0.$$

The normalization constant N_0 assures that at $\Delta t = 0$, $\bar{P} = 1$. The constants A and B are uniquely fixed by the two boundary conditions, and the

4.8 Application: Diffusive Wave Spectroscopy

results for reflection measurement are given by

$$g_1(\Delta t) = \bar{P}(\Delta t, z = 0)$$

$$= \frac{3L + [(1 + \rho)/(1 - \rho)]4l^*}{3(L - z_0) + [(1 + \rho)/(1 - \rho)]2l^*}$$

$$\times \left\{ \frac{\sinh\left[\sqrt{\frac{6\Delta t}{t_0}} \frac{L - z_0}{l^*}\right] + \frac{1 + \rho}{1 - \rho}\sqrt{\frac{8\Delta t}{3t_0}} \cosh\left[\sqrt{\frac{6\Delta t}{t_0}} \frac{L - z_0}{l^*}\right]}{\left[1 + \left(\frac{1 + \rho}{1 - \rho}\right)^2 \frac{8\Delta t}{3t_0}\right] \sinh\left[\sqrt{\frac{6\Delta t}{t_0}} \frac{L}{l^*}\right] + \left(2\frac{1 + \rho}{1 - \rho}\sqrt{\frac{8\Delta t}{3t_0}}\right) \cosh\left[\sqrt{\frac{6\Delta t}{t_0}} \frac{L}{l^*}\right]} \right\}. \quad (4.142)$$

It is easily checked that $g_1(0) = 1$, as it should be. For a sample of infinite thickness, $L = \infty$, the expression simplifies to

$$g_1(\Delta t) = \frac{\exp\left[-\sqrt{\frac{6\Delta t}{t_0}} \frac{z_0}{l^*}\right]}{1 + \frac{1 + \rho}{1 - \rho}\sqrt{\frac{8\Delta t}{3t_0}}}. \quad (4.143)$$

For small $\Delta t/t_0$, the time dependence in the denominator means that $g_1(\Delta t)$ behaves as $\exp[-c_0\sqrt{6\Delta t/t_0}]$, where

$$c_0 \cong \frac{z_0}{l^*} + \frac{2}{3}\frac{1 + \rho}{1 - \rho}. \quad (4.144)$$

The value of c_0 is known to be on the order of 2. For example, if reflectively $\rho \cong 0.4$ and $z_0 \cong l^*$, then $c_0 \cong 2.5$. The value of $\rho \cong 0.4$ is not unreasonable for a ratio of indices of refraction of ~ 1.4 at the two sides of a sample interface. By knowing c_0, t_0 (and hence D_p) is uniquely determined from the reflection measurements. In Figure 4.9 it is shown that the experimentally measured $g_2(\Delta t)$ follows very well the predicted behavior.

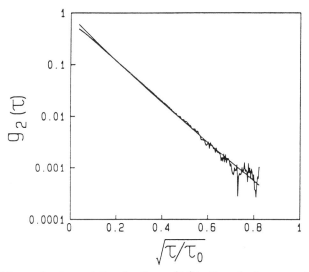

Figure 4.9 Measured autocorrelation function $g_2(\Delta t)$ in the reflection geometry. It is seen that the data are well described by an exponential $\sqrt{\Delta t}$ decay, as predicted. The slope directly gives the value of c_0. The figure is reproduced from Pine *et al.* (1990b). In the figure, Δt is denoted by τ, t_0 by τ_0. The sample is an L = 5-mm-thick cell with 5% 0.412-μm-diameter polystyrene spheres. The solid line is the theoretical fit to the data.

One can also measure the transmitted intensity of the diffusing light. The predicted form of $g_1(\Delta t)$ is

$$g_1(\Delta t) = \bar{P}(\Delta t, z = L) = \frac{L/l^* + (4/3)(1+\rho)/(1-\rho)}{z_0/l^* + (2/3)(1+\rho)/(1-\rho)}$$

$$\times \left\{ \frac{\sinh\left[\sqrt{\frac{6\Delta t}{t_0}}\frac{z_0}{l^*}\right] + \frac{1+\rho}{1-\rho}\sqrt{\frac{8\Delta t}{3t_0}}\cosh\left[\sqrt{\frac{6\Delta t}{t_0}}\frac{z_0}{l^*}\right]}{\left[\left(\frac{1+\rho}{1-\rho}\right)^2\frac{8\Delta t}{3t_0}+1\right]\sinh\left[\sqrt{\frac{6\Delta t}{t_0}}\frac{L}{l^*}\right] + 2\frac{1+\rho}{1-\rho}\sqrt{\frac{8\Delta t}{3t_0}}\cosh\left[\sqrt{\frac{6\Delta t}{t_0}}\frac{z_0}{l^*}\right]} \right\}. \quad (4.145)$$

In this case L/l^* is a sensitive parameter of the solution. Therefore l^* may be obtained by performing experiments on samples of different thicknesses. For light scattering in a solution with a 1% concentration of 0.5-μm-diameter polystyrene spheres, l^* has been measured to be $\cong 180$ μm, implying a light diffusion constant of $D^{(B)} \cong 2 \times 10^8$ cm^2/sec. In strong-scattering media, values of $D^{(B)}$ as low as 10^5–10^6 cm^2/sec have been observed (Genack, 1990).

Problems and Solutions

4.1. Derive an expression for the spatial spectral decomposition of the configurationally averaged $\langle G^+ G^- \rangle_c$.

From definitions,

$$G^+\left(\omega + \frac{\Delta\omega}{2}, \mathbf{r}, \mathbf{r}'\right) = \frac{L^{2d}}{N^2(2\pi)^{2d}} \iint d\mathbf{k}\, d\mathbf{k}'$$
$$\times \exp[i(\mathbf{k}\cdot\mathbf{r}) - i(\mathbf{k}'\cdot\mathbf{r}')]\langle \mathbf{k}|G^+(\omega^+)|\mathbf{k}'\rangle, \quad (P4.1)$$

$$G^-\left(\omega - \frac{\Delta\omega}{2}, \mathbf{r}', \mathbf{r}\right) = \frac{L^{2d}}{N^2(2\pi)^{2d}} \iint d\mathbf{k}_1\, d\mathbf{k}'_1$$
$$\times \exp[i(\mathbf{k}'_1\cdot\mathbf{r}') - i(\mathbf{k}_1\cdot\mathbf{r})]\langle \mathbf{k}'_1|G^-(\omega^-)|\mathbf{k}_1\rangle,$$
$$(P4.2)$$

Let $\mathbf{R}_c = (\mathbf{r} + \mathbf{r}')/2$ so that

$$\mathbf{r} = \mathbf{R}_c + \frac{(\mathbf{r} - \mathbf{r}')}{2},$$

$$\mathbf{r}' = \mathbf{R}_c - \frac{(\mathbf{r} - \mathbf{r}')}{2}.$$

As seen from Problem 3.1, configurational averaging may be written as

$$\langle G^+ G^- \rangle_\mathbf{r} = \frac{L^{4d}}{N^3(2\pi)^{4d}} \cdot \iiiint d\mathbf{k}\, d\mathbf{k}'\, d\mathbf{k}'_1\, d\mathbf{k}_1$$
$$\times \frac{1}{L^d} \int d\mathbf{R}_c \langle G^+ G^- \rangle_\mathbf{k} \exp[i(\mathbf{k} - \mathbf{k}' + \mathbf{k}'_1 - \mathbf{k}_1)\cdot \mathbf{R}_c]$$
$$\times \exp[i(\mathbf{k} + \mathbf{k}' - \mathbf{k}'_1 - \mathbf{k}_1)\cdot(\mathbf{r} - \mathbf{r}')/2], \quad (P4.3)$$

where $\langle\ \rangle_\mathbf{r}$ denotes the configurational average of the product of the left-hand sides of (P4.1) and (P4.2) in real space, $\langle\ \rangle_\mathbf{k}$ denotes the average of the product of the two corresponding quantities in the integrals of (P4.1) and (P4.2) in **k** space. Integration over $\mathbf{R}_c$ yields $(2\pi)^d \delta(\mathbf{k} - \mathbf{k}' + \mathbf{k}'_1 - \mathbf{k}_1)$. Integration with respect to $\mathbf{k}_1$ then gives

$$\langle G^+G^-\rangle_{\mathbf{r}} = \frac{L^{3d}}{N^3(2\pi)^{3d}} \cdot \iiint d\mathbf{k}\, d\mathbf{k}'\, d\mathbf{k}'_1$$
$$\times \langle G^+G^-\rangle_{\mathbf{k}} \exp[i(\mathbf{k}' - \mathbf{k}'_1) \cdot (\mathbf{r} - \mathbf{r}')]. \quad (P4.4)$$

One can introduce a new set of variables that satisfy the delta function constraint automatically. That is to call $\mathbf{k}' - \mathbf{k}'_1 = \mathbf{k} - \mathbf{k}_1 = \Delta\mathbf{k}$, then

$$\mathbf{k} \to \mathbf{k} + \frac{\Delta\mathbf{k}}{2} = \mathbf{k}_+$$

$$\mathbf{k}' \to \mathbf{k}' + \frac{\Delta\mathbf{k}}{2} = \mathbf{k}'_+$$

$$\mathbf{k}_1 \to \mathbf{k} - \frac{\Delta\mathbf{k}}{2} = \mathbf{k}_-$$

$$\mathbf{k}'_1 \to \mathbf{k}' - \frac{\Delta\mathbf{k}}{2} = \mathbf{k}'_-, \quad (P4.5)$$

and

$$\langle G^+G^-\rangle_{\mathbf{r}} = \frac{L^{3d}}{N^3(2\pi)^{3d}} \cdot \iiint \langle G^+G^-\rangle_{\mathbf{k}}$$
$$\times \exp[i(\Delta\mathbf{k} \cdot (\mathbf{r} - \mathbf{r}'))]\, d\mathbf{k}\, d\mathbf{k}'\, d\Delta\mathbf{k}. \quad (P4.6)$$

It is noted that whereas only $\Delta\mathbf{k}$ appears in the exponent, $\langle G^+G^-\rangle_{\mathbf{k}}$ in general can depend on $\mathbf{k}$, $\mathbf{k}'$, as well as $\Delta\mathbf{k}$. However, $\langle G^+G^-\rangle_{\mathbf{r}}$ depends only on $\mathbf{r} - \mathbf{r}'$ because for intensity transport, G^+ and G^- have the same source and detection points [see Eq. (4.1)]. However, if the quantity to be averaged is $G^+(\omega^+, \mathbf{r}_+, \mathbf{r}'_+)G^-(\omega^-, \mathbf{r}'_-, \mathbf{r}_-)$, where $\mathbf{r}'_+ \neq \mathbf{r}'_-$, $\mathbf{r}_+ \neq \mathbf{r}_-$, then substitution of these **r**'s into Eqs. (P4.1) and (P4.2) gives for the general case:

$$\langle G^+G^-\rangle_{\mathbf{r}} = \frac{L^{3d}}{N^3(2\pi)^{3d}} \cdot \iiint \langle G^+G^-\rangle_{\mathbf{k}} \exp[i(\Delta\mathbf{k} \cdot \mathbf{R}_m)$$
$$+ i\mathbf{k} \cdot (\mathbf{r}_+ - \mathbf{r}_-) + i\mathbf{k}' \cdot (\mathbf{r}'_+ - \mathbf{r}'_-)]\, d\mathbf{k}\, d\mathbf{k}'\, d\Delta\mathbf{k}. \quad (P4.7)$$

Here $\mathbf{R}_m = [(\mathbf{r}_+ - \mathbf{r}'_+) + (\mathbf{r}_- - \mathbf{r}'_-)]/2$ is the "average" vector that bisects the two vectors linking the input and output sites of the + and − channels.

4.2. Show that $\nabla_k e(\mathbf{k})$ is an odd function of $\mathbf{k}$.
From the definition of $e(\mathbf{k})$, Eq. (2.37), we have

$$\nabla_k e(\mathbf{k}) = \frac{2}{a}\left(\sin k_x a\,\hat{\mathbf{i}} + \sin k_y a\,\hat{\mathbf{j}} + \sin k_z a\,\hat{\mathbf{k}}\right), \tag{P4.8}$$

where $\hat{\mathbf{i}}, \hat{\mathbf{j}}, \hat{\mathbf{k}}$ are unit vectors along the x, y, z directions, respectively. Each component of the vector is an odd function of its argument. Therefore, if $\mathbf{k} \to -\mathbf{k}$, we have $\nabla_k e(-\mathbf{k}) = -\nabla_k e(\mathbf{k})$, i.e., $\nabla_k e(\mathbf{k})$ is an odd function of $\mathbf{k}$.

4.3. Show that the configurationally averaged value of $\mathbf{L}^\pm$ is given by $\tilde{\mathbf{G}}_e^\pm$, where $\mathbf{L}^\pm$ is defined by Eq. (4.79), and $\tilde{\mathbf{G}}_e^\pm$ is $\mathbf{G}_e^\pm$ restricted on the sites of one particle only.
From the CPA condition,

$$\left\langle (\mathbf{v}_i - \tilde{\Sigma})\left[\mathbf{I} - \tilde{\mathbf{G}}_e(\mathbf{v}_i - \tilde{\Sigma})\right]^{-1}\right\rangle_c = 0,$$

one can factor out $\tilde{\mathbf{G}}_e^{-1}$ to get

$$\left\langle (\mathbf{v}_i - \tilde{\Sigma})\left[\tilde{\mathbf{G}}_e^{-1} - (\mathbf{v}_i - \tilde{\Sigma})\right]^{-1}\right\rangle_c = 0, \tag{P4.9}$$

where the $\pm$ superscripts have been suppressed. Equation (P4.9) may be alternatively written as

$$\left\langle -\mathbf{I} + \tilde{\mathbf{G}}_e^{-1}\left[\tilde{\mathbf{G}}_e^{-1} - (\mathbf{v}_i - \tilde{\Sigma})\right]^{-1}\right\rangle_c = 0, \tag{P4.10}$$

or

$$\langle \mathbf{L}^\pm \rangle_c = \tilde{\mathbf{G}}_e^\pm$$

as asserted.

4.4. Derive the Siegert relation

$$\frac{\langle I(\Delta t)I(0)\rangle_c}{\langle I\rangle_c^2} = 1 + g_2(\Delta t),$$

where $g_2(\Delta t) = |g_1(\Delta t)|^2$.
Expressed in terms of amplitudes,

$$\langle I(\Delta t)I(0)\rangle_c = \langle \phi(\Delta t)\phi^*(\Delta t)\phi(0)\phi^*(0)\rangle_c. \tag{P4.11}$$

174   Diffusive Waves 4

Each $\phi(\Delta t)$ may be expressed as the sum of amplitudes resulting from different scattering paths. For example,

$$\phi(\Delta t) = \sum_i \phi_i(\Delta t), \tag{4.12}$$

where the index i denotes a particular scattering path. Therefore,

$$\langle I(\Delta t)I(0)\rangle_c = \sum_{i,j,k,l} \langle \phi_i(\Delta t)\phi_j^*(\Delta t)\phi_k(0)\phi_l^*(0)\rangle_c. \tag{P4.13}$$

The sum is zero if one index is different from the other three, since that term may be averaged independently, and $\langle \phi_i\rangle_c = 0$ because of random phase addition. Therefore, we are left with the possibility of $i = j$, $k = l$, or $i = l$, $j = k$, or $i = k$, $j = l$. The last case still gives zero because the phases do not cancel unless ϕ is paired with ϕ^*. Also, there is the term $i = j = k = l$. However, compared with the number of other terms the contribution of this term is negligible. Therefore,

$$\langle I(\Delta t)I(0)\rangle_c \cong \langle I(\Delta t)\rangle_c \langle I(0)\rangle_c$$
$$+ \langle \phi(\Delta t)\phi^*(0)\rangle_c \langle \phi^*(\Delta t)\phi(0)\rangle_c$$
$$= \langle I\rangle_c^2 + |\langle \phi(\Delta t)\phi^*(0)\rangle_c|^2. \tag{P4.14}$$

Dividing by $\langle I\rangle_c^2$ on both sides and recalling that

$$g_1(\Delta t) = \frac{\langle \phi(\Delta t)\phi^*(0)\rangle_c}{\langle I\rangle_c},$$

we get the Siegert relation directly from Eq. (P4.14).

4.5. Derive an expression for the diffusive particle flux across the surface defined by $z = 0$.

Consider the flux in the $-z$ direction across a surface element in the neighborhood of the origin. If the number density of diffusive particles at $\mathbf{r}$ is denoted by $P(\mathbf{r})$, then the contribution to the flux J from a small volume element centered at $\mathbf{r}$ is given by

$$P(\mathbf{r})d\Omega(v_t \cos\theta)\exp(-r/l^*)/4\pi l^*.$$

Here r is the distance of the volume element to the surface element, $\exp(-r/l^*)/l^*$ accounts for the probability of no scattering in traversing the distance r, $v_t \cos\theta$ is the projected z component velocity, and $d\Omega/4\pi$

gives the probability for a velocity in that particular direction, $d\Omega$ being the solid angle differential $d\phi \sin\theta\, d\theta$, where θ is defined relative to the surface normal. The total flux is therefore given by

$$J_- = \frac{v_t}{4\pi l^*} \int_0^{2\pi} d\phi \int_0^{\pi/2} d\theta \int_0^\infty dr\, P(\mathbf{r}) \cos\theta \sin\theta \exp(-r/l^*)$$

$$= \frac{v_t}{2} \int_0^{\pi/2} d\theta \int_0^\infty dr\, [P(0) + \mathbf{r}\cdot(\nabla P(\mathbf{r}))_{\mathbf{r}=0}] \cos\theta \sin\theta \exp(-r/l^*)$$

$$= \frac{v_t}{4} P(0) + \frac{v_t l^*}{6} \left.\frac{\partial P}{\partial z}\right|_{z=0}$$

$$= \frac{v_t}{4} P(0) + \frac{D^{(B)}}{2} \left.\frac{\partial P}{\partial z}\right|_{z=0}. \tag{P4.15}$$

In the equation above, it should be noted that ∇P has only the z-component. For J_+, the only difference is in the second term, where $\mathbf{r} \to -\mathbf{r}$ (due to the fact that the volume contributing to the flux has to be on the other side of the surface) results in a negative sign. Therefore

$$J_+ = \frac{v_t P(0)}{4} - \frac{D^{(B)}}{2} \left.\frac{\partial P}{\partial z}\right|_{z=0}. \tag{P4.16}$$

References

Genack, A. Z. (1990). In "Scattering and Localization of Classical Waves in Random Media," (P. Sheng, ed.), p. 207. World Scientific Publishing.
Greenwood, A. D. (1958). *Proc. Phys. Soc.* **71**, 585.
Kubo, R. (1956). *Can. J. Phys.* **34**, 1274.
Mahan, G. D. (1981). "Many-Particle Physics," Plenum, New York.
Pine, D. J., Weitz, D. A., Zhu, J. X., and Herbolzheimer, E. (1990a), *J. Phys. France* **18**, 2101.
Pine, D. J., Weitz, D. A., Maret, G., Wolf, P. W., Herbolzheimer, E., and Chaiken, P. M. (1990b). In "Scattering and Localization of Classical Waves in Random Media," (P. Sheng, ed.), p. 312. World Scientific Publishing.
van Albada, M. P., van Tiggelen, B. A., Lagendijk, A., and Tip, A. (1991). *Phys. Rev. Lett.* **66**, 3132.
Ward, J. C. (1950). *Phys. Rev.* **78**, 182.

5
The Coherent Backscattering Effect

5.1 Wave Diffusion versus Classical Diffusion

In the last chapter it was argued that wave intensity transport is diffusive in character after multiple scattering. The purpose of this and subsequent chapters is to point out the inadequacies of this description and the interesting and rich phenomena that can arise from the nonclassical character of wave transport.

The difference between classical diffusion and wave diffusion may be easily demonstrated. Take a slab of strong-scattering material and illuminate its front surface by a coherent laser beam. If only the total transmitted light is measured, then the measured value and its dependence on sample thickness should be well described by the solution to the diffusion equation with appropriate boundary conditions. However, whereas the diffusion equation predicts the transmitted intensity to have smooth spatial variation, in reality the transmitted intensity is not smooth but consists of interlaced bright and dark spots visible when the transmitted light is projected onto a screen, e.g., a sheet of paper. These intensity variations are called the speckle pattern, which is a result of wave interference in the absence of inelastic scattering. The essential point is, therefore, that the diffusive character and the interference effect are not mutually exclusive in the absence of inelastic scattering. In the case of the speckle pattern, its effect shows up after configurational averaging as increased magnitude of *intensity fluctuation* at a given spatial point, when compared to classical diffusion.

The coherent backscattering is another interference effect, sometimes also known as the weak localization effect, that manifests itself at the level of configurationally averaged *mean intensity*. It can therefore directly affect the diffusive behavior when its dynamic ramifications are taken into account. The purpose of this chapter is to introduce the effect in its static version.

5.2 Coherence in the Backscattering Direction

Figure 5.1 is a schematic illustration showing two ray paths in a scattering medium, composed of fixed, randomly placed scatterers. Both rays have the same frequency. Ray path A has amplitude A_i at some far-away source point $\mathbf{R}_0$ and wave vector $\mathbf{k}_i$. After n scatterings, ray path A exits from the front surface with wave vector $\mathbf{k}_f$. The spatial locations of the n scatterers are labeled consecutively as $\mathbf{r} = \mathbf{r}_1, \mathbf{r}_2, \mathbf{r}_3, \ldots, \mathbf{r}' = \mathbf{r}_n$. Ray path B has amplitude $B_i = A_i$ at the same source point $\mathbf{R}_0$ and the same incident and exit wave vectors $\mathbf{k}_i$, $\mathbf{k}_f$ as A. It also experiences n scatterings, but in exactly the reverse order to ray path A. Namely, ray path B starts with $\mathbf{r}' = \mathbf{r}_n$ and ends with $\mathbf{r} = \mathbf{r}_1$. Since all scatterings are elastic, the complex amplitude A_0 of ray path A at some far-away

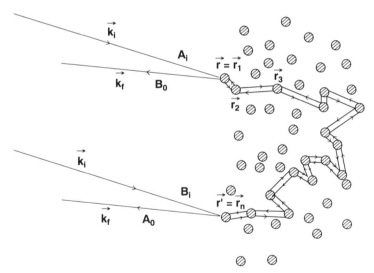

Figure 5.1 A schematic illustration of two interfering rays A and B. Both rays have the same incident wave vector $\mathbf{k}_i$ and the exit wave vector $\mathbf{k}_f$. The scattering sites of ray path A are labeled consecutively as $\mathbf{r} = \mathbf{r}_1, \mathbf{r}_2, \ldots \mathbf{r}_n = \mathbf{r}'$. Ray path B has exactly the reverse scattering order. This choice is made purposely by following the time reversal rule in which ray path B may be regarded as ray path A propagating backward in time between $\mathbf{r}_1$ and $\mathbf{r}_n$. If the system is time reversal invariant, A and B are perfectly coherent when $\mathbf{k}_f = -\mathbf{k}_i$. It should be noted that the backscattering direction is different from the direction of specular reflection. The two coincide only at normal incidence.

5.2 Coherence in the Backscattering Direction

observation point $\mathbf{r}_0$ may be related to A_i as

$$A_0 = A_i G_A \exp[i\mathbf{k}_i \cdot (\mathbf{r}_1 - \mathbf{R}_0) + i\mathbf{k}_{1,2} \cdot (\mathbf{r}_2 - \mathbf{r}_1) + \cdots$$
$$+ i\mathbf{k}_{n-1,n} \cdot (\mathbf{r}_n - \mathbf{r}_{n-1}) + i\mathbf{k}_f \cdot (\mathbf{r}_0 - \mathbf{r}_n)]. \quad (5.1)$$

Here $\mathbf{k}_{n-1,n}$ is the wave vector of ray propagation between scatterers $n-1$ and n, and G_A is a factor that accounts for the scattering strength of each scatterer as well as the other effects (such as geometric spreading) associated with the fact that we have only written out explicitly the phase factors of the Green's functions, which are the mathematically accurate wave propagators between successive scatterers. Similarly, for ray path B, amplitude B_0 at the same far-away observation point $\mathbf{r}_0$ may be related to $B_i = A_i$ as

$$B_0 = A_i G_B \exp[i\mathbf{k}_i \cdot (\mathbf{r}_n - \mathbf{R}_0) + i\mathbf{k}_{n,n-1} \cdot (\mathbf{r}_{n-1} - \mathbf{r}_n) + \cdots$$
$$+ i\mathbf{k}_{2,1} \cdot (\mathbf{r}_1 - \mathbf{r}_2) + i\mathbf{k}_f \cdot (\mathbf{r}_0 - \mathbf{r}_1)], \quad (5.2)$$

where G_B is a factor similar to G_A. Since $\mathbf{k}_{n,n-1} = -\mathbf{k}_{n-1,n}$, it is clear that the sum of all the intervening phase factors, between the first term and the last term, is identical to that of Eq. (5.1). Also, $G_A = G_B$ because identical scatters are involved. Therefore,

$$\frac{A_0}{B_0} = \exp[i(\mathbf{k}_i + \mathbf{k}_f) \cdot (\mathbf{r} - \mathbf{r}')]. \quad (5.3)$$

From Fig. 5.1, Eq. (5.3) has the simple physical interpretation that B_0 has the extra phase $-\mathbf{k}_i \cdot (\mathbf{r} - \mathbf{r}')$ due to the extra path length at incidence, whereas A_0 has the extra phase $\mathbf{k}_f \cdot (\mathbf{r} - \mathbf{r}')$ due to the extra path length at exit. The two cancels if $\mathbf{k}_i = -\mathbf{k}_f$, and the two outgoing rays are exactly coherent in the direction opposite to the incident direction.

In actual scattering, the outgoing amplitude A_0 should be the sum of an infinite number of scattering paths, of which the one shown in Figure 5.1 is just a single example. For fixed $\mathbf{r}_1 = \mathbf{r}$, $\mathbf{r}_n = \mathbf{r}'$, and $\mathbf{k}_i$, $\mathbf{k}_f$, A_0 and B_0 should be

$$A_0 = \exp[i\mathbf{k}_i \cdot (\mathbf{r} - \mathbf{R}_0) + i\mathbf{k}_f \cdot (\mathbf{r}_0 - \mathbf{r}')] \sum_{\{\alpha\}} A_\alpha, \quad (5.4)$$

$$B_0 = \exp[i\mathbf{k}_i \cdot (\mathbf{r}' - \mathbf{R}_0) + i\mathbf{k}_f \cdot (\mathbf{r}_0 - \mathbf{r})] \sum_{\{\beta\}} B_\beta, \quad (5.5)$$

where A_i is set equal to 1, and α, β are the indices for the scattering paths. Under a certain assumption, it is possible to show that there is a one-to-one correspondence between the terms of the two sums, $\sum_\alpha A_\alpha$ and $\sum_\beta B_\beta$. This correspondence is already clear in the choice of ray paths A and B shown in Figure 5.1. Namely, for every path α, it is possible to find a *time-reversed* path β which has exactly the reverse order of scattering to α. Also, for every path β one can find a corresponding path α with the reverse order. Therefore, time reversal defines a one-to-one mapping between the paths in the set $\{\alpha\}$ and those in the set $\{\beta\}$. As a result, *if the system is assumed to be invariant under time reversal*, then

$$\sum_{\{\alpha\}} A_\alpha = \sum_{\{\beta\}} B_\beta,$$

and A_0/B_0 is given exactly by Eq. (5.3). The assumption of time reversal invariance requires, for example, that there be no uniform velocity for the scatterers so that the time-reversed path sees the same scatterers' positions. But since the wave speed is usually fairly high, the uniform velocity has to be comparable to that in order to make any appreciable difference. A more serious requirement is the absence of magnetic field in electronic systems, since the vector potential associated with the magnetic field can give different phases to a scattering path and its time-reversed counterpart. When the time reversal invariance is broken, the coherence effect in the backscattering direction is not totally destroyed. It is only diminished in proportion to the degree of time reversal symmetry breaking.

It is interesting to note that the coherent backscattering effect was "discovered" independently in two different fields: as the basis of weak localization of electrons in dirty metals in solid-state physics (Langer and Neal, 1966) and in optics as a reflection enhancement effect for electromagnetic wave propagation in a turbulent atmosphere (deWolf, 1971). The word "discovered" was put in quotation marks to mean within the realm of recent literature. In any case, the parallel discoveries do reflect the common wave character of the coherent backscattering effect, regardless of whether the wave is quantum or classical in nature. This similarity, however, does not carry over to the ease of actual demonstration. In the quantum case, the difficulty in observing an electronic wave function means there are so far only indirect demonstrations of the effect. But in the case of the electromagnetic wave there have been several experiments involving laser light reflection from disordered samples in which the coherent backscattering effect was directly observed and quantitatively measured (Tsang and Ishimaru, 1984; van Albada and Lagendijk, 1985; Wolf and Maret, 1985). Below we specialize to the classical scalar wave

5.3 Angular Profile of the Coherent Backscattering

case for a more explicit calculation of the effect in a fixed sample geometry.

5.3 Angular Profile of the Coherent Backscattering

Equation (5.3) implies that if a coherent beam is incident on a random media, its backscattered intensity from the two rays A and B is given by

$$|A_0 + B_0|^2 = |A_0|^2 + |B_0|^2 + A_0 B_0^* + A_0^* B_0$$
$$= |A_0|^2 |1 + \exp[-i(\mathbf{k}_i + \mathbf{k}_f) \cdot (\mathbf{r} - \mathbf{r}')]|^2$$
$$= 2\{1 + \cos[(\mathbf{k}_i + \mathbf{k}_f) \cdot (\mathbf{r} - \mathbf{r}')]\} |A_0|^2 \qquad (5.6)$$

Inside the curly brackets, the constant term is due to incoherent scattering, $|A_0|^2 + |B_0|^2$, and the $\cos[(\mathbf{k}_i + \mathbf{k}_f) \cdot (\mathbf{r} - \mathbf{r}')]$ term is due to the interference terms, $A_0 B_0^* + A_0 B_0$. It is seen that the coherence decreases as $\mathbf{k}_f$ moves away from $-\mathbf{k}_i$ and as $|\mathbf{r} - \mathbf{r}'|$ becomes large. At its maximum, i.e., $\mathbf{k}_i + \mathbf{k}_f = 0$, the coherence enhances the backscattering intensity by a factor of 2.

Equation (5.6) suggests a new way of looking at the reflection intensity. For three points, $\mathbf{r}$, $\mathbf{r}'$, and $\mathbf{r}''$, we have pairs of rays $A_0(\mathbf{r} \to \mathbf{r}')$ and $B_0(\mathbf{r}' \to \mathbf{r})$, $C_0(\mathbf{r} \to \mathbf{r}'')$ and $D_0(\mathbf{r}'' \to \mathbf{r})$, plus $E_0(\mathbf{r}' \to \mathbf{r}'')$ and $F_0(\mathbf{r}'' \to \mathbf{r}')$. The net intensity in the direction of $\mathbf{k}_f$ is obtained by the addition of the six ray amplitudes, squared. That means

$$|A_0 + B_0 + C_0 + D_0 + E_0 + F_0|^2 = |A_0 + B_0|^2 + |C_0 + D_0|^2$$
$$+ |E_0 + F_0|^2 + (A_0 + B_0)(C_0 + D_0)^* + (A_0 + B_0)^*(C_0 + D_0)$$
$$+ (A_0 + B_0)^*(E_0 + F_0) + (A_0 + B_0)(E_0 + F_0)^*$$
$$+ (C_0 + D_0)^*(E_0 + F_0) + (C_0 + D_0)(E_0 + F_0)^*.$$

There are two types of terms. The first type consists of square of the paired amplitudes, each giving a result similar to Eq. (5.6). Then there are the interference terms between the paired amplitudes, which would yield a reflected intensity profile that is particular to the configuration. In other words, the phase between any two paired amplitudes, e.g., $(C_0 + D_0)$ and $(E_0 + F_0)$, is expected to vary randomly from one configuration to the next because there is no reason for the scattering paths of two pairs to have any fixed relationship. Therefore, if configurational averaging is applied, only the first type of terms is expected to remain, whereas the second type would vanish. This configurational averaging effect has been demonstrated experimentally (Etemad *et al.*, 1986). By using light scattering from fixed scatterer configurations, it was found that for each configuration, there were may intensity peaks occurring at arbitrary reflection angles. This is exactly the speckle pattern mentioned earlier (in reflection in this case). However, after configurational averaging only the backscattering peak emerged, and all the other peaks vanished. The reason is that while there may be a large peak at $\theta' \neq 0$ for a given configuration, the intensity at θ' is likely to be small for most other configurations. At $\theta = 0$, however, the backscattering peak may be smaller than the other peaks for any given configuration, but it is always present in every configuration. Therefore, upon configurational averaging, whereas the other peaks diminish, the coherent backscattering peak persists. In other words, the speckle pattern gets smoothed over at the level of mean intensity, but the coherent backscattering peak does not. This fact implies that if it is the configurationally averaged result one desires (which may be realized physically if the scatterers are moving randomly, such as in a colloid due to the Brownian motion, and the measured quantity is the time-averaged intensity), then the sum of the amplitudes squared may be replaced by the sum of the *paired intensities*, where each pair sees a enhancement factor $[1 + \cos(\mathbf{k}_i + \mathbf{k}_f) \cdot (\mathbf{r} - \mathbf{r}')]$.

This simplification enables us to calculate the coherent backscattering angular profile for light reflected from a slab of disordered scattering medium, extending from $z = 0$ to $z = L$. Since the coherent enhancement depends on $\mathbf{r} - \mathbf{r}'$, the intensity addition has to take into account the amount $P(\mathbf{r}, \mathbf{r}')$ of the source intensity that a ray starting at $\mathbf{r}$ would end up at $\mathbf{r}'$. The quantity $P(\mathbf{r}, \mathbf{r}')$ may be approximated by the solution to the diffusion equation, just as in the case of DWS analyzed in the last chapter. The difference here is that $P(\mathbf{r}, \mathbf{r}')$ is not concerned with a given path *length*. Rather, it is the sum of *all* paths that end at $\mathbf{r}'$ and begin at $\mathbf{r}$. Since path length is proportional to time, the solution needed here is the time-dependent solution [Eq. (4.126)] *integrated over time*, i.e., the time-independent, *static* solutions to the diffusion equation. If the boundary conditions are ignored for the moment, then

$$P(\mathbf{r}, \mathbf{r}') = \int_0^\infty \frac{\exp\left[-(\mathbf{r} - \mathbf{r}')^2 v_t / 4 D^{(B)} s\right]}{\left(4\pi D^{(B)} s / v_t\right)^{3/2}} ds$$

$$= \frac{v_t}{4\pi D^{(B)} |\mathbf{r} - \mathbf{r}'|}. \tag{5.7}$$

5.3 Angular Profile of the Coherent Backscattering

$P(\mathbf{r},\mathbf{r}')$ is recognized to be the solution to the 3D Laplace equation (diffusion equation with the $\partial/\partial t$ term dropped) with a point source. Because of the requirement that the source be a *diffusive* wave source, $\mathbf{r}$ cannot be located on the sample surface but must be on a plane $z = z_0 \cong l^*$ inside the sample where the incident wave is likely to have experienced at least one scattering. In order to get a physical feeling first, let us simplify to the case of $L \to \infty$, with the boundary conditions at $z = 0$ [Eq. (4.135b)] given by

$$P - C\frac{\partial P}{\partial z} = 0,$$

where $C = (2l^*/3)(1 + \rho)/(1 - \rho)$, ρ being the reflection coefficient at the sample surface. C is also called the extrapolation length because if the solution is linearly extrapolated in the negative z direction, then P vanishes at $z = -C$. An approximate solution can be obtained from Eq. (5.7) by the method of images:

$$P(\mathbf{r},\mathbf{r}') = \frac{v_t}{4\pi D^{(B)}} \left\{ \frac{1}{\left[(x-x')^2(y-y')^2 + (z_0-z')^2\right]^{1/2}} \right.$$

$$\left. - \frac{1}{\left[(x-x')^2(y-y')^2 + (z'+z_0+2C)^2\right]^{1/2}} \right\}, \quad (5.8)$$

where it is seen that at $z' = -C$ the condition of $P = 0$ is guaranteed by the subtraction of an "image" source term located at $\mathbf{r} = (x, y, -2C - z_0)$. Since $\mathbf{r}'$ represents the position of the last scattering, it is reasonable to assume that $z' \cong z_0$. Then

$$P(\mathbf{r},\mathbf{r}') = \frac{3}{4\pi l^*} \left\{ \frac{1}{r_\parallel} - \frac{1}{\left[r_\parallel^2 + 4(z_0 + C)^2\right]^{1/2}} \right\}, \quad (5.9)$$

where $r_\parallel = [(x-x')^2 + (y-y')^2]^{1/2}$. By adding the paired intensities for those paths that start at $\mathbf{r}$ and end at all possible $\mathbf{r}'$, the total scattered intensity in the direction of $\mathbf{k}_f$ is obtained as

$$\frac{I(\mathbf{k}_i,\mathbf{k}_f)}{I_{\text{inc.}}} \cong \frac{3}{(4\pi)^2 l^*} \int d\mathbf{r}_\parallel \left\{ \frac{1}{r_\parallel} - \frac{1}{\left[r_\parallel^2 + 4(z_0 + C)^2\right]^{1/2}} \right\}$$

$$\times \{1 + \cos[(\mathbf{k}_i + \mathbf{k}_f) \cdot (\mathbf{r} - \mathbf{r}')]\}. \quad (5.10)$$

The result is noted to be independent of the source position $\mathbf{r}$. Here the extra factor of $1/4\pi$ gives the fraction of intensity scattered into the direction of $\mathbf{k}_f$ at $\mathbf{r}'$, $I_{inc.}$ is the incident intensity, and $\mathbf{r}_\parallel = (x - x', y - y')$. In the solution to Problem 5.1 it is shown that the integral yields

$$\frac{I(\mathbf{k}_i, \mathbf{k}_f)}{I_{inc.}} \cong \frac{3(z_0 + C)}{4\pi l^*} \left\{ 1 + \frac{1 - \exp[-2|\mathbf{k}_i + \mathbf{k}_f|(z_0 + C)]}{2|\mathbf{k}_i + \mathbf{k}_f|(z_0 + C)} \right\}. \quad (5.11)$$

Several interesting features of Eq. (5.11) should be noted. First, $z_0 + C$ is just some constant times l^*. Because $l^*|\mathbf{k}_i|$ is generally $\gg 1$, it follows that the factor $2|\mathbf{k}_i + \mathbf{k}_f|(z_0 + C)$ can vary from 0 (at the backscattering direction $\mathbf{k}_f = -\mathbf{k}_i$) to some value that is $\gg 1$. Thus Eq. (5.11) predicts an enhancement factor of 2 at the backscattering direction over the incoherent baseline, as anticipated.

Second, the coherent enhancement exists only over a fairly narrow angular cone whose width may be evaluated from Eq. (5.11). If the width is defined by the condition

$$2|\mathbf{k}_i + \mathbf{k}_f|(z_0 + C) = 1, \quad (5.12)$$

and the deviation of $\mathbf{k}_f$ from $-\mathbf{k}_i$ is denoted by an angle θ so that $\mathbf{k}_f + \mathbf{k}_i \approx 2\pi\theta/\lambda$, where λ is the wavelength, then the condition (5.12) implies an angular width of

$$\Delta\theta \cong \frac{\lambda}{4\pi l^*} \left(1 + \frac{2}{3} \frac{1+\rho}{1-\rho} \right)^{-1}, \quad (5.13)$$

where it is assumed that $z_0 = l^*$. The inverse relationship between $\Delta\theta$ and l^* is not surprising, because once $\mathbf{k}_f \neq -\mathbf{k}_i$, the dephasing becomes a function of separation between $\mathbf{r}$ and $\mathbf{r}'$, and the (incoherent) diffusive behavior is expected to be dominant for $|\mathbf{r} - \mathbf{r}'|$ larger than l^*. Thus one can add amplitudes coherently only for $|\mathbf{r}_\parallel - \mathbf{r}'_\parallel| \leq l^*$. The path difference between two rays emerging at an angle θ from the backscattering direction but with the separation $|\mathbf{r}_\parallel - \mathbf{r}'_\parallel| = l^*$ is roughly θl^*. Since constructive interference may be expected only when $\theta l^* \lesssim \lambda$, it follows that $\Delta\theta l^* \cong \lambda$ should be the condition for the angular width $\Delta\theta$ of the coherent enhancement cone, as given by Eq. (5.13). In general, for light scattering experiments $\Delta\theta$ is on the order of a fraction of a degree. The small magnitude of $\Delta\theta$ could be the reason that the definitive verification of the effect is only a relatively recent event.

The third feature to be noted from Eq. (5.11) is that at small θ, the decrease from the peak maximum is linear in θ, independent of the

5.4 Sample Size (Path Length) Dependence

azimuthal angle. Therefore, the enhancement angular profile has a cusp, i.e., a derivative discontinuity, at $\theta = 0$.

To make the calculation more accurate means relaxing the condition that $\mathbf{r}$ and $\mathbf{r}'$ must be at $z = z_0$, and using a general solution $P(\mathbf{r}, \mathbf{r}')$ for finite L. In the solution to Problem 5.2 it is shown that if the boundary conditions of Eq. (4.135) are approximated by the conditions that P vanishes at $z = -C$ and $z = L + C$, then

$$P(\mathbf{r}, \mathbf{r}') = \frac{3}{\pi l^*(L + 2C)} \sum_{n=1}^{\infty} \sin \frac{n\pi(z + C)}{L + 2C} \sin \frac{n\pi(z' + C)}{L + 2C}$$
$$\times K_0 \left(\frac{n\pi}{L + 2C} r_\parallel \right), \quad (5.14)$$

where K_0 denotes the zeroth order modified Bessel function of the second kind. Since $\mathbf{r}$ and $\mathbf{r}'$ are the sites of the first scattering and the last scattering, respectively, the probability for this to happen is given by

$$\frac{1}{l^*} \exp\left(\frac{-z}{l^*} \right)$$

for $\mathbf{r}$ being the first scattering position under normal incidence, and

$$\frac{1}{l^*} \exp\left(\frac{-z'}{l^*} \right)$$

for having no further scattering after $\mathbf{r}'$. Here we have assumed $\mathbf{k}_i$, $\mathbf{k}_f$ to be close to the normal to the front surface so that the distance projection factor $\cos\theta \cong 1$. By combining all the factors, including $1/4\pi$ for the fraction scattered into the direction $\mathbf{k}_f$, the final result is

$$\frac{I(\mathbf{k}_i, \mathbf{k}_f)}{I_{\text{inc.}}} = \frac{1}{4\pi(l^*)^2} \int d\mathbf{r}_\parallel \int_0^L dz \int_0^L dz' \exp\left(\frac{-z}{l^*} \right)$$
$$\times [1 + \cos(\mathbf{k}_i + \mathbf{k}_f) \cdot (\mathbf{r} - \mathbf{r}')] P(\mathbf{r}, \mathbf{r}') \exp\left(\frac{-z'}{l^*} \right). \quad (5.15)$$

Here $P(\mathbf{r}, \mathbf{r}')$ is understood to be given by Eq. (5.14).

5.4 Sample Size (Path Length) Dependence

The coherent backscattering angular profiles for $L = 1.4l^*$, $4l^*$, and $100l^*$, calculated from Eqs. (5.14) and (5.15), are shown in Figure 5.2. It is

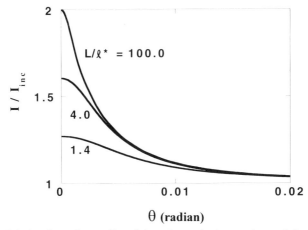

Figure 5.2 Calculated angular profiles of the coherent backscattering peak for three sample thicknesses $L/l^* = 1.4$, 4, and 100. $\theta = 0$ is the 180° backscattering direction. The wavelength λ is assumed to be that of the He-Ne laser light, 633 nm, and l^* is taken to be 11 μm. The coherent peak is seen to decrease in magnitude as the sample thickness decreases. The baseline of the profile, here denoted as 1, represents the incoherent component of the scattering from an $L/l^* = \infty$ sample. The incoherent baselines of the finite-thickness samples are exactly a factor of 2 lower than their respective coherent backscattering peaks.

seen that in the case of $L = 100l^*$, the calculated angular profile indeed has a sharp peak which decays quickly to the incoherent background, which is half the peak value. Here the incoherent background value 1 is noted to correspond to that for a semi-infinite sample. Figure 5.3 shows a comparison between a theoretically calculated profile and the experimental result for a semi-infinite sample of colloidal suspension (Akkermans *et al.*, 1986), where the random movement of the scattering particles and the time average of the measured intensity assures that the measured result is configurationally averaged. Here the value of $l^* = 19$ μm was measured independently, so the fit has no adjustable parameter (apart from making the baseline agree at the value 1). The fact that the peak value in Figure 5.3 is not 2 is due to the convolution of the theoretical result with a known instrumental resolution profile. The agreement is seen to be very good indeed.

For $L = 1.4l^*$ and $4l^*$, the theoretical profiles for the coherent enhancement are seen to have lower peak values when they are normalized by the incoherent background for a semi-infinite sample. This is a consequence of the path length dependence, or sample size dependence, of the coherent backscattering effect. That is, as the sample becomes thinner, the long backscattering paths are cut off from the effect. To have a better feel

5.4 Sample Size (Path Length) Dependence

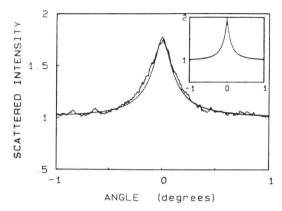

Figure 5.3 Comparison of the experimentally measured profile (wiggly line) and the theoretical prediction. The sample used in the experiment is a colloidal suspension with $l^* = 19\ \mu$m and a wavelength $\lambda = 515$ nm. The theoretical profile shown in the inset is convolved with a known instrumental resolution function to obtain the profile shown by the smooth solid line. The incoherent baseline of the profile is normalized to 1. The figure is from Akkermans et al. (1986).

for this path length dependence, let us refer to Eq. (5.7), where the static diffusion solution is expressed in terms of an integral over the solution for a given path length s. The integrand is a Gaussian where $|\mathbf{r} - \mathbf{r}'|$ decays over the distance $\sim \sqrt{l^* s}$ (which can be seen by expressing $D^{(B)} = v_t l^*/3$). The decay length is the same for the image solution, Eq. (5.9), if it were also expressed as the superposition of solutions for different path lengths. By substituting the integral form of the point source solution, Eq. (5.7), into Eq. (5.10), there would be an integrand involving the product

$$\exp\left[-3(\mathbf{r} - \mathbf{r}')^2/4l^*s\right]\cos[(\mathbf{k}_i + \mathbf{k}_f)\cdot(\mathbf{r} - \mathbf{r}')].$$

The path length dependence of the coherence enhancement effect is clear from this product. If $(\mathbf{k}_i + \mathbf{k}_f)$ is expressed as $2\pi\theta/\lambda$, then by using the decay length of $|\mathbf{r} - \mathbf{r}'|$, $\sqrt{l^*s}$, from the Gaussian factor, the condition for coherence is seen to be

$$\frac{2\pi\theta}{\lambda}\sqrt{l^*s} \leq \pi, \tag{5.16}$$

or

$$s \leq \frac{\lambda^2}{4\theta^2 l^*}. \tag{5.17}$$

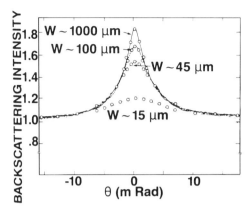

Figure 5.4 Experimentally measured angular profiles of the coherent backscattering peak for four sample thicknesses (denoted by W in the figure) of $L = 15$ μm, 45 μm, 100 μm, and 1000 μm. The samples used are colloidal suspensions with $l^* = 11$ μm, and the light wavelength is 633 nm (He-Ne laser). The three cases of $L = 15$ μm, 45 μm, and 1000 μm should be compared with the theoretical profiles shown in Figure 5.2 with similar L/l^* ratios. The figure is from Etemad *et al.* (1987).

Equation (5.17) tells us that at any given angle θ away from the backscattering direction, there is an upper limit to the path length which can contribute to the coherent backscattering effect. As θ approaches zero this upper limit diverges. The angular profiles shown in Figure 5.2 are entirely consistent with this interpretation. The thinner samples are seen to have lower peak values, but far away from the peak the profiles remain unchanged. The angles below which the profile of the finite-thickness samples deviate from the semi-infinite case may be estimated from Eq. (5.17). Since the number of scatterings for a diffusive ray to travel a net distance L is $3(L/l^*)^2$, where the factor 3 arises from the three directions, it follows that the path lengths being cut off are those with $s \geq 3(L/l^*)^2 l^* = 3L^2/l^*$. From Eq. (5.17) it follows that a sample of thickness L should have a backscattering profile which differs from the semi-infinite case at

$$\theta < \frac{\lambda}{2\sqrt{3}\,L}.$$

For $\lambda = 633$ nm (He-Ne laser) and $L = 50$ μm, the difference occurs for $\theta \leq 2.5$ milliradians from the backscattering direction. In Figure 5.4 the measured coherent backscattering profiles (Etemand *et al.*, 1987) are plotted for four sample thicknesses. They correspond remarkably well with what are expected theoretically.

An important point to be noted is that the incoherent background for a thin sample is always decreased by the *same* factor as the coherent peak, so that the ratio 2 between the coherent peak and the incoherent background is preserved for any sample thickness. As noted before, the decrease in the coherent peak seen in Figure 5.4 is relative to the incoherent background of the semi-infinite sample case.

Another means of eliminating the long scattering paths is through absorption which can be characterized by an inelastic scattering length l_{in}. However, there is a subtle difference between the sample size limitation and the limitation through absorption because absorption is a probabilistic process, whereas sample size limitation is deterministic. Therefore, for a finite sample size L the paths with lengths shorter than L are not affected. But if the sample has absorption, there is a finite probability for all paths to be affected, regardless of their lengths (of course, paths longer than l_{in} are expected to be affected more than those shorter than l_{in}). However, the difference between the two is actually fairly small, and the two mechanisms are roughly equivalent. If we regard paths with $s > l_{in}$ as cut off in the absorption case and those with $s > 3L^2/l^*$ as cut off in the finite sample thickness case, then a conversion between the two mechanisms may be achieved by using the relation

$$L = \sqrt{l_{in} l^*/3}. \tag{5.18}$$

The effect of inelastic scattering is addressed in more detail in Chapter 9.

The sample size dependence of the coherent backscattering effect foreshadows many interesting developments, including the scaling theory of localization. Together with the requirement of time reversal invariance (and the consequent degradation of the effect when the requirement is not satisfied), these two aspects of the coherent backscattering effect play an important role in providing indirect evidence of weak localization in dirty electronic systems. At the same time, they also provide the means for manipulating the wave localization effect through external control. These points are addressed in Chapters 9 and 10.

Problems and Solutions

5.1. Show that the integral of Eq. (5.10) gives the result of Eq. (5.11). The integral to be performed may be written as

$$\frac{3}{4\pi^2 l^*} \left\{ \int_0^{2\pi} d\phi \int_0^{\infty} \left(1 - \frac{r}{\sqrt{r^2 + a^2}}\right) dr + \int_0^{\infty} \int_0^{2\pi} \cos(\mathbf{q} \cdot \mathbf{r}) \right.$$

$$\left. \times \left[1 - \frac{r}{\sqrt{r^2 + a^2}}\right] d\phi \, dr \right\}.$$

where $\mathbf{q} = \mathbf{k}_i + \mathbf{k}_f$, $a = 2(z_0 + C)$, the two-dimensional $d\mathbf{r}_\|$ integral is written as $d\phi r\, dr$, where the subscript $\|$ on r has been dropped, and ϕ is the angle between $\mathbf{q}$ and $\mathbf{r}$. For the first term inside the bracket, a change of variable,

$$r = a \tan\theta,$$

converts the r integration into a simple θ integration:

$$\text{First term} = 2\pi a \int_0^{\pi/2} (\sec^2\theta - \sec\theta \tan\theta)\, d\theta$$

$$= 2\pi a \left[\frac{\sin\theta - 1}{\cos\theta}\right]_0^{\pi/2}$$

$$= 2\pi a. \tag{P5.1}$$

For the second term inside the bracket, the $d\phi$ integration yields $2\pi J_0(qr)$. As a result,

$$\text{Second term} = 2\pi \int_0^\infty J_0(qr)\, dr - 2\pi \int_0^\infty J_0(qr) \frac{r}{\sqrt{r^2 + a^2}}\, dr. \tag{P5.2}$$

Since the Laplace transform of $J_0(t)$ is $1/\sqrt{1+s^2}$, it follows that the first term of (P5.2) can be obtained from the Laplace transform with $s = 0$, which yields $2\pi/q$. The second term of (P5.2) can be obtained from a table of Hankel transforms. It has the value

$$\frac{2\pi}{q} \exp(-aq).$$

Collecting terms, we get

$$\frac{3}{(4\pi)^2 l^*} 2\pi a \left[1 + \frac{1 - \exp(-aq)}{aq}\right],$$

which is exactly the result shown by Eq. (5.11). The value of the expression inside the square brackets approaches 2 as $aq \to 0$ and approaches 1 as $aq \to \infty$.

5.2. Find the solution to the equation

$$\nabla_\mathbf{r}^2 P(\mathbf{r}, \mathbf{r}') = -\frac{v_t}{D^{(B)}} \delta(\mathbf{r} - \mathbf{r}')$$

with the boundary conditions that $P(\mathbf{r}, \mathbf{r}') = 0$ at $z = -C, L + C$.

Without the boundary conditions, $P(\mathbf{r},\mathbf{r}')$ is given by Eq. (5.7). With the boundary conditions, it is advantageous to solve the problem in cylindrical coordinates, where the equation may be cast in the form

$$\nabla_{\mathbf{r}}^2 P(\mathbf{r},\mathbf{r}') = -\frac{v_t}{D^{(B)}} \frac{1}{r_\|} \delta(r_\| - r_\|') \delta(\phi - \phi') \delta(z - z'). \quad (P5.3)$$

The boundary conditions at $z = -C, L + C$ suggest that $\delta(z - z')$ should be expanded in terms of the complete orthonormal functions of the form

$$\sqrt{\frac{2}{L + 2C}} \sin \frac{n\pi(z + C)}{L + 2C}, \quad n = 1, 2, 3, \ldots.$$

In other words,

$$\delta(z - z') = \frac{2}{L + 2C} \sum_{n=1}^{\infty} \sin \frac{n\pi(z + C)}{L + 2C} \sin \frac{n\pi(z' + C)}{L + 2C}. \quad (P5.4)$$

Equation (P5.4) is easily verified by integration with $\sin m\pi(z' + C)/(L + 2C)$, for example. It is a statement of the completeness of the expansion functions. Similarly,

$$\delta(\phi - \phi') = \frac{1}{2\pi} \sum_{m=-\infty}^{\infty} \exp[im(\phi - \phi')]. \quad (P5.5)$$

By writing

$$P(\mathbf{r},\mathbf{r}') = \frac{v_t}{D^{(B)}\pi(L + 2C)} \sum_{m=-\infty}^{\infty} \sum_{n=1}^{\infty} \sin \frac{n\pi(z + C)}{L + 2C}$$

$$\times \sin \frac{n\pi(z' + C)}{L + 2C} \exp[im(\phi - \phi')] g_m(r_\|), \quad (P5.6)$$

and substituting it into Eq. (P5.3), with $\delta(z - z')$ and $\phi(\phi - \phi')$ on the right-hand side expanded in accordance with Eqs. (P5.4) and (P5.5), the resulting equation for g_m is

$$\frac{d^2 g_m}{dr_\|^2} + \frac{1}{r_\|} \frac{dg_m}{dr_\|} - \left[\frac{m^2}{r_\|^2} + \frac{n^2\pi^2}{(L + 2C)^2}\right] g_m = -\frac{1}{r_\|} \delta(r_\| - r_\|'). \quad (P5.7)$$

This is the equation for modified Bessel function of order m. One can verify that the solution is given by

$$g_m = I_m\left(\frac{n\pi}{L+2C}r_\|\right) K_m\left(\frac{n\pi}{L+2C}r'_\|\right) \quad \text{if } r'_\| > r_\|$$

$$= I_m\left(\frac{n\pi}{L+2C}r'_\|\right) K_m\left(\frac{n\pi}{L+2C}r_\|\right) \quad \text{if } r'_\| < r_\|. \quad \text{(P5.8)}$$

Here I_m and K_m are the modified Bessel functions of the first kind and the second kind, respectively. By letting $r'_\| = 0$, the problem becomes isotropic with respect to ϕ, so only the $m = 0$ term remains, and $I_0(0) = 1$. It follows

$$P(\mathbf{r},\mathbf{r}') = \frac{3}{\pi l^*(L+2C)} \sum_{n=1}^{\infty} \sin\frac{n\pi(z+C)}{L+2C}$$

$$\times \sin\frac{n\pi(z'+C)}{L+2C} K_0\left(\frac{n\pi}{L+2C}r_\|\right),$$

which is exactly Eq. (5.14).

References

Akkermans, E., Wolf, P. E., and Maynard, R. (1986). *Phys. Rev. Lett.* **56**, 1471.
deWolf, D. A. (1971). *IEEE Trans. Antennas Propag.* **19**, 254.
Etemad, S., Thompson, R., and Anderson, M. J. (1986). *Phys. Rev. Lett.* **57** 575.
Etemad, S., Thompson, R., Andrejco, M. J., John, S., and MacKintosh, F. C. (1987). *Phys. Rev. Lett.* **59**, 1420.
Langer, J. L., and Neal, T. (1966). *Phys. Rev. Lett.* **16**, 984.
Tsang, L., and Ishimaru, A. (1984). *J. Opt. Soc. Am.* **A1**, 836.
van Albada, M. P., and Lagendijk, A. (1985). *Phys. Rev. Lett.* **55**, 2692.
Wolf, P. E., and Maret, G. (1985). *Phys. Rev. Lett.* **55**, 2696.
Wolf, P. E., Maret, G., Akkermans, E., and Maynard, R. (1988). *J. Phys. France* **49**, 63.

6

Renormalized Diffusion

6.1 Coherent Backscattering Effect in the Diagrammatic Representation

In the last chapter the coherent backscattering effect was seen as an enhancement of scattered intensity in the backward direction. The purpose of this chapter is to examine the *dynamic* implications of this effect in the context of diffusive transport behavior as described in Chapter 4. To quantify such implications, it is necessary to represent the coherent backscattering effect in terms of the Green's functions. A first step in this direction is the association of the ray paths in Figure 5.1 with products of Green's functions and $\bar{t}_i$'s. For example, for the ray path A,

$$A_0 = A_i G_0^+(\omega, \mathbf{r}_1, \mathbf{R}_0) \bar{t}_1^+ G_e^+(\omega, \mathbf{r}_2, \mathbf{r}_1) \bar{t}_2^+ G_e(\omega, \mathbf{r}_3, \mathbf{r}_2) \cdots$$
$$\times \bar{t}_n^+ G_0^+(\omega, \mathbf{r}_0, \mathbf{r}_n). \tag{6.1}$$

This association is based on the physical picture in which each scatterer acts as a secondary wave source, so that the amplitude of reaching the next scatterer is given by the Green's function. Since each scattering occurs in an environment of other scatterers, G_e^+ is used so that the presence of the random environment is taken into account in an averaged sense. Similarly, $\bar{t}_i^+$ gives the scattering amplitude for the ith scatter in an effective medium. For each Green's function, $G_e^+(\omega, \mathbf{r}_j, \mathbf{r}_i)$, $\mathbf{r}_i$ is noted to be the source point and $\mathbf{r}_j$ the detection point. Therefore, the line that represents the Green's function in a diagram should have an arrow that points from $\mathbf{r}_i$ toward $\mathbf{r}_j$. $\mathbf{r}_0$ ($\mathbf{R}_0$) is defined to be the observation (source) point far from the sample. Comparison with Eq. (5.1) shows the factor G_A in that equation to be the product of $\bar{t}_1^+ \cdots \bar{t}_n^+$ with the magnitude part of the Green's functions.

Since we are interested in the behavior of intensity transport, it is necessary to examine the quantity $A_0 A_0^*$, where

$$A_0^* = G_0^-(\omega, \mathbf{r}_n, \mathbf{r}_0) \bar{t}_n^- \cdots$$
$$\times G_e^-(\omega, \mathbf{r}_2, \mathbf{r}_3) \bar{t}_2^- G_e^-(\omega, \mathbf{r}_1, \mathbf{r}_2) \bar{t}_1^- G_0^-(\omega, \mathbf{R}_0, \mathbf{r}_1) A_i^*. \quad (6.2)$$

Here we have used the fact that $[G^+(\omega, \mathbf{r}_i, \mathbf{r}_j)]^* = G^-(\omega, \mathbf{r}_j, \mathbf{r}_i)$, and $(\bar{t}_i^+)^* = \bar{t}_i^-$. The product $A_0 A_0^*$ is thus the outer product of the right-hand sides of Eqs. (6.1) and (6.2). In accordance with the rules of drawing diagrams from outer products as explained in Chapter 4, the diagram corresponding to $A_0 A_0^*$, for the part between the first and the nth scatterer, is given by Figure 6.1, where we have made the slight change of replacing $\bar{t}_i^+ \otimes \bar{t}_i^-$ by the CPA vertex $U_i^{(B)}$. The net effect of this replacement is recalled to be the removal of the restriction on the indices of the scatterers. In addition, the dynamics of the multiply scattered wave is introduced by letting A_0 and A_0^* have frequencies $\omega^+ = \omega + (\Delta\omega/2)$ and $\omega^- = \omega - (\Delta\omega/2)$, respectively. The Green's functions are also labeled by their momenta, with the beginning and ending momenta $\mathbf{k}_i^+ = \mathbf{k}_i + (\Delta\mathbf{k}/2)$, $\mathbf{k}_f^+ = \mathbf{k}_f + (\Delta\mathbf{k}/2)$ for A_0, and $\mathbf{k}_i^- = \mathbf{k}_i - (\Delta\mathbf{k}/2)$, $\mathbf{k}_f^- = \mathbf{k}_f - (\Delta\mathbf{k}/2)$ for A_0^*. It should be noted that although the direction of the arrows on the A_0^* path is opposite to that of A_0, yet an advanced Green's function is also time reversed compared with a retarded Green's function. As a result, despite the arrows' directions, the momenta of the $+$ and $-$ channels are the same except for the sign of $\Delta\mathbf{k}$, which is introduced here to delineate the dynamics of diffusive waves. Figure 6.1 is immediately recognized to be one of the ladder diagrams responsible for the wave diffusion constant evaluation given in Chapter 4. Since $A_0 A_0^*$ and $B_0 B_0^*$ constitute the

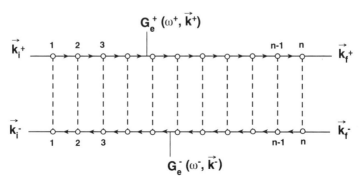

Figure 6.1 Representation of $A_0 A_0^*$ (or $B_0 B_0^*$) as a ladder diagram.

6.2 Evaluation of the Maximally Crossed Diagrams

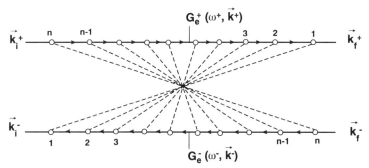

Figure 6.2 Representation of $B_0 A_0^*$ (or $A_0 B_0^*$) as a maximally crossed diagram.

incoherent scattering background as seen in the last chapter, we have thus established a correspondence between the incoherent scattering and the ladder diagrams.

For the coherent interference part of the scattering, it is necessary to look at terms such as $A_0 B_0^*$ and $B_0 A_0^*$. From Figure 5.1, it is straightforward to write down for the ray path B,

$$B_0 = A_i G_0^+(\omega, \mathbf{r}_n, \mathbf{R}_0) \bar{\iota}_n^+ G_e^+(\omega, \mathbf{r}_{n-1}, \mathbf{r}_n) \bar{\iota}_{n-1}^+ G_e^+(\omega, \mathbf{r}_{n-2}, \mathbf{r}_{n-1})$$
$$\times \cdots \bar{\iota}_1^+ G_0^+(\omega, \mathbf{r}_0, \mathbf{r}_1). \tag{6.3}$$

The product $B_0 A_0^*$ is shown in Figure 6.2. The difference from Figure 6.1 is clear. Because the order in which the scatterings occur in B_0 is reversed from that in A_0^*, the dashed interaction lines that link the upper (G^+) line and the lower (G^-) line have to cross each other. Figure 6.2 represents one of the so-called maximally crossed diagrams. The evaluation of the maximally crossed diagrams and their dynamic implications is now the task at hand.

6.2 Evaluation of the Maximally Crossed Diagrams

To evaluate the maximally crossed (MC) diagrams, let us consider for the moment the simplest MC diagram shown in Figure 6.3a for two scatterings. For the evaluation purpose, it makes no difference if we redraw it as in Figure 3.6b, which can be obtained from 6.3a by regarding the dashed interaction lines as elastic rubber bands connected to two parallel broomsticks (the Green's functions), and by turning the bottom broomstick 180° to untangle the two dashed lines. The only change, which is immaterial to the actual evaluation of the diagram, is that now the − input channel is on the right side and its output is on the left, reversed

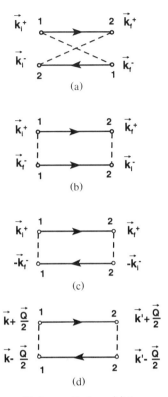

Figure 6.3 (a) An MC diagram with two scatterings. (b) Same as in (a), but with the bottom line turned 180°. This operation should have no effect on the evaluation of the diagram. (c) The sign of the momenta in the − channel is reversed from that of (b). This operation would leave the diagram the same only if the system is time-reversal invariant. (d) The momenta are relabeled in accordance with the rule $\mathbf{k}_f + \mathbf{k}_i = \mathbf{Q}$, and $\mathbf{k}, \mathbf{k}'$ as defined by Eq. (6.4). The result is a proper-looking ladder diagram.

from the + channels. Because the order of the two scatterings is preserved and the momenta remain unchanged, diagram 6.3b is identical to the original 6.3a from which it is obtained. Figure 6.3b looks very much like a ladder diagram. The one difference from a proper ladder diagram is in the sign of the momenta in the − channel. In general, the sign of momenta is physically significant. For example, in the presence of a magnetic field two electrons travelling a loop in opposite directions can have different phases that depend on the vector potential (Schiff, 1968). However, if the system is time reversal invariant, then reversing the momenta would yield the same physical state. Thus, by *assuming* the system to be time reversal invariant, the sign of momenta in the − channel may be reversed as shown

6.2 Evaluation of the Maximally Crossed Diagrams

in Figure 6.3c. The − output and input channels are also reversed back to the same order as the + channels. Now the diagram is almost like a proper ladder diagram. To make it completely proper, it is noted that the difference in the + and − channel momenta, which is $\Delta \mathbf{k}$ in the ladder diagrams, is now given by $\mathbf{k}_i^+ - (-\mathbf{k}_f^-) = \mathbf{k}_i^+ + \mathbf{k}_f^- = [\mathbf{k}_i + (\Delta\mathbf{k}/2)] + [\mathbf{k}_f - (\Delta\mathbf{k}/2)] = \mathbf{k}_i + \mathbf{k}_f$. Therefore, one can write

$$\mathbf{k}_i^+ = \left(\frac{\mathbf{k}_i - \mathbf{k}_f}{2} + \frac{\Delta\mathbf{k}}{2}\right) + \frac{\mathbf{k}_i + \mathbf{k}_f}{2}, \tag{6.4a}$$

$$\mathbf{k}_f^- = \left(\frac{\mathbf{k}_i - \mathbf{k}_f}{2} + \frac{\Delta\mathbf{k}}{2}\right) - \frac{\mathbf{k}_i + \mathbf{k}_f}{2}, \tag{6.4b}$$

$$\mathbf{k}_f^+ = \left(\frac{\mathbf{k}_f - \mathbf{k}_i}{2} + \frac{\Delta\mathbf{k}}{2}\right) + \frac{\mathbf{k}_i + \mathbf{k}_f}{2}, \tag{6.4c}$$

$$\mathbf{k}_i^- = \left(\frac{\mathbf{k}_f - \mathbf{k}_i}{2} + \frac{\Delta\mathbf{k}}{2}\right) - \frac{\mathbf{k}_i + \mathbf{k}_f}{2}. \tag{6.4d}$$

By relabeling

$$\frac{\mathbf{k}_i - \mathbf{k}_f}{2} + \frac{\Delta\mathbf{k}}{2} \quad \text{as } \mathbf{k},$$

$$\frac{\mathbf{k}_f - \mathbf{k}_i}{2} + \frac{\Delta\mathbf{k}}{2} \quad \text{as } \mathbf{k}',$$

$$\mathbf{k}_i + \mathbf{k}_f \quad \text{as } \mathbf{Q},$$

the transformed diagram is now just the ladder diagram with the three $\mathbf{k}$ variables transformed from $\mathbf{k}_i$, $\mathbf{k}_f$, $\Delta\mathbf{k}$ to $\mathbf{k}$, $\mathbf{k}'$, and $\mathbf{Q}$. The final transformed diagram is shown in Figure 6.3d. The net outcome of this whole exercise may be summarized as follows. In a time reversal invariant system, an MC diagram can be converted into a ladder diagram with the provision that

$$\mathbf{Q} = \mathbf{k}_i + \mathbf{k}_f \tag{6.5}$$

plays the role of $\Delta\mathbf{k}$ in the ladder diagrams. On the other hand, the initial and final momenta of the transformed ladder diagram are related by $\mathbf{k} = -\mathbf{k}' + \Delta\mathbf{k}$.

The MC diagrams modify the vertex function beyond its Boltzmann approximation. That is,

$$\mathbf{U} = \mathbf{U}^{(B)} + \mathbf{U}^{(MC)}. \tag{6.6}$$

We have seen from Chapter 4 and the discussion above that $\mathbf{U}^{(B)}$ is composed of ladder diagrams. Here $\mathbf{U}^{(MC)}$ is defined to be the sum of all

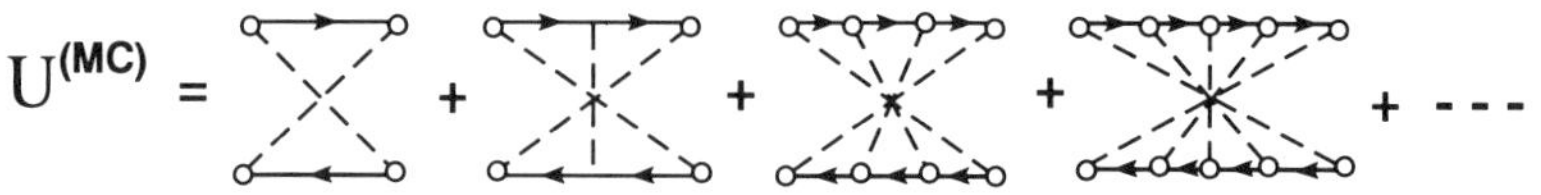

Figure 6.4 The definition of $\mathbf{U}^{(MC)}$ as the sum of all the maximally crossed diagrams.

MC diagrams as shown in Figure 6.4. By following the rule of transformation, every one of the diagrams can be made into a corresponding ladder diagram with the condition of Eq. (6.5). In operator form, that means

$$\begin{aligned}\mathbf{U}^{(MC)} &= \mathbf{U}^{(B)}{:}(\mathbf{G}_e^+ \otimes \mathbf{G}_e^-){:}\mathbf{U}^{(B)} + \mathbf{U}^{(B)}{:}(\mathbf{G}_e^+ \otimes \mathbf{G}_e^-){:}\mathbf{U}^{(B)}{:}(\mathbf{G}_e^+ \otimes \mathbf{G}_e^-){:}\mathbf{U}^{(B)} \\ &+ \mathbf{U}^{(B)}{:}(\mathbf{G}_e^+ \otimes \mathbf{G}_e^-){:}\mathbf{U}^{(B)}{:}(\mathbf{G}_e^+ \otimes \mathbf{G}_e^-){:}\mathbf{U}^{(B)}{:}(\mathbf{G}_e^+ \otimes \mathbf{G}_e^-){:}\mathbf{U}^{(B)} + \cdots \\ &= \mathbf{U}^{(B)}{:}\left[\mathbf{G}_e^+ \otimes \mathbf{G}_e^- + (\mathbf{G}_e^+ \otimes \mathbf{G}_e^-){:}\mathbf{U}^{(B)}{:}(\mathbf{G}_e^+ \otimes \mathbf{G}_e^-) + \cdots \right]{:}\mathbf{U}^{(B)} \\ &= \mathbf{U}^{(B)}{:}\langle\mathbf{G}^+ \otimes \mathbf{G}^-\rangle_c^{(B)}{:}\mathbf{U}^{(B)}. \end{aligned} \quad (6.7)$$

The last equality of (6.7) follows from Eq. (4.19). What Eq. (6.7) tells us is that by summing over all the MC diagrams, the task of evaluating $U^{(MC)}$ actually becomes easier because one can rely on what is already known. Since we can make the correspondence

$$\langle\mathbf{G}^+ \otimes \mathbf{G}^-\rangle_c^{(B)} \Leftrightarrow \phi_{\mathbf{k},\mathbf{k}'}^{(B)}(\Delta\omega, \mathbf{Q}|\omega), \quad (6.8)$$

which follows from Eq. (4.11) with $\Delta\mathbf{k}$ replaced by $\mathbf{Q} = \mathbf{k}_i + \mathbf{k}_f$, the MC vertex contribution, Eq. (6.7), may be written in the form

$$\left(\frac{N}{L^d}\right)^2 U^{(MC)} = \frac{1}{N^2}\sum_{\mathbf{k}}\sum_{\mathbf{k}'}\left[\frac{d\kappa_e^2}{d\omega}\frac{1}{\pi\rho_e(\omega)}\right]^2(-\operatorname{Im}\Sigma^+)^2\phi_{\mathbf{k},\mathbf{k}'}^{(B)}, \quad (6.9)$$

where the Ward identity, Eqs. (4.96c), has been used to convert $U^{(B)}$ into an expression that involved $-\operatorname{Im}\Sigma^+$. If Σ^+ is assumed to be $\mathbf{k}$-independent (as in CPA), then for the evaluation of $U^{(MC)}$ all those factors besides $\phi_{\mathbf{k},\mathbf{k}'}^{(B)}$ can be moved to the outside of the double-$\mathbf{k}$ summation. The summations over $\mathbf{k}, \mathbf{k}'$ then yield $S^{(B)}(\Delta\omega, \mathbf{Q}|\omega)$ by definition [Eq. (4.10)]. From Eq. (4.45) it is known that

$$S^{(B)}(\Delta\omega, \mathbf{Q}|\omega) = \frac{N}{L^d}\frac{2\pi\rho_e(\omega)(d\omega/d\kappa_e^2)(d\omega/d\kappa_0^2)}{-i\Delta\omega + D^{(B)}(\omega)|\mathbf{k}_i + \mathbf{k}_f|^2}, \quad (6.10)$$

where we have imposed the MC diagram condition of $\mathbf{Q} = \mathbf{k}_i + \mathbf{k}_f$. Combining terms yields

$$\frac{N}{L^d} U^{(MC)}_{\mathbf{k}_i, \mathbf{k}_f} = \frac{2}{\rho_e(\omega)\pi} \frac{d\kappa_e^2}{d\kappa_0^2} \frac{(-\text{Im}\,\Sigma^+)^2}{-i\Delta\omega + D^{(B)}(\omega)|\mathbf{k}_i + \mathbf{k}_f|^2}. \quad (6.11)$$

Since $U^{(MC)}$ peaks at $\mathbf{k}_i = -\mathbf{k}_f$, the imprint of the coherent backscattering effect is very much evident in the foregoing form.

6.3 Renormalized Diffusion Constant

The contribution of the MC vertex function to the diffusion constant is expressed through the integral M_0, Eq. (4.42a). By denoting the MC component of the contribution as $M_0^{(MC)}$, one obtains

$$M_0^{(MC)} = i \frac{L^d}{N^3} \sum_{\mathbf{k}_i} \sum_{\mathbf{k}_f} (\Delta G_e)_{\mathbf{k}_i} (\Delta G_e)^2_{\mathbf{k}_f}$$

$$\times \left[\nabla_{\mathbf{k}_i} e(\mathbf{k}_i) \cdot \Delta\hat{\mathbf{k}} \right] \frac{N}{L^d} U^{(MC)}_{\mathbf{k}_i, \mathbf{k}_f} \left[\nabla_{\mathbf{k}_f} e(\mathbf{k}_f) \cdot \Delta\hat{\mathbf{k}} \right]. \quad (6.12)$$

For the weak scattering limit, Eq. (6.12) may be reduced to the form

$$M_0^{(MC)} = \frac{1}{N^2} \frac{1}{-\text{Im}\,\Sigma^+} \sum_{\mathbf{k}_i} \sum_{\mathbf{k}_f} (\Delta G_e)_{\mathbf{k}_i} (\Delta G_e)_{\mathbf{k}_f}$$

$$\times \left[\nabla_{\mathbf{k}_i} e(\mathbf{k}_i) \cdot \Delta\hat{\mathbf{k}} \right] \frac{N}{L^d} U^{(MC)}_{\mathbf{k}_i, \mathbf{k}_f} \left[\nabla_{\mathbf{k}_f} e(\mathbf{k}_f) \cdot \Delta\hat{\mathbf{k}} \right]. \quad (6.13)$$

In contrast to the case of $M_0^{(B)}$ for isotropic scattering, $M_0^{(MC)}$ is clearly nonzero because the two summations are coupled through the $|\mathbf{k}_i + \mathbf{k}_f|$ dependence of $U^{(MC)}$. Another way of saying the same thing is that the MC diagram is inherently related to anisotropic scattering (in fact, it is peaked in the backward direction), so its total contribution to the diffusion constant does not vanish. In Chapter 8 the integrals in (6.12) will be treated with more care. In this chapter the focus is to get a general idea about the dynamic implications of coherent backscattering. For this purpose an approximate evaluation of (6.13) suffices. To this end, the double integral in Eq. (6.13) is separated into two single integrals by observing that the diffusive pole of $U^{(MB)}_{\mathbf{k}_i, \mathbf{k}_f}$ makes it peaked at $\mathbf{k}_i + \mathbf{k}_f = 0$. Therefore, the $\mathbf{k}_i$ integral of $(\Delta G_e)_{\mathbf{k}_i}(\Delta G_e)_{\mathbf{k}_f}[\nabla_{k_i}e(\mathbf{k}_i)\cdot\Delta\hat{\mathbf{k}}][\nabla_{k_f}e(\mathbf{k}_f)\cdot\Delta\hat{\mathbf{k}}]$ may be evaluated approximately with the condition $\mathbf{k}_i = -\mathbf{k}_f$. That is, as

$$-\frac{L^d}{N}\int\frac{d\mathbf{k}_i}{(2\pi)^d}(\Delta G_e)_{\mathbf{k}_i}(\Delta G_e)_{-\mathbf{k}_i}\left[\nabla_{k_i}e(\mathbf{k}_i)\cdot\Delta\hat{\mathbf{k}}\right]^2$$

$$= -\frac{L^d}{N}\int\frac{d\mathbf{k}}{(2\pi)^d}(\Delta G_e)_{\mathbf{k}}^2\left[\nabla_k e(\mathbf{k})\cdot\Delta\hat{\mathbf{k}}\right]^2,$$

$$= 4K_0\frac{N}{L^d}, \quad (6.14)$$

where the minus sign is due to the fact that $\nabla_k e(\mathbf{k})$ is an odd function of $\mathbf{k}$, whereas $(\Delta G_e)_{\mathbf{k}}$ is an even function. The remaining $\mathbf{k}_f$ integral may be converted to an $\mathbf{Q}$ integral over the MC vertex function. That is,

$$M_0^{(MC)} = \frac{L^d}{N}\frac{1}{-\mathrm{Im}\,\Sigma^+}\int\frac{d\mathbf{Q}}{(2\pi)^d}4K_0\frac{N}{L^d}\frac{N}{L^d}U_\mathbf{Q}^{(MC)}. \quad (6.15)$$

Substitution of Eq. (6.11) into Eq. (6.15) gives

$$M_0^{(MC)} \cong \frac{N}{L^d}\frac{8K_0}{\rho_e(\omega)\pi}\frac{d\kappa_e^2}{d\kappa_0^2}(-\mathrm{Im}\,\Sigma^+)\frac{L^d}{N}\int\frac{d\mathbf{Q}}{(2\pi)^d}\frac{1}{-i\Delta\omega + D^{(B)}(\omega)|\mathbf{Q}|^2}$$

$$= 8D^{(B)}(\omega)(-\mathrm{Im}\,\Sigma^+)\int\frac{d\mathbf{Q}}{(2\pi)^d}\frac{1}{-i\Delta\omega + D^{(B)}(\omega)|\mathbf{Q}|^2}, \quad (6.16)$$

where Eq. (4.99a) is used for $D^{(B)}(\omega)$, which is also correct for the classical scalar wave if the scattering is isotropic, and $e(\mathbf{k}) = k^2$.

From Eqs. (4.46) and (4.99) it follows that

$$\frac{1}{D(\omega)} = \frac{1}{D^{(B)}(\omega)} + \frac{M_0^{(MC)}\pi\rho_e(\omega)}{8K_0^2(d\kappa_e^2/d\kappa_0^2)(-\mathrm{Im}\,\Sigma^+)}$$

$$= \frac{1}{D^{(B)}(\omega)}\left[1 + \frac{(d\kappa_e^2/d\kappa_0^2)}{\pi\rho_e(\omega)}\int\frac{d\mathbf{Q}}{(2\pi)^d}\frac{1}{-i\Delta\omega + D^{(B)}(\omega)|\mathbf{Q}|^2}\right], \quad (6.17)$$

where $D(\omega)$ represents the diffusion constant renormalized by the coherent backscattering effect. The $(1 + \delta)$ factor for classical scalar waves is neglected in this chapter. A more detailed discussion of its effect is presented in Chapter 8.

In order to get a better idea of the consequences of the correction term due to the MC diagrams, let us set $\Delta\omega = 0$ so as to focus on the infinite travel-time limit. The correction term is clearly positive, which means $D(\omega) < D^{(B)}(\omega)$. By multiplying both sides by $D(\omega)D^{(B)}(\omega)$ and assuming the correction term to be small so that $D(\omega) \cong D^{(B)}(\omega)$, Eq. (6.17) (with $\Delta\omega = 0$) becomes

$$D(\omega) \cong D^{(B)}(\omega) - \frac{(d\kappa_e^2/d\kappa_0^2)}{\pi\rho_e(\omega)} \int \frac{d\mathbf{Q}}{(2\pi)^d} \frac{1}{|\mathbf{Q}|^2}. \quad (6.18)$$

Some striking implications of the correction term are shown below.

6.4 Sample Size and Spatial Dimensionality Dependences of Wave Diffusion

The first step in the approximate evaluation of $D(\omega)$ is to set limits on the $\mathbf{Q}$ integral in (6.18). The upper limit of the integral cannot be infinity because the diffusive behavior has a minimum length scale, given by the mean free path l. That means the upper limit of $|\mathbf{Q}|$ should be $\sim l^{-1}$. On the other hand, the lower limit of $|\mathbf{Q}|$ should be $\sim L^{-1}$, the inverse of the sample size. The resulting correction factor is more conveniently expressed in terms of the electrical conductivity in the electronic case:

$$\sigma_E(\omega) = \sigma_E^{(B)}(\omega) - \delta\sigma_E, \quad (6.19)$$

with

$$\delta\sigma_E = \frac{e^2}{2\hbar\pi^2} \begin{cases} \frac{1}{\pi}\left(\frac{1}{l} - \frac{1}{L}\right) & \text{3D} \\ \ln\frac{L}{l} & \text{2D} \\ 2(L - l) & \text{1D,} \end{cases} \quad (6.20)$$

where we have used the Einstein relation, Eq. (4.104), and approximated $d\kappa_e^2/d\kappa_0^2 \cong 1$. The conductivity here does not account for electron's spin degeneracy, which would increase both $\sigma_E^{(B)}$ and $\delta\sigma_E$ by a factor of 2. For the classical wave case we have

$$D(\omega) = D^{(B)}(\omega) - \delta D(\omega), \quad (6.21)$$

with

$$\delta D(\omega) = \frac{1}{\pi} v_t \kappa_c^{1-d} \begin{cases} \dfrac{1}{l} - \dfrac{1}{L} & \text{3D} \\ \ln \dfrac{L}{l} & \text{2D} \\ L - l & \text{1D.} \end{cases} \qquad (6.22)$$

What is remarkable about the correlation to the diffusion constant is that the sample size L matters, which means the renormalized diffusion constant is no longer an intensive physical quantity. The physical origin of this L dependence may be traced to the sample size dependence of the coherent backscattering effect. However, what is most significant is the fact that in 1D and 2D, the (negative) correction term diverges as $L \to \infty$. In other words, renormalized diffusion implies the absence of diffusion over infinite distances in 1D and 2D. Since the conductivity or the diffusion constant cannot be negative, the condition $D(\omega) = 0$ defines a length, the localization length ξ, beyond which diffusive transport is no longer an accurate description. While Eqs. (6.20) and (6.22) are obtained on the basis of $D(\omega) \cong D^{(B)}(\omega)$ so that the correction term is small, we may nevertheless obtain a rough estimate of the 1D and 2D localization length by setting $\delta\sigma_E = \sigma_E^{(B)}(\omega)$, $\delta D(\omega) = D^{(B)}(\omega)$ at $L = \xi$. That means in the electronic case

$$\xi \cong l \exp(\rho_e(\omega) v_t l \pi^2) \quad \text{2D}$$
$$\cong (1 + \pi^2 \rho_e(\omega) v_t) l \quad \text{1D}, \qquad (6.23)$$

and for the classical wave case

$$\xi \cong l \exp\left(\frac{\pi}{2} \kappa_e l\right) \quad \text{2D}$$
$$\cong (1 + \pi) l \quad \text{1D}, \qquad (6.24)$$

where we have substituted $v_t l/d$ for $D^{(B)}$ ($l^* = l$ for isotropic scattering). Although the numerical factors in Eqs. (6.23) and (6.24) may not be accurate, they nevertheless tell us that in 1D, the localization length is directly proportional to the mean free path, whereas in 2D the localization length is a transcendental function of l and therefore can be extremely large in the weak scattering limit.

The fact that the localization tendency always prevails in 1D and 2D does not invalidate what have been derived before about diffusive transport in these dimensions. The existence of a localization length, however, does imply that the diffusive behavior is restricted to a wave transport regime which is intermediate between l and ξ.

In 3D, the correction term does not diverge as $L \to \infty$. Therefore whether $\sigma(\omega)$ or $D(\omega)$ can vanish depends on the magnitude of the coefficient. In the electronic case, the condition for localization is obtained by setting $\sigma_E^{(B)} = \delta\sigma_E$ with $L = \infty$. That gives the localization condition

$$1 \gtrsim \frac{2\pi^3}{3} \rho_e v_l l^2. \quad (6.25)$$

Since ρ_e can become vanishingly small near the band edge and $v_t l^2$ generally does not diverge, the condition expressed by Eq. (6.25) is easily satisfied for the Anderson model near the band edges. We therefore expect localization of the electronic band-edge states in the presence of disorder. By using the spherical approximation for the Brillouin zone, i.e., $e(\mathbf{k}) = k^2$, then

$$\rho_e = \frac{m}{\hbar} \frac{\kappa_e}{2\pi^2},$$

whereas $v_t = \hbar\kappa_e/m$ [see Eqs. (4.102) and (4.103)], whereupon the criterion for localization becomes

$$\kappa_e l \lesssim \sqrt{3/\pi} \cong 1. \quad (6.26)$$

In the classical wave case, the condition obtained by setting $D^{(B)} = \delta D$ with $L = \infty$ has exactly the same form as Eq. (6.26). If the constant on the right-hand side of (6.26) is taken to be 1, i.e.,

$$\kappa_e l \lesssim 1, \quad (6.27)$$

then the inequality is known as the Ioffe–Regel criterion for localization (Ioffe and Regel, 1960; Mott, 1974). The original proposition of this criterion is based on the simple physical notion that if the mean free path becomes comparable to the effective wavelength, then a wave can no longer be plane wave–like, and wave localization is postulated to occur. A more formal derivation of this criterion is given in Chapter 8 for classical waves.

6.5 Localization in One Dimension: The Herbert–Jones–Thouless Formula

The divergence of the MC diagrams' contribution in 1D means that the diffusive behavior is valid only within a limited transport range. In order to have a better idea about what happens beyond the diffusive regime, it is instructive to examine the 1D case with a different perspective from the one pursued so far.

In 1D, it is easy to see heuristically why a wave cannot propagate indefinitely in the presence of disorder. Consider a 1D sample as composed of many segments, each of length $\sim l$. Since over that distance the phase is randomized, the wave intensity transmission coefficient, $|\tau|^2$, over a sample of length $L \gg l$ may be written approximately as *the product of intensity transmission coefficients*, i.e.,

$$|\tau|^2 = \prod_i |\tau_i|^2, \qquad (6.28)$$

where τ_i is the amplitude transmission coefficient for the ith segment. The error made in this approximation is to be noted below. But for the moment let us assume its validity. By writing

$$|\tau|^2 = \exp(-2L/\xi), \qquad (6.29)$$

the localization length ξ is obtained as

$$\xi = -2lN \bigg/ \sum_{i=1}^{N} \ln|\tau_i|^2, \qquad (6.30)$$

where $N = L/l$. Equation (6.30) clearly demonstrates that $\xi \propto l$ in 1D, in agreement with Eq. (6.24). It also shows that the wave intensity should decay exponentially inside a random medium, and ξ is that decay length. However, the decay cannot be a smooth exponential because $|\tau|^2$ is expected to be randomly varying, both as a function of frequency ω and as a function of sample length L. The range of such variations can extend from nearly 0 to 1 because at special frequencies and sample lengths, Eq. (6.28) breaks down due to the fact that in the absence of inelastic scattering, there can be constructive or destructive interference effects over the whole sample, so that the transmission may reach the limiting value of 1 or nearly 0. For a fixed sample length L, the fact that special frequencies exist may be inferred from the well-known resonant transmission phenomenon of periodic dielectric stacks. When a periodic stack becomes disordered, these frequencies would shift but do not disappear. As $L \to \infty$, the frequency width of each total transmission peak shrinks to zero, so it becomes difficult to achieve such resonant conditions. For a fixed frequency, the same effect can also appear as L is varied.

What are the statistics of such fluctuations? Since the total transmission $|\tau|$ can vary by orders of magnitude as a function of L, it indicates that

6.5 Localization in One Dimension: The Herbert–Jones–Thouless Formula

$\ln |\tau|$ (rather than $|\tau|$) is likely to be Gaussian distributed. Physically, the rationale for $\ln |\tau|$ to have Gaussian statistics is based on the fact that wave scattering in a 1D system is *successive* in nature; i.e., the transmission probability of the whole sample is the product of transmission coefficients of the segments. That means if each scattering is independent, then $\ln |\tau_i|$'s are additive, independent random variables. In fact, if $\ln|\tau_i|$'s have Gaussian statistics, then for a fixed L, the fluctuation in $1/\xi$, which is determined by the mean of all the $\ln |\tau_i|$'s, can jump randomly. As $L \to \infty$, however, the evaluation of $1/\xi$ becomes accurate because the central limit theorem of the Gaussian statistics requires it to be so. In other words, $1/\xi$ is a *self-averaging* quantity.

In the preceding discussion it is clear that the multiplicative nature of 1D wave transport is the basis of the 1D localization phenomenon. This multiplicative property can be expressed rigorously if transfer matrices were used in place of transmission coefficients. A transfer matrix is defined as the matrix which relates the two amplitudes of the left-going and right-going waves in one homogeneous segment of the sample (if the material property is continuously varying, the segment can be made infinitesimal in thickness so as to approximate homogeneity) to those of the next segment. For a piecewise homogeneous sample, the transfer matrix simply expresses the boundary conditions at the interface of the two segments. In terms of transfer matrices, Furstenberg (1963) proved that the multiplication of *random* transfer matrices always results in localization. However, care has to be exercised in equating random transfer matrices to random media because a random medium does not always imply random transfer matrices. A trivial example of this nonequivalence may be found in the case of a random binary system where component 1 has a fixed segment thickness, and there is a resonant transmission phenomenon at a set of discrete frequencies for one such segment of component 1 embedded in an otherwise homogeneous component 2. At those special frequencies, the segment of component 1 is invisible to the wave because resonant transmission implies an identity transfer matrix for the segment. It follows that if an identical second segment is put in, it would still be invisible at the same frequencies. One can easily deduce that a randomly placed collection of such segments would always be invisible to the wave at those particular frequencies because if transfer matrices corresponding to the segments of components 1 are identity matrices, then the overall transfer matrix must correspond to a uniform medium. Wave delocalization can thus occur at this discrete set of resonant transmission frequencies. Another possibility for nonlocalization in 1D is for the transfer matrix at *each interface* to be an identity matrix. This is the case, for example, for P-polarized electromagnetic wave propagation in a binary randomly layered medium at the *Brewster angle* (Sipe et al., 1988). At that particular angle (which has different values in the two components), there is total transmission through the stack, regardless of whether the medium is random or ordered.

The existence of a localization length directly implies a basic change in the character of the wave eigenfunction. Whereas a plane wave extends throughout the sample and has an intrinsic magnitude N, in the presence of disorder every eigenfunction in 1D and 2D exists in a finite spatial domain. Thus, their normalization does not involve N as for the plane wave case. Also, localization causes transport properties of a system to become sample size dependent, because whereas for $l < L \ll \xi$ the localization effect is not manifest so that the wave transport behavior is still diffusive, for $L \gg \xi$ such is not the case because the wave cannot be transported beyond that scale. The localization length is thus the second length scale, aside from the mean free path, which marks a transition in the character of wave transport behavior. In 1D, it turns out that there is a simple approach to the accurate evaluation of the localization length, known as the Herbert–Jones–Thouless formula. This is presented below.

A special property of the 1D eigenfunctions is that they can be indexed sequentially in accordance with the number of nodes in each eigenfunction. For example, if the eigenfunction on a sample of length L (extending from $x = 0$ to $x = L$) is determined by the condition $\phi(L) = 0$, with $\phi(0) = 0$, $\partial \phi(0)/\partial x = 1$ being the initial conditions, then the lowest-frequency eigenfunction would have no node inside the interval, the next eigenfunction one node, etc. This is true regardless of the nature of disorder in the interval. With this knowledge, let us define a quantity,

$$\Omega_\omega = \lim_{L \to \infty} \frac{1}{L} \ln \phi_\omega(L), \qquad (6.31)$$

which is known as the characteristic function. The real part of Ω_ω yields the inverse of the localization length, ξ^{-1}, with a positive sign. That is because in a disordered medium, if one of the two solutions to the wave equation is known to decay with the exponent ξ^{-1}, then the other solution must *grow* with exactly the same exponent. This is shown in the solution to Problem 6.1. For arbitrary initial condition, the growing solution always dominates, which accounts for the expression Ω_ω giving ξ^{-1} as its real part.

For the imaginary part of Ω_ω, it is instructive to focus on what happens to $\phi_\omega(L)$ as ω varies across an eigenfrequency. Since $\phi_\omega(L) = 0$ at an eigenfrequency, it is clear that $\phi_\omega(L)$ must change sign as ω varies across it so as to increase the number of nodes by 1. As a result, $\ln \phi_\omega(L)$ jumps

6.5 Localization in One Dimension: The Herbert–Jones–Thouless Formula

by an amount $i\pi$ every time ω hits an eigenfrequency. If the imaginary part of Ω_ω is specified to be a monotonic function of ω, then $\text{Im}[\ln \phi_\omega(L)/\pi L]$ gives the *accumulated* density of states. In other words,

$$\frac{1}{\pi} \text{Im}\left[\frac{d\Omega_\omega}{d\omega}\right] = \bar{\rho}(\omega) = -\frac{1}{\pi}\frac{d\kappa_0^2}{d\omega} \text{Im}\langle G^+(\omega, x = x')\rangle_c. \quad (6.32)$$

The casual nature of the wave equation demands that the imaginary part of $d\Omega_\omega/d\omega$ be related to its real part through the Kramers–Kronig relation, as expressed in the solution to Problem 2.3 [Eq. (P2.16)]:

$$\text{Re}\left[\frac{d\Omega_\omega}{d\omega}\right] = \frac{d\kappa_0^2(\omega)}{d\omega} P\int d\omega' \frac{\bar{\rho}(\omega')}{\kappa_0^2(\omega) - \kappa_0^2(\omega')}$$

$$= \text{Re}\left\{\frac{d\kappa_0^2(\omega)}{d\omega}\langle G^+(\omega, x = x')\rangle_c\right\}, \quad (6.33)$$

or

$$\text{Re}\left[\frac{d\Omega_\omega}{d\kappa_0^2(\omega)}\right] = P\int d\omega' \frac{\bar{\rho}(\omega')}{\kappa_0^2(\omega) - \kappa_0^2(\omega')}. \quad (6.34)$$

By integrating both sides with respect to $\kappa_0^2(\omega)$, a relation between $\bar{\rho}(\omega)$ and $\xi^{-1}(\omega)$ is obtained:

$$\text{Re}\, \Omega_\omega = \xi^{-1}(\omega) = \text{constant} + \int d\omega' \ln|\kappa_0^2(\omega) - \kappa_0^2(\omega')|\bar{\rho}(\omega'). \quad (6.35)$$

This equation was first derived by Herbert and Jones (1971) and by Thouless (1972). In the case of the Anderson model, one can write

$$a\xi^{-1}(u) = \text{constant} + \frac{\beta\hbar}{ma}\int_{-\infty}^{\infty} dv \ln|u - v|\bar{\rho}(v), \quad (6.36a)$$

where $v, u = (\hbar\omega - \epsilon_0)/(\beta\hbar^2/ma^2)$. The constant of integration is uniquely determined by the condition that in a uniform lattice, $\xi^{-1} = 0$. Therefore,

$$\text{constant} = -\frac{2}{\pi}\int_0^1 \frac{\ln v}{\sqrt{1-v^2}} dv = \ln 2. \quad (6.36b)$$

It is noted that we have set $u = 0$ for the evaluation of the integral in (6.36b). This is possible because the integral has to be independent of u, a

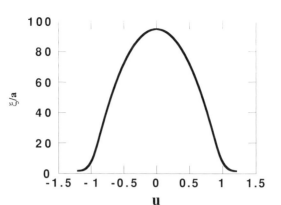

Figure 6.5 Localization length for the Anderson model with $a^2\sigma_w = 0.5$. This figure should be compared with Figure 3.4 for the 1D mean free path, with the same amount of randomness. ξ is almost exactly $4l$.

fact which may be deduced by noting that if $F(u)$ is the value of the integral, then $dF(u)/du$ gives the real part of the 1D (diagonal) Green's function, which is identically zero inside the band [see Eq. (2.65)]. Consequently $F(u)$ is independent of u. In Figure 6.5, ξ/a as calculated from Eqs. (6.36) is plotted as a function of u for $a^2\sigma_w = 0.5$. In Figure 6.6 the σ_w dependence of $a\xi^{-1}$ is shown for $u = 0$ and $u = 1$. It is seen that at $u = 0$, $\xi^{-1} \propto \sigma_w^2$ for small σ_w. At the band edge, however, ξ^{-1} is seen to vary as $\sigma_w^{2/3}$ at small σ_w.

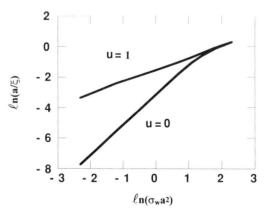

Figure 6.6 Dependence of $a\xi^{-1}$ as a function of randomness $a^2\sigma_w$ at two frequencies: one at the band center and one at the band edge. On the log–log scale, the straight line implies a power-law dependence, with the slope giving the power. At small $\sigma_w a^2$ values, the band center case shows a σ_w^2 dependence, whereas the band edge case shows a $\sigma_w^{2/3}$ dependence.

6.5 Localization in One Dimension: The Herbert–Jones–Thouless Formula

For classical scalar waves in continuum, the constant of integration in Eq. (6.35) is given by the condition that as $\omega \to 0$, scattering becomes weak, and $\xi^{-1} \to 0$ independent of $\bar{\rho}(\omega)$. That means

$$\xi^{-1}(\omega) = \int_0^\omega du \ln\left|1 - \frac{\omega^2}{u^2}\right| \bar{\rho}(u) \tag{6.37}$$

Here u has the dimension of frequency. In the solution to Problem 6.2 it is shown that for a uniform medium, $\xi^{-1}(\omega) = 0$ identically, as it should. For the purpose of numerical calculation, however, it is more convenient to integrate Eq. (6.33) directly:

$$\xi^{-1}(\omega) = \text{Re} \int_0^\omega d\omega' \, \frac{2\omega'}{v_0^2} \langle G^+(\omega', x = x') \rangle_c. \tag{6.38a}$$

In the low-frequency regime where the effective medium description is valid, one can write Eq. (6.38a) as

$$\xi^{-1}(\omega) = \text{Re} \int_0^\omega d\omega' \, \frac{2\omega'}{v_e^2} G_e^+(\omega', x = x'), \tag{6.38b}$$

so that everything is consistently defined relative to the effective medium. From Eq. (2.33),

$$G_e^+(\omega', x = x') = \frac{1}{2i\kappa_e^*}, \tag{6.39}$$

where $\kappa_e^* = \kappa_e + i/2l$. To be consistent, one has to write $v_e^2 = (\omega')^2/(\kappa_e^*)^2$. Therefore

$$\xi^{-1}(\omega) = \text{Re} \int_0^\omega d\omega' \, \frac{2\omega'}{(\omega')^2} \frac{\kappa_e^*}{2i}$$

$$= \int_0^\omega d\omega' \, \frac{2\omega'}{4(\omega')^2 l(\omega')}. \tag{6.40a}$$

In Chapter 3, it has been shown that at low frequencies l has the Rayleigh frequency dependence of ω^{-2} [see Eq. (3.77)]. Therefore, for $\omega \to 0$ the denominator of the integrand in Eq. (6.40a) is frequency independent. One can thus rewrite $\xi^{-1}(\omega)$ in the $\omega \to 0$ limit as

$$\lim_{\omega \to 0} \xi^{-1}(\omega) = \frac{1}{4\omega^2 l(\omega)} \int_0^\omega d\omega' \, 2\omega', \tag{6.40b}$$

which means

$$\xi^{-1}(\omega) = \tfrac{1}{4}l^{-1}(\omega). \tag{6.41}$$

This equation clearly demonstrates the proportionality relation between ξ and l in 1D. The factor 4 is valid at low frequencies. This factor turns out to be valid for the Anderson model as well (see Figures 6.5 and 3.4). It is noted that in Eq. (6.24), a very rough estimate of the proportionality constant gives $1 + \pi$, which is close to what is obtained.

At high frequencies, the behavior of $\xi^{-1}(\omega)$ is expected to depend on the details of each model. However, some general comments are possible based on the expression given by Eq. (6.37). For a 1D uniform medium, $\bar{\rho}(\omega) = 1/\pi v_0$ is a constant. For an inhomogeneous medium it can be argued that heuristically, $\bar{\rho}(\omega) \cong 1/\pi v(\omega)$. The value of $v(\omega)$ is known in the low- and high-frequency limits. At low frequencies, the effective-medium wave speed $\bar{v}$ is given by

$$\frac{1}{\bar{v}^2} = \frac{p}{v_1^2} + \frac{1-p}{v_2^2}.$$

At the high-frequency geometric optics limit, the effective speed is given by the travel time average, i.e.,

$$\frac{1}{\bar{v}_H} = \frac{p}{v_1} + \frac{1-p}{v_2}.$$

Mathematically, $\bar{v}_H > \bar{v}$ always, so from the expression $\bar{\rho}(\omega) \cong 1/\pi v(\omega)$ the low-frequency density of states should always be higher than the high-frequency density of states. In Eq. (6.37), the $\ln|1 - (\omega^2/u^2)|$ expression is positive for $u \le \omega/\sqrt{2}$ and negative for $u > \omega/\sqrt{2}$. Since the positive area of the expression exactly cancels the negative area [so for $\bar{\rho}(\omega) =$ constant, $\xi^{-1} = 0$], any increase in the low-frequency density of states gives the right-hand side of Eq. (6.37) a finite positive value. Moreover, as long as the low-frequency density of states is bounded, $\xi^{-1}(\omega)$ cannot diverge. As a result, there are two possibilities when $\omega \to \infty$: $\xi^{-1}(\omega)$ either approaches a constant or decreases to zero. It turns out that a more detailed investigation of the high-frequency behavior, using a very different theoretical technique, shows the first possibility applies when the interfaces in a medium are abrupt and sharp. The second possibility applies when the medium is smoothly varying (Sheng *et al.*, 1986). Since $\xi^{-1}(\omega) \sim \omega^2$ at low frequencies, it follows that in the second case $\xi^{-1}(\omega)$ has a maximum.

A large amount of work has been done on 1D localization behavior, ranging from rigorous proof of the various localization properties to numerical simulations to experimental measurements, because the one-dimensionality allows the application of numerous theoretical techniques as well as detailed experiments. Once the problem deviates from 1D, however, the situation becomes much more complicated. The next chapter addresses a fruitful phenomenological approach to the localization problem in higher spatial dimensionalities.

Problems and Solutions

6.1. Show that for the two solutions to the 1D wave equation at a fixed frequency ω, if one decays exponentially with an exponent ξ^{-1}, then the other one must grow exponentially with the same exponent.

The wave equation,

$$\frac{d^2}{dx^2}\phi + \kappa^2(\omega, x)\phi = 0, \tag{P6.1}$$

may be rewritten as a 2×2 first-order system:

$$\frac{d}{dx}\begin{pmatrix} \phi \\ \phi' \end{pmatrix} = \begin{pmatrix} 0 & 1 \\ -\kappa^2 & 0 \end{pmatrix}\begin{pmatrix} \phi \\ \phi' \end{pmatrix}, \tag{P6.2}$$

where $\phi' = d\phi/dx$. The important feature of Eq. (P6.2) is that the 2×2 transfer matrix on the right-hand side has zero trace. By defining

$$\boldsymbol{\Phi} = \begin{pmatrix} \phi & \phi_1 \\ \phi' & \phi_1' \end{pmatrix}, \tag{P6.3}$$

where ϕ and ϕ_1 are the two eigensolutions of the system, Eq. (P6.2) can be put in the discrete matrix form as

$$\boldsymbol{\Phi}(x + \Delta x) = (\mathbf{I} + \mathbf{M}\,\Delta x)\boldsymbol{\Phi}(x), \tag{P6.4}$$

where

$$\mathbf{M} = \begin{pmatrix} 0 & 1 \\ -\kappa^2 & 0 \end{pmatrix}. \tag{P6.5}$$

By taking the determinant of both sides of Eq. (P6.4), one gets

$$\det[\boldsymbol{\Phi}(x + \Delta x)] = \det(\mathbf{I} + \mathbf{M}\,\Delta x)\det[\boldsymbol{\Phi}(x)]$$
$$\cong [1 + (\mathrm{Tr}\,\mathbf{M})\Delta x]\det[\boldsymbol{\Phi}(x)], \tag{P6.6}$$

where only terms linear in Δx are retained. Here Tr **M** means the trace of the matrix **M**. As $\Delta x \to 0$, Eq. (P6.6) means

$$\frac{d}{dx} \det \mathbf{\Phi} = (\text{Tr}\,\mathbf{M}) \det \mathbf{\Phi} = 0, \tag{P6.7}$$

since **M** has zero trace. Therefore, the determinant of the solution matrix is a constant, independent of x. It follows that if one eigensolution decays like $\exp(-x/\xi)$, then the other eigensolution must grow exponentially as $\exp(x/\xi)$ so that the determinant of the solution matrix can maintain its constancy as a function of x. In particular, the growth rate must accurately equal the decay rate.

6.2. Show that Eq. (6.37) gives $\xi^{-1} = 0$ for the 1D uniform continuum.

In a uniform 1D medium, $\bar{\rho}$ is a constant independent of the frequency. From the equation,

$$\xi^{-1} = \int_0^\infty \ln\left|1 - \frac{\omega^2}{u^2}\right| \bar{\rho}(u)\, du,$$

it is clear that if $\bar{\rho}(\omega)$ is constant, then the Lyapunov exponent ξ^{-1} is proportional to a simple integral,

$$\xi^{-1} \propto \int_0^\infty \ln\left|1 - \frac{1}{v^2}\right| dv,$$

where $v = u/\omega$. The integral may be rewritten as

$$\int_0^\infty \ln\left|1 - \frac{1}{v^2}\right| dv$$

$$= \int_0^1 \ln\left|\frac{1}{v} - 1\right| dv + \int_1^\infty \ln\left|1 - \frac{1}{v}\right| dv + \int_0^\infty \ln\left|1 + \frac{1}{v}\right| dv$$

$$= \int_0^1 \ln|1 - v|\, dv - \int_0^1 \ln v\, dv + \int_0^\infty \ln\frac{z}{z+1}\, dz + \int_0^\infty \ln\frac{v+1}{v}\, dv,$$

where $z = v - 1$ from the previous line. From the last line of this equation it is clear that the first two terms cancel exactly because by setting $z' = 1 - v$ the first integral becomes exactly the second one. The third and fourth integrals also cancel because their integrands are the negatives of each other. Therefore $\xi^{-1} = 0$ exactly in a 1D uniform medium, as it should. Equation (6.37) tells us that a constant density of states is intimately related to this fact. Since $\xi^{-1}(\omega)$ must be positive, Eq. (6.37) is also a constraint on the physically allowable form for $\bar{\rho}(\omega)$.

References

Furstenberg, H. (1963). *Trans. Am. Math. Soc.* **108**, 377.
Herbert, D. C., and Jones, R. (1971). *J. Phys.* **C4**, 1145.
Ioffe, A. F., and Regel, A. R. (1960). *Progr. Semiconductors* **4**, 237.
Mott, N. F. (1974). "Metal-Insulator Transitions," Taylor & Francis, London.
Schiff, L. I. (1968). "Quantum Mechanics," 3rd Ed. p. 179. McGraw-Hill, New York.
Sheng, P., White, B., Zhang, Z. Q., and Papanicolaou, G. (1986). *Phys. Rev.* **B34**, 4757.
Sipe, J. E., Sheng, P., White, B. S., and Cohen, M. H. (1988). *Phys. Rev. Lett.* **60**, 108.
Thouless, D. J. (1972). *J. Phys.* **C5**, 53.

7

The Scaling Theory of Localization

7.1 Distinguishing a Localized State from an Extended State

The basis of a phenomenological theory of localization is the simple observation that a localized state has a length scale beyond which it is not mobile. To build upon this observation, one can imagine the following thought experiment. Consider a sample whose wave transport property is to be determined in the absence of inelastic scattering. By dividing the sample into many smaller cubic samples of varying sizes, the experiment consists of introducing a single-frequency excitation source at the center of each sample and detecting for the response at the samples' surfaces (perimeter). By plotting the response as a function of sample size L, it is possible to distinguish a localized state, which is expected to exhibit a decreasing response as sample size increases, from an extended state, for which the integrated response around the sample surface should stay constant.

To quantify this approach, this argument may be turned around and focused on the sensitivity of the wave function inside the sample to the boundary condition(s) at the surfaces (Edward and Thouless, 1972; Licciardello and Thouless, 1975; Thouless, 1974). The rationale here is that a localized eigenfunction should become insensitive to the boundary conditions as the sample size increases beyond its localization length, whereas an extended state would always be sensitive, regardless of the sample size. One way to characterize this sensitivity is to define a frequency shift $\delta\omega$ that is the difference in the eigenvalues when the boundary condition is changed from the symmetric, periodic boundary condition to the antisymmetric boundary condition. If the sample is homogeneous, this change of boundary condition is equivalent to adding or subtracting an extra half of a wavelength, resulting in a change in the wavevector by the amount

$$\delta k = \pi/L. \tag{7.1}$$

For the electronic case, $\omega = \hbar k^2/2m$, so $\delta\omega = (\hbar k/m)\delta k = v\pi/L$. For the classical scalar wave, $\omega = v_0 k$, so $\delta\omega = v_0\pi/L$, also. In both cases it is seen that $\delta\omega$ corresponds to the inverse of a time scale, L/v, required for the wave to travel from one side of the sample to the other. If the sample is inhomogeneous and the eigenfunctions are extended, the calculation of $\delta\omega$ is not trivial. Thouless has argued that the relevant time scale should correspond to the *diffusive* transport time to cover the length L, which means

$$\delta\omega \propto \frac{1}{L^2}. \tag{7.2}$$

$\delta\omega$ may be interpreted physically as the frequency width arising from the (stochastic) time scale required for a change in the boundary condition to be communicated to the wave function, since it can be argued that in changing the boundary condition abruptly at one end of a sample, the wave function cannot fully adjust to it until the change signal is propagated to the other end. If the nature of signal propagation is diffusive, then the uncertainty principle, $\delta\omega \sim (\delta t)^{-1}$, gives the L dependence of (7.2). To obtain a quantitative measure of the sensitivity to a boundary condition change, the frequency shift $\delta\omega$ has to be compared with the average frequency separation $\Delta\omega$ between two neighboring eigenvalues, and the dimensionless ratio $\delta\omega/\Delta\omega$ is denoted as

$$\gamma = \frac{\delta\omega}{\Delta\omega}. \tag{7.3}$$

Since $\Delta\omega \propto L^{-d}$, the parameter γ is noted to increase as L for the 3D extended states and $\gamma \propto L^{d-2}$ in general.

For a localized state $\delta\omega$ must decrease exponentially when L increases beyond ξ, since it is sensitive to the boundary condition only to the degree that it can "feel" the boundary through its exponential tail. However, $\Delta\omega$ should have the same variation with L regardless of whether the states are extended or localized. Therefore, γ always decreases as a function of L for localized states.

A fruitful way to examine the sample size variation of γ is through the consideration of joining smaller samples to form successively larger ones, i.e., by *scaling* the sample size. Let us first consider the $d = 3$ case. When

7.1 Distinguishing a Localized State from an Extended State

$\delta\omega \gg \Delta\omega$, an eigenstate in the smaller sample can easily couple onto other eigenstates in the neighboring samples that overlap with its eigenfrequency, thus resulting in a new wave function that can extend throughout the larger sample. Since in the larger sample the ratio $\delta\omega/\Delta\omega$ increases, the renormalization to even larger samples guarantees the formation of new extended states. On the other hand, if $\delta\omega \ll \Delta\omega$, then the eigenstates in neighboring samples would find it difficult to overlap with each other through their finite level width. Because they cannot couple effectively, the states would stay localized in the smaller samples. That means in the larger sample $\delta\omega$ is smaller by an exponential factor, since a localized state can feel the boundary only through its tail. Consequently, although $\Delta\omega$ is smaller in the larger sample, the ratio $\delta\omega/\Delta\omega$ still decreases, thus guaranteeing localization in the infinite sample size limit. The lesson of this scaling argument, due to Thouless (1979), is that a large $\delta\omega/\Delta\omega$ starting ratio tends to renormalize to $\gamma = \infty$, and a small $\delta\omega/\Delta\omega$ starting ratio tends to renormalize to $\gamma = 0$. It is thus natural to infer heuristically that there is a unique value of $\gamma = \gamma_c$ at which one behavior crosses into another. This plausible assumption of a critical γ_c is an essential element for the scaling theory of localization.

For the case of $d = 1$, on the other hand, the same argument implies that as the sample size increases, γ would continuously decrease, first as $1/L$ and then as $\exp(-2L/\xi)$. The case of $d = 2$ is interesting because $\delta\omega/\Delta\omega \sim$ constant in the "extended state" limit, so it can go either way. Thus, the scaling argument indicates that the localization phenomenon is sensitive to the spatial dimensionality of the sample, with $d = 2$ being the "marginal" dimension for localization. From Eqs. (7.2) and (7.3) the origin of the marginal spatial dimension can be traced to the diffusive character of the multiply scattered wave.

The quantity γ may be heuristically interpreted as the dimensionless conductance in electronic systems. From the Einstein relation and the fact that conductance $\Gamma_E = \sigma_E L^{d-2}$, it follows that

$$\Gamma_E = \frac{e^2}{\hbar} D\rho_e(\omega) L^{d-2}.$$

By using the Thouless argument that $\delta\nu L^2 = D$, where $\delta\nu = \delta\omega/2\pi$, we have

$$\begin{aligned}\Gamma_E &= \frac{e^2}{\hbar} \frac{\delta\omega}{2\pi} L^2 \frac{1}{\Delta\omega L^d} L^{d-2} \\ &= \frac{e^2}{h} \gamma,\end{aligned} \quad (7.4)$$

where $\gamma = \delta\omega/\Delta\omega$. From Eqs. (7.2) and (7.3), it is seen that

$$\gamma(L) \propto L^{d-2} \qquad (7.5)$$

in the extended regime, just what is expected for the conductance of an electric conductor with cubic (square) geometry. In the localized regime, it follows from the arguments presented above that γ should have the behavior

$$\gamma(L) \propto \exp(-2L/\xi) \qquad (7.6)$$

for $L \gg \xi$.

In the context of a critical γ_c, an interesting phenomenon, first pointed out by Mott, is that the electrically conducting materials all have 3D conductivities larger than a certain minimum value (Mott *et al.*, 1975), given approximately by $0.026 - 0.1 e^2/(\hbar a)$, where a is the atomic unit cell size. In terms of conductance, that translates into the condition of $\gamma > \gamma_c \simeq 0.16 - 0.6$ for conducting behavior, since a is the smallest physical length scale in the problem. Due to the electron's spin degeneracy, the actual γ_c to be compared with the nondegenerate case is $0.08 - 0.3$. While there is certainly the additional physics of electron–electron interaction underlying this phenomenon, Mott's observation nevertheless has an important influence on the development of the scaling theory of Anderson localization.

7.2 The Scaling Hypothesis and Its Consequences

One fact that has not been mentioned so far is that γ for a finite sample is a highly configuration-dependent quantity (see Chapter 10). Moreover, the measured γ in an electronic system is dependent on how the leads are attached to it. Suppose the latter problem is solved by defining a uniform way of measuring γ, e.g., by attaching leads of width L to two opposite sides of the sample; then there is still the generic problem of how to define γ for a finite sample. Since we want to associate a single value of γ to a given sample size, that value must be the configurational average of the γ values. The question is: What is the correct average to take? It turns out that averaging γ and averaging $\ln \gamma$ can give quite different answers. In order to define the proper average for γ, it is necessary first to know the distribution of γ. Ideally, one would like to have a Gaussian distribution for γ so that the central limit theorem applies to its mean. Evidence from numerical simulations in 2D indicated that, as with the 1D transmission coefficient, whereas γ displays Gaussian

7.2 The Scaling Hypothesis and Its Consequences

statistics when it is large in value (delocalized limit), but in the limit of small γ, $\ln \gamma$ is a random variable that displays Gaussian statistics (Zhang and Sheng 1991a). Therefore an attractive choice is to define $\langle \gamma \rangle_c$ or $\langle \ln \gamma \rangle_c$ as the quantity to be associated with a given sample size in the extended or the localized regime, respectively. What is meant by γ in the localized regime is then defined by $\exp\langle \ln \gamma \rangle_c$. Of course, a Gaussian distribution has two parameters, the mean and the variance. The basis of the one-parameter scaling theory, to be discussed below, is that the mean alone is sufficient to describe most of the important features of the localization phenomena in different spatial dimensionalities. For simplicity, the configurational averaging brackets $\langle \rangle_c$ will be dropped in the following so that γ and $\ln \gamma$ should be understood as their configurationally averaged values in their respective regimes.

Let us summarize what has been discussed in Section 7.1 as follows. In terms of $\ln \gamma(L)$, it is known that

$$\ln \gamma(L) \propto \begin{cases} (d-2)\ln L & \text{extended states} \\ -\exp(\ln L) & \text{localized states.} \end{cases} \quad (7.7)$$

Note that as L increases, whereas $\ln \gamma$ increases for an extended states in 3D, it decreases for a localized state. In other words, the derivative $d[\ln \gamma(L)]/(d \ln L)$ is positive for 3D extended states but is negative for localized states. When combined with the fact that in 3D there is a critical point at γ_c, it is clear that if the transition from the localized to the extended state is continuous, then $d[\ln \gamma(L)]/(d \ln L) = 0$ at $\gamma = \gamma_c$. In other words, $\ln \gamma_c$ is a *fixed point* in the scaling variation of $\ln \gamma$ at which sample size variation has no effect on its value. If we define

$$\beta = \frac{d \ln \gamma}{d \ln L}, \quad (7.8)$$

(the β notation is used to be consistent with the literature notation; it should be distinguished from the effective-mass β factor introduced in Chapter 2), then

$$\beta \propto \begin{cases} d-2 & \text{large } \gamma \\ \ln \gamma & \text{small } \gamma \\ 0 & \text{at } \gamma = \gamma_c. \end{cases} \quad (7.9)$$

In terms of β, *the scaling hypothesis states that β is a function of γ only*. This hypothesis, first proposed by Abrahams *et al.* (1979), is consistent with the right-hand side of Eq. (7.9), and it is automatically satisfied if γ can be expressed as a function of L only, so that if $\beta = f[\gamma(L), L]$, then L may be solved in terms of γ. Since γ must also depend on the amount of randomness in the system, the scaling hypothesis thus implies that the *effect of increasing or decreasing the randomness may be completely compensated by varying L*. One way this can happen is that γ is a function of only *one variable* which combines both randomness and L, and the domain of values for that one variable can be completely covered by varying L. A plausible realization of this variable is ξ/L in the localized domain and ζ/L in the extended domain. The physical significance of ζ, denoted the correlation length, will be clarified later. Both ξ and ζ are functions of randomness, but if γ is a function only of the ratio ξ/L or ζ/L, then varying L completely compensates for the effect of randomness.

What would happen if the scaling hypothesis were not valid? Then $\beta = 0$ would define a γ_c which is either sample size dependent or randomness dependent. In this context the fact that γ_c should be independent of L, which is based on the physically plausible renormalization picture presented earlier, argues in favor of the scaling hypothesis.

With the scaling hypothesis, a schematic picture of $\beta(\ln \gamma)$ is shown in Figure 7.1, where the *continuity* of the β function is assumed. Some interesting behavior can be immediately deduced. First, for $d = 3$, if one starts at a point $\ln \gamma > \ln \gamma_c$ so that β is positive, then the definition of $\beta = (d \ln \gamma)/(d \ln L)$ implies that as L increases, $\ln \gamma$ increases also, which would make β even larger. The arrow on the curve thus indicates the direction of variation of $\ln \gamma$ as L increases. If β is negative, on the other hand, then as L increases $\ln \gamma$ would decrease, which makes β even

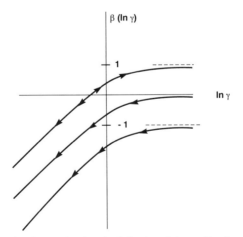

Figure 7.1 A schematic picture showing the behavior of the scaling function. The picture is constructed from the known asymptotic behavior of $\beta(\ln \gamma)$ in different spatial dimensionalities, plus the assumption of continuity. The arrows indicate the directions of variation as the sample size L increases.

7.2 The Scaling Hypothesis and Its Consequences

more negative. Therefore γ_c is an unstable fixed point where the direction of variation of $\beta(\ln \gamma)$ points away from γ_c. It is seen that $\beta(\ln \gamma)$ embodies the scaling property of γ discussed earlier.

For the $d = 2$ case, Eq. (7.9) states that $\beta = 0$ to the leading order in the large-γ limit. The sign of the next-order term is therefore important in deciding whether there is a fixed point in 2D. In drawing Figure 7.1, which shows no mobility edge for 2D, use has been made of the information given by Eqs. (6.19) and (6.20), which show that in 2D, $\sigma_E = \Gamma_E$,

$$\gamma = \gamma^{(B)} - \delta\gamma(L), \tag{7.10}$$

with

$$\delta\gamma(L) = \frac{1}{\pi} \ln\left(\frac{L}{l}\right). \tag{7.11}$$

The constant here, $1/\pi$, is for a γ that has no spin degeneracy. It follows from (7.11) that

$$\frac{d\gamma}{d(\ln L)} = -\frac{1}{\pi}. \tag{7.12}$$

By dividing both sides by γ, we get

$$\frac{d(\ln \gamma)}{d(\ln L)} \cong -\frac{1}{\pi\gamma} \tag{7.13}$$

in 2D. That means the direction of variation for β with increasing L is as indicated in Figure 7.1, so that all states are localized in 2D, no matter how weak the randomness. The same conclusion can be reached for the 1D case.

The significant feature of the 1D or 2D scaling function $\beta(\ln \gamma)$ is the switch-over from the linear behavior at negative $\ln \gamma$ to the constant behavior at large $\ln \gamma$. This switch-over is a manifestation of the effect of the localization length; for $L < \xi$ the measured behavior should approximate that of a conductor, but as L increases beyond the localization length the insulating behavior prevails. This understanding may be used to estimate the variation of the localization length with randomness. For example, from the integration of Eq. (7.12) one gets,

$$\gamma = \gamma_0 - \frac{2}{\pi} \ln\left(\frac{L}{l}\right) \tag{7.14}$$

as the 2D behavior of γ in the regime $L < \xi$, where γ_0 is the value of γ at $L = l$ and may be regarded as a measure of randomness. The value of $\gamma = \gamma_s$ at the switch-over is defined by setting $L = \xi$:

$$\gamma_s = \gamma_0 - \frac{2}{\pi} \ln\left(\frac{\xi}{l}\right),$$

or

$$\frac{\xi}{l} = \exp\left[\frac{\pi}{2}\left(\frac{\gamma_0}{\gamma_s} - 1\right)\right]. \quad (7.15)$$

As γ_0 becomes larger than γ_s in the weak-scattering regime, the localization length grows exponentially. This result reproduces what was deduced in the last chapter, but here the continuity of the β function into the strong-scattering regime shows that the whole phenomenon can be viewed in a unified framework through the variation of the sample size L.

There is an interesting issue about taking the limit of $\lambda \to \infty$, $L \to \infty$ for the classical wave in 1D and 2D. It is known that as $\lambda \to \infty$ the classical wave scattering strength approaches zero as $\lambda^{-(d+1)}$. Therefore, by fixing L and letting $\lambda \to \infty$, the conducting limit is always attained. However, if λ is fixed and L is increased $\to \infty$, then the renormalization arrow tells us that the limiting behavior is the localized state. The noncommuting limits are seen to be caused by the existence of *three length scales* in the problem: R, λ, and L. As R/λ determines ξ, taking the $\lambda \to \infty$ limit first means $\xi \to \infty$, so $\xi/L \to \infty$ always. However, if the limit $L \to \infty$ is taken first, then $\xi/L \to 0$.

In 3D, the most intriguing feature of the β function is the existence of a critical γ_c. Near γ_c,

$$\beta(\gamma) \cong s[\ln \gamma - \ln \gamma_c] = s \ln \frac{\gamma}{\gamma_c}, \quad (7.16)$$

where s is the slope of the β function when it crosses zero at $\ln \gamma_c$. From the definition of $\beta(\ln \gamma)$, one gets

$$\frac{d \ln \gamma}{\ln(\gamma/\gamma_c)} = s\, d(\ln L), \quad (7.17)$$

Integration of (7.17) yields

$$\frac{\ln(\gamma/\gamma_c)}{\ln(\gamma_0/\gamma_c)} = \left(\frac{L}{l}\right)^s, \quad (7.18)$$

7.2 The Scaling Hypothesis and Its Consequences

where γ_0 is the value of γ at $L = l$, determined solely by the amount of randomness. One can define

$$\epsilon = \ln \gamma_0 - \ln \gamma_c = \ln \frac{\gamma_0}{\gamma_c} = \ln\left(\frac{\gamma_0 - \gamma_c}{\gamma_c} + 1\right) \cong \frac{\gamma_0 - \gamma_c}{\gamma_c} \quad (7.19)$$

as the *distance to the mobility edge in terms of the randomness*. Then

$$\ln\left(\frac{\gamma}{\gamma_c}\right) = \epsilon \left(\frac{L}{l}\right)^s. \quad (7.20)$$

When ϵ is large enough such that

$$\beta \cong s \ln \frac{\gamma}{\gamma_c} \cong 1, \quad (7.21)$$

the same switch-over in the behavior of β occurs in the positive β region. From Eqs. (7.20) and (7.21), this condition is satisfied at a length scale ζ such that $\ln(\gamma/\gamma_c) \cong 1/s$, or

$$\frac{\zeta}{l} = (s\epsilon)^{-1/s}. \quad (7.22)$$

Therefore, in 3D there is a length scale, ζ, denoted the correlation length, which defines the property of a strongly scattered wave field in the delocalized regime. ζ is seen to diverge at the mobility edge where $\epsilon \to 0$. From the arrow attached to the β curve in the positive β region, one can deduce that for $L > \zeta$ so that $\beta \cong 1$, the classical behavior prevails. For $L < \zeta$, on the other hand, the dimensionless conductance follows the behavior given by Eq. (7.18), or

$$\gamma = \gamma_c \left(1 + \frac{\gamma_0 - \gamma_c}{\gamma_c}\right)^{(L/l)^s}$$

$$\cong \gamma_c + \left(\frac{L}{l}\right)^s (\gamma_0 - \gamma_c). \quad (7.23)$$

As γ_0 approaches γ_c, it is seen that γ becomes *independent of L*. In other words, the conductivity, which is given by γ/L, is no longer intensive in nature, even though the wave is still delocalized. Since $D \propto \gamma/L$, it follows from Eq. (7.23) that

$$\frac{\gamma}{L} \cong \frac{\gamma_c}{L} + \frac{1}{L}\left(\frac{L}{l}\right)^s (\gamma_0 - \gamma_c)$$

$$\cong \frac{\gamma_0}{L} \qquad (7.24)$$

when $\gamma_0 \cong \gamma_c$. Observing that $\gamma_0/l \propto D^{(B)}$ from the definition of γ_0, one obtains from (7.24)

$$D = D^{(B)} \frac{l}{L} \qquad (7.25a)$$

for $l < L < \zeta$. Continuity and the requirement that D be L independent for $L > \zeta$ means

$$D = D^{(B)} \frac{l}{\zeta} \qquad (7.25b)$$

in the $L \to \infty$ limit. For weak scattering, ϵ is large and from (7.22) ζ is expected to be on the order of l, so that means $D \cong D^{(B)}$. For strong scattering, however, $\zeta \gg l$, and the renormalized D can be significantly smaller than its Boltzmann value. A natural formula for interpolation between the two limits is

$$D = D^{(B)} l \left(\frac{1}{\zeta} + \frac{1}{L}\right). \qquad (7.26)$$

It is interesting to observe that near the mobility edge, i.e., when $L \ll \zeta$, the diffusion is anomalous in its time dependence. That is, from the diffusion relation

$$L^2 = Dt,$$

the L dependence of D as given by Eq. (7.25a) directly implies

$$L^2 = D^{(B)} \frac{l}{L} t,$$

or

$$L \propto t^{1/3}. \qquad (7.27)$$

7.2 The Scaling Hypothesis and Its Consequences

Equation (7.25a) also enables us to estimate the value of γ_c, a dimensionless constant, in terms of the Ioffe–Regel criterion. In 3D, we have the relation

$$\gamma = 2\pi \rho_e(\omega) DL. \qquad (7.28)$$

The 3D density of states at a frequency ω is given by

$$\rho_e(\omega) = \frac{4\pi \kappa_e^2 \Delta \kappa_e}{(2\pi)^3 \Delta \omega} = \frac{4\pi \kappa_e^2}{(2\pi)^3 v} \qquad (7.29)$$

in the approximation where $e(\mathbf{k})$ is replaced by k^2, and $\Delta\omega/\Delta\kappa_e$ is identified as v. By writing D as

$$D = \frac{1}{3} vl \frac{l}{L} \qquad (7.30)$$

for D close to γ_c, Eqs. (7.29) and (7.30) imply

$$\gamma \cong \frac{(\kappa_e l)^2}{3\pi}, \qquad \gamma \equiv \gamma_c. \qquad (7.31)$$

In particular,

$$\gamma_c = \frac{1}{3\pi} \simeq 0.106 \qquad (7.32)$$

when $\kappa_e l$ is set equal to 1 at γ_c as dictated by the Ioffe–Regel criterion. This value of γ_c is noted to be on the same order as what has been deduced from Mott's minimum metallic conductivity criterion.

When β is negative in the 3D case, i.e., in the localized regime, Eq. (7.20) may be written as

$$\ln\left(\frac{\gamma}{\gamma_c}\right) = -|\epsilon|\left(\frac{L}{l}\right)^s = -\left(\frac{L}{l|\epsilon|^{-1/s}}\right)^s. \qquad (7.33)$$

Here, however, one has to remember that for γ to be close to γ_c so that Eq. (7.33) is valid, L cannot be very large since increasing L would move $\ln \gamma$ away from $\ln \gamma_c$ as indicated by the arrow on the β curve shown in

Figure 7.1. If we let L become large, $\ln \gamma$ has to be proportional to $-L$ as specified by Eq. (7.7). In that case

$$\ln\left(\frac{\gamma}{\gamma_c}\right) = -\frac{L}{l|\epsilon|^{-1/s}} = -\frac{L}{\xi}, \qquad (7.34)$$

so that the localization length is seen to depend on $|\epsilon|$ as

$$\xi \propto |\epsilon|^{-1/s}. \qquad (7.35)$$

One way to alter the randomness as seen by a wave is to change its frequency. For example, in the Rayleigh scattering regime the strength of scattering by the random inhomogeneities increases with frequency. Therefore γ_0 can also be regarded as a function of frequency. In that case if ω_c is defined by the conditions $\gamma_c = \gamma_0(\omega_c)$, then

$$\epsilon \cong [\gamma_0(\omega) - \gamma_0(\omega_c)]/\gamma_0(\omega_c) \propto (\omega - \omega_c) \qquad (7.36)$$

so that

$$\xi \propto |\omega - \omega_c|^{-1/s} \qquad (7.37)$$

near the mobility edge. Similarly, from Eq. (7.22) it may be concluded that

$$\zeta \propto |\omega - \omega_c|^{-1/s}. \qquad (7.38)$$

Therefore, both the correlation length and the localization length diverge at the mobility edge with the exponent $1/s$.

The above discussion shows that γ is a function of $L/l|\epsilon|^{-1/s}$. From (7.20) and (7.34) it is seen that $l|\epsilon|^{-1/s}$ is proportional to either ζ or ξ, depending on whether the regime is localized or delocalized. Therefore γ is always a function of only one variable, L/ξ or L/ζ, where ξ and ζ are functions of randomness. This feature embodies the anticipated scaling property of γ.

7.3 Finite-Size Scaling Calculation of $\beta(\ln \gamma)$

The scaling hypothesis is a powerful assumption from which many localization properties may be deduced. One way to check its validity is via numerical simulation, which has gained acceptance over the past few decades as a means of verifying theoretical hypothesis and predicting the consequences of theories where analytical results are impossible or not yet available. Since numerical simulation can deal only with finite samples, the

7.3 Finite-Size Scaling Calculation of $\beta(\ln \gamma)$

variation of γ with a finite sample size L is a near-perfect case for numerical simulation. In fact, it is shown below that in 1D, $\beta(\ln \gamma)$ may be written down explicitly, and numerical simulations in 2D and 3D can be carried out by a scheme, commonly known as the finite-size scaling approach (MacKinnon and Kramer, 1981, 1983), which is a generalization of the 1D problem to systems of finite cross sections. This approach is numerically efficient and can provide the necessary accuracy for checking the validity of the scaling hypothesis, as well as giving the actual functional form of $\beta(\ln \gamma)$.

In the finite-size scaling approach, one considers a strip of length N and cross-sectional dimension M, where $N \gg M$ are both in units of the grid size. In the Anderson model, the grid coincides with the lattice. The condition that $N \gg M$ is crucial here because in that limit a unique localization length $\xi_M(W)$, independent of N, may be associated with a given cross-sectional dimension and randomness. Here W is used as a general parameter to denote the amount of randomness. The central dimensionless quantity considered in the finite-size scaling approach is $\Lambda = \xi_M(W)/M$, i.e., the localization length of the strip in units of its cross-sectional dimension. Just as in the scaling hypothesis, here the main finite-size scaling assumption, which is to be verified numerically, is that as M varies, the behavior of $\Lambda = \xi_M(W)/M$ depends only on Λ, not on W or M separately. That is,

$$\frac{d(\ln \Lambda)}{d(\ln M)} = \chi(\ln \Lambda). \tag{7.39}$$

Equation (7.39) can be integrated to get

$$h(\Lambda) = \int \frac{d \ln \Lambda}{\chi(\ln \Lambda)} = \ln M - \ln(\text{constant}) = -\ln \frac{\text{constant}}{M}, \tag{7.40}$$

where the "constant" means independence from M, so it can depend only on W. By denoting f as the inverse function of h, one obtains

$$\Lambda = f\left[\frac{\text{some function }(W)}{M}\right]. \tag{7.41}$$

When Λ is small, i.e., $M \gg \xi_M(W)$, the system is essentially in the 2D or 3D limit, and $\xi_M(W)$ should be the true localization length in 2D or 3D, $\xi(W)$, where $\xi(W)$ is independent of M. That gives

$$\frac{d(\ln \Lambda)}{d(\ln M)} = \frac{d\{\ln[\xi(W)/M]\}}{d(\ln M)} = -1 = \chi(\ln \Lambda). \tag{7.42}$$

The integration of Eq. (7.42) gives back

$$\Lambda = \frac{\xi(W)}{M}. \tag{7.43}$$

That is, $f(x) = x$ if $\xi(W)$ is identified as the W-dependent constant in Eq. (7.40).

For Λ large, $\xi_M(W) \gg M$, the problem in effectively 1D; and the wave amplitude is expected to be evenly distributed across the cross section of the strip, so that the *effective* 1D disorder W' as seen by the wave is averaged over the width of the strip. By assuming Gaussian statistics, that means

$$W'^2 = W^2/M^{d-1}. \tag{7.44}$$

Since the 1D localization length in the Anderson model is proportional to σ_w^{-2} (for σ_w not too large and for energy not close to be band edge), and $\sigma_w \propto W$, it means

$$\xi_M \sim (W')^{-2} = W^{-2}M^{d-1}, \tag{7.45}$$

so that

$$\Lambda = \frac{\xi_M}{M} \propto W^{-2}M^{d-2}.$$

For classical waves, ξ_M is proportional to the mean free path in 1D, which in turn is proportional to $(\kappa_e R)^{-2}$ (see Chapter III). That means $W \sim$ frequency ω for classical waves. In any case,

$$\chi(\ln \Lambda) = d - 2, \tag{7.46}$$

and the integration of Eq. (7.39) yields

$$\Lambda = \left[\frac{\xi_\infty(W)}{M}\right]^{2-d}, \tag{7.47}$$

i.e., $f(x) = x^{2-d}$. Here $\xi_\infty(W)$ is a length that depends only on the randomness. In 2D, $\xi_\infty(W) = \xi(W)$. In 3D, $\xi_\infty(W) = \zeta(W)$, the correlation length, in the extended regime and $\xi_\infty(W) = \xi(W)$ in the localized regime. For the 1D case, where $M = 1$ and $\Lambda > 1$ always, $\chi(\ln \Lambda) = -1$ exactly and (7.43) always holds. From Eqs. (7.43) and (7.47), it is seen that in 3D, Λ behaves as M^{-1} when Λ is small but as M when Λ is large. In 1D and 2D, however, Λ behaves either as M^{-1} or is a constant in the two limits.

By combining the behaviors of Λ at the two limits, it can be concluded that $\Lambda = f(x)$, with $x = \xi(W)/M$ in the localized regime and $x =$

7.3 Finite-Size Scaling Calculation of $\beta(\ln \gamma)$

$\zeta(W)/M$ in the delocalized regime. Λ is clearly a quantity very similar to γ. The scaling variable $x = \xi_\infty(W)/M$ is a combination of both randomness W and size M, and the domain of positive x values is completely covered by varying M, no matter what the value of W [and consequently $\xi_\infty(W)$]. Therefore, Λ may be regarded as a function of only one variable, which ensures the validity of the scaling relation (7.39). It is also seen that Λ's behavior is the same as that of γ when both are large. $\chi(\ln \Lambda)$ is therefore similar to $\beta(\ln \gamma)$. To relate the two so that $\beta(\ln \gamma)$ may be explicitly calculated, we let $M = L$ and write

$$\beta = \frac{d(\ln \gamma)}{d(\ln L)} = \frac{d(\ln \gamma)}{d(\ln \Lambda)} \frac{d(\ln \Lambda)}{d(\ln L)} = -\beta_1(\ln \gamma)\chi(\ln \Lambda). \quad (7.48)$$

Since in the 1D case the value of $\chi(\ln \Lambda)$ is identically -1, $\beta_1(\ln \gamma)$ denotes the β function in 1D. In 2D and 3D the β function may be obtained from $\beta_1(\ln \gamma)$ through numerical simulation of $\chi(\ln \Lambda)$.

To calculate $\beta_1(\ln \gamma)$, it is necessary to have an expression for γ. For a conductor, γ should be ∞ in the absence of inelastic scattering. However, for an insulator γ is proportional to $\exp(-2L/\xi) = \exp(-2/\Lambda)$, remembering that $L = M$. Based on an analogy with the tunnel junction, Landauer (1970) proposed that γ may be expressed in terms of the 1D transmission coefficient, $|\tau|^2 = \exp(-2/\Lambda)$, as

$$\gamma = \frac{|\tau|^2}{1 - |\tau|^2}. \quad (7.49)$$

This definition of γ is seen to satisfy the conditions for both the conducting limit and the insulating limit, since $|\tau|^2 = 1$ for a conductor so that $\gamma \to \infty$, and $|\tau|^2 \to 0$ as $L \to \infty$ for an insulator, so $\gamma \sim |\tau|^2$. A more thorough discussion of the Landauer formula, including its generalization to 2D and 3D, is given in Chapter X. Here it is easy to see that γ as given by Eq. (7.49) is a function of Λ. Therefore $\beta_1(\ln \gamma)$ may be easily calculated to be:

$$\beta_1 = -\frac{d \ln \gamma}{d \ln \Lambda} = -\frac{\Lambda}{\gamma}\frac{d\gamma}{d\Lambda} = -\frac{\Lambda}{\gamma}\frac{d\gamma}{d\tau}\frac{d\tau}{d\Lambda}$$

$$= (1 + \gamma)\ln\left(\frac{\gamma}{1 + \gamma}\right). \quad (7.50)$$

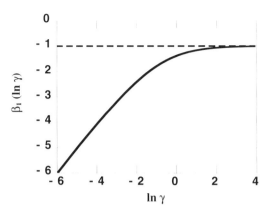

Figure 7.2 The 1D scaling function $\beta_1(\ln \gamma)$.

Since $\gamma = \exp(\ln \gamma)$, β_1 can thus be plotted as a function of $\ln \gamma$ as shown in Figure 7.2. For γ large,

$$\beta_1 \simeq -\gamma \ln\left(1 - \frac{1}{\gamma}\right) \to -1, \qquad (7.51)$$

and for γ small, $\beta_1 \sim \ln \gamma$, which are exactly the expected limits.

Figures 7.3a and b give the numerically calculated values of Λ as a function of $\xi_\infty(W)/M$ for the 2D and 3D Anderson models. A numerical approach for the evaluation of ξ_M, called the recursive Green's function method, is detailed in the solution to Problem 7.1. Data for many different values of W are included. Here the value of $\xi_\infty(W)$ is chosen to fit all the data onto a single curve. The fitted values of $\xi_\infty(W)$ are shown in the inset as a function of W, expressed in units of $V = \hbar^2/2ma^2(W/V = 2\sigma_w a^2)$. The data show convincingly that the scaling hypothesis is valid at least to the first approximation (numerically it is not possible to verify the accuracy of the scaling hypothesis to all orders). For the 2D case Λ is seen to be a monotonically increasing function of $x = \xi_\infty/M$. Therefore, $\chi = d(\ln \Lambda)/d(\ln M)$ is always negative, since $d \ln M = -d \ln x$. As a result, from Eq. (7.48) $\beta(\ln \gamma)$ for the 2D case always has the same sign as $\beta_1(\ln \gamma)$. For the 3D case, the function Λ is composed of two different segments with opposite trends as a function of increasing x. χ can therefore be both positive and negative. In this case the upper segment corresponds to $\sigma_w a^2 \lesssim 8$, whereas the lower segment is for $\sigma_w a^2 \gtrsim 8$. The fitted values of $\xi(W)$ and $\zeta(W)$ are both seen to increase dramatically at $\sigma_w a^2 \cong 8$. It follows that $\sigma_w a^2 \cong 8$ defines the mobility edge for the 3D Anderson model. By evaluating χ numerically for the 2D and 3D cases, the resulting 2D and 3D scaling functions may be obtained from Eqs.

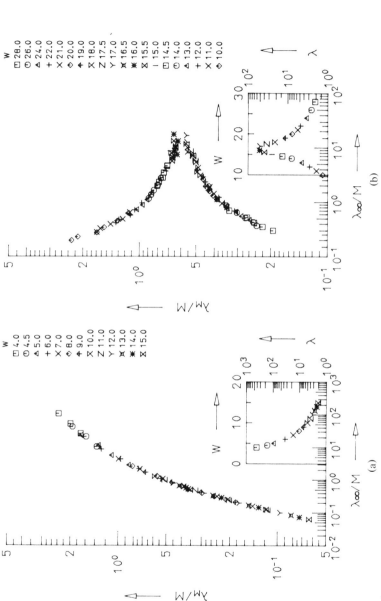

Figure 7.3 (a) Numerically calculated ξ_M/M (shown as λ_M/M) plotted as a function of ξ/M (shown as λ_∞/M) for the 2D Anderson model at the band center. The value of $\xi(W)$, determined by fitting all data with different W's onto a single curve, is shown in the inset. Here $W = 2\sigma_w a^2$. (b) Same as (a) for the 3D Anderson model at the band center. There are now two segments. The top segment corresponds to delocalized states and is obtained for $\sigma_w^2 a > 8$. The fitted value of ξ_∞ in this case is the correlation length $\zeta(W)$. The lower segment corresponds to localized states with $\sigma_w^2 a < 8$. The fitted values of ξ_∞ in this case corresponds to $\xi(\omega)$. Both ζ and ξ are shown as a function of W in the inset. Both figures are reproduced from MacKinnon and Kramer (1981).

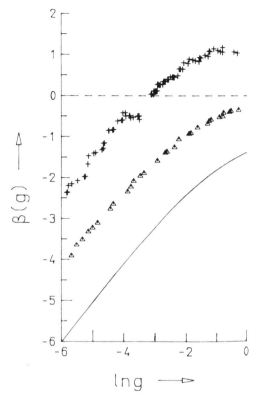

Figure 7.4 The numerically calculated scaling function from the data shown in Figure 7.3a and b. Note the similarity with Figure 7.1. Here γ is shown as g. The figure is reproduced from MacKinnon and Kramer (1981).

(7.48) and (7.50). They are plotted in Figure 7.4. Similarity with the expected behavior, shown in Figure 7.1, is noted. In 3D, $\gamma_c \cong \exp(-3.4) \cong 0.03$, which is on the same order as but smaller than that predicted by the Ioffe–Regel criterion. The slope s at which the 3D β function crosses zero may also be evaluated. It is approximately 1–1.5. From Eq. (7.37), that directly gives the exponent for the divergence of the localization length at the mobility edge.

7.4 Universality and Limitations of the Scaling Theory Results

While the finite-size scaling calculations have been carried out by using the Anderson model, the scaling function should be the same for classical scalar waves because the basis of the scaling argument—that the effect of

7.4 Universality and Limitations of the Scaling Theory Results

randomness (or frequency) is manifest only through a length scale, and that the dimensionless scaling parameter $\ln \gamma$ is dependent only on the ratio x between that length scale and the sample size—is wave-type as well as lattice-type independent. In other words, $\ln \gamma(x)$ may be regarded as the composition of two functions. The outer function depends only on the sample's spatial dimensionality, but the inner variable x is a model-dependent function of W, ω, wave type, lattice type, etc. The scaling theory conclusions are based on the *universality of the outer function*.

While the scaling theory should hold for both quantum and classical waves, it is important to point out that in 3D, it is entirely possible for classical scalar waves not to have the $\beta < 0$ branch of the scaling function, i.e., no localized states, unless some other condition(s) is satisfied, such as the ratio between the dielectric constants of the scatterers and the matrix exceeding a critical value (see Chapter 8 for more discussion on this point). This is because, whereas in a lattice model wave localization is aided by the presence of the band edges, where the group velocity vanishes and the wave is literally "waiting" to be localized by the addition of randomness, for waves in the continuum the absence of an upper band edge means that the coherent backscattering effect must work by itself, thus making localization more difficult. One is reminded in this regard that classical waves can never be localized at the lower band edge, i.e., $\omega \to 0$, because the effect of randomness is always seen by the wave as $\omega^2[\epsilon(\mathbf{r}) - 1]/v_0^2$, so the scattering vanishes in the zero-frequency limit. This point should also be obvious from the frequency dependence of Rayleigh scattering.

Although the localization of classical scalar waves in 3D is potentially more difficult, but if localization *does occur*, it is expected to have the same $\beta(\ln \gamma)$ function as shown in Figure 7.3b; that is, universality is not affected.

Before leaving this chapter, some implicit assumptions of the scaling theory should be made explicit so that we are aware of the theory's limitations. The three most important assumptions of the scaling theory are homogeneity, isotropy, and the absence of inelastic scattering. Homogeneity means that the density of the scatterers is constant beyond a finite sampling scale. Most of the disordered media are expected to be homogeneous. An example of an *inhomogeneous* system is one in which the wave scatterers are distributed as a fractal; i.e., the scatterer density decays as a power law function of the sampling scale. In such a system a wave can delocalize even in 2D, because whereas the effect of the coherent backscattering varies as $\ln L$ in the weak scattering limit [as derived in the last chapter; see Eqs. (6.19) and (6.20)], the density of the scatterers decays even faster (as a power law), thus making it possible for the wave to delocalize (Zhang and Sheng, 1991b).

Isotropy is another important assumption. In the Anderson model, there are two ways to introduce anisotropy. One is to make each scatterer anisotropic, e.g., by making the effective mass anisotropic so that the scattering and consequently the mean free paths become anisotropic as well. This model proves to have the same scaling behavior as the isotropic case because one can scale the sample size in different directions by its respective anisotropic mean free paths. After scaling, the model returns to being isotropic (Li *et al.*, 1989). Therefore this type of anisotropy does not invalidate the scaling property. Another way to introduce anisotropy is to make the diagonal terms have both an isotropic and an anisotropic component. For example, the anisotropic component can be a randomly layered system where the randomness is correlated over infinite distances in the directions parallel to the layers. Therefore, while each scatterer is isotropic, the anisotropy resides in its (infinite-range) spatial correlation. In fact, such models are relatively realistic for most of the Earth's subsurface, which is known to be predominantly layered. The coherent backscattering effect then has two components, one isotropic and the other one anisotropic. It can be shown that scaling breaks down in this case, and novel localization behavior can emerge (Xue *et al.*, 1989).

Another implicit assumption of the scaling theory is the absence of inelastic scattering. The effect of inelastic scattering will be discussed in Chapter 9, in relation to the physical manifestations of the localization effect.

Problem and Solution

7.1. Give a formulation of the numerical approach for the calculation of ξ_M.

In order to illustrate the numerical scheme, let us restrict ourselves to the 2D Anderson model defined by Eqs. (2.51), (2.53), and (2.54). Consider a strip of width M and length $N \gg M$. Both N and M are in units of the lattice (discretization) constant. This strip is the disordered material whose ξ_M we wish to measure. In order to do so, the strip is connected on both the left and the right to two perfect leads of the same width but infinite length. The perfect leads have no disorder, i.e., $\sigma(\mathbf{l}) = 0$. In analogy with a 1D system where ξ is calculated as $-2Na/\ln\langle|\tau|^2\rangle_c$, ξ_M will be calculated similarly. However, the calculation of τ is more complicated here because there is more than one plane wave for a given frequency. To see that, we recall that in a 2D perfect lead, $e(\mathbf{k}) = 2(2 - \cos k_x a - \cos k_z a)/a^2$ so that at a fixed frequency, $\omega = \hbar e(\mathbf{k})/2m$, there can be multiple (k_x, k_z) combinations that satisfy the dispersion relation. Here x

is defined as the width direction and z the length direction of the strip. We shall impose the periodic boundary condition in the width direction, so that there is only a discrete set of allowed k_x values, given by

$$k_x^{(i)} = \frac{2\pi}{aM} i, \quad i = 1, \cdots M. \quad (P7.1)$$

$k_z^{(i)}$ is then defined as the value of k_z that satisfies $e(\mathbf{k})|_{k_x^{(i)}} = 2m\omega/\hbar$. However, for fixed ω and $k_x^{(i)}$ there is not always a real solution for $k_z^{(i)}$. These cases correspond to k_z being an imaginary, i.e., evanescent, mode. It follows that the number of propagating modes, M_0, is less than M in general. For the propagating modes on both the left and right sides, a transmission coefficient τ_{ij} may be defined, where i corresponds to the left-hand side $k_z^{(i)}$ and j corresponds to the right-hand side $k_z^{(j)}$. The total transmission $|\tau|^2$ is given by

$$|\tau|^2 = \sum_i^{M_0} \sum_j^{M_0} \tau_{ij} \tau_{ij}^*$$

$$= \text{Tr}\, \tau\tau^\dagger. \quad (P7.2)$$

where $\dagger$ means Hermitian conjugate. In getting to the last line, the j summation is seen to be just the matrix multiplication, and Tr (the operation of taking trace of a matrix) is the summation over i. This $|\tau|^2$ is used to calculate ξ_M as $-2aN/\langle \ln|\tau|^2 \rangle_c$.

To calculate τ_{ij}, Lee and Fisher (1981) had pioneered a recursive Green's function method that is accurate and numerically stable. This method is presented below.

First, τ_{ij} has to be related to the Green's function. On physical ground,

$$\tau_{ij} \sim G_{ij}^+(0, N+1), \quad (P7.3)$$

where $G_{ij}^+(0, N+1)$ stands for the Green's function that gives the amplitude of the wave in momentum channel j on the right side, at $z = N+1$, that started as a plane wave in momentum channel i on the left side. The Green's function of (P7.3) is a bit unusual in that the indices i, j indicate the momentum channels of the perfect lead, whereas $0, N+1$ indicate the real space positions of the incoming and the outgoing waves. The question now is what is the proportionality factor. In Chapter 2, the 1D lattice Green's function is shown to exhibit the form [Eq. (2.65b)]

$$G_0^+(\omega, l - l') = -\frac{i}{\partial \beta e(k)/\partial k|_{k_0}} \exp(ik_0|l - l'|),$$

where k_0 is the solution to the equation $\kappa_0^2 = \beta e(k_0)$, $e(k)$ being the 1D dispersion function $2(1 - \cos ka)/a^2$. From this 1D expression, we see that since the τ_{ij} is defined relative to the incoming *flux* in the z direction, G_{ij}^+ first has to be multiplied by $i\bar{k}_z(i)$ so that the incident plane wave has unit amplitude (apart from a phase factor), and then it has to be multiplied by $\sqrt{\bar{k}_z(j)/\bar{k}_z(i)}$ in order to normalize the outgoing z flux relative to the incoming z flux. Here

$$\bar{k}_z(i) = \left.\frac{\partial \beta e(\mathbf{k})}{\partial k_z}\right|_{k_x^{(i)}, k_z^{(i)}}$$

is proportional to the *group velocity* in the z direction, and $e(\mathbf{k})$ here is the 2D expression $2(2 - \cos k k_x a - \cos k_z a)/a^2$. The net result is

$$|\tau_{ij}| = a\sqrt{\bar{k}_z(i)\bar{k}_z(j)}\,|G_{ij}^+(0, N+1)|. \tag{P7.4}$$

It is noted that $G_{ij}^+(0, N+1)$ is a 2D Green's function, which is dimensionless in the notations of this book. The task now is to calculate the Green's function. To do that, we show first how to calculate $G_{ij}^L(0,0)$, $G_{ij}^L(1,1),\ldots,G_{ij}^L(N,N)$, which are the half-space (the left half) Green's functions built up from the perfect-lead Green's function from the left side, from which $G_{ij}(0, N+1)$ may be obtained (superscript + is suppressed here and in the following).

First consider $G_{ij}^L(0,0)$. This is the value of the Green's function at the end of the perfect lead on the left side. It should be emphasized that $G_{ij}^L(0,0) \neq G_{ij}(0,0)$, which is the true Green's function of the full system at $(0,0)$. $G_{ij}^L(0,0)$ has two properties. First, it is diagonal in the i,j indices, i.e.,

$$G_{ij}^L(0,0) = \delta_{i,j}g_i^{-1}, \tag{P7.5}$$

because i and j are the momentum eigenstates of the perfect lead, so they must be orthogonal. Second, a new ending to the perfect lead may be generated by joining onto it a strip with $\sigma(\mathbf{l}) = 0$, where $\mathbf{l} = (0, 1, \ldots M-1)$ in the x direction. The new Green's function, $[G_{ij}^L(0,0)]'$, is expressible by the perturbation series

$$[G_{ij}^L(0,0)]' = G_{ij}^0 + G_{ij}^0 v_0 G_{ij}^L(0,0) v_0 G_{ij}^0$$
$$+ G_{ij}^0 v_0 G_{ij}^L(0,0) v_0 G_{ij}^0 v_0 G_{ij}^L(0,0)$$
$$\times v_0 G_{ij}^0 + \cdots, \tag{P7.6}$$

which should be *identical* to $G_{ij}^L(0,0)$. Here $v_0 = -(\beta/a^2)a^d = -\beta$ is the constant off-diagonal matrix element [see Eq. (2.51)], and

$$G_{ij}^0 = \frac{\delta_{i,j}}{\kappa_0^2 a^2 - 2\beta(1 - \cos k_x^{(i)} a)} \tag{P7.7}$$

is the Green's function for the (isolated) ordered 1D strip that has been added. The meaning of this series is fairly straightforward: It involves G_{ij}^0 plus all the interaction paths with the rest of the perfect lead, which is completely embodied in $G_{ij}^L(0,0)$. The interaction in the z direction involves just the multiplication of G_{ij}^0 of the isolated strip by v_0 in linking it to the perfect lead, and the same in the other direction. The perturbation series may be summed to get

$$G_{ij}^L(0,0) = G_{ij}^0 \left[1 - v_0^2 G_{ij}^L(0,0) G_{ij}^0\right]^{-1}$$
$$= \left[\left(G_{ij}^0\right)^{-1} - v_0^2 G_{ij}^L(0,0)\right]^{-1}. \quad \text{(P7.8)}$$

Since everything is diagonal in i, j, (P7.8) represents a simple algebraic equation to get g_i and hence $G_{ij}^L(0,0)$:

$$g_i^{-1} = \frac{1}{m_0 a^2 - 2v_0 \cos k_x^{(i)} a - v_0^2 g_i^{-1}}, \quad \text{(P7.9)}$$

where $m_0 a^2 = \kappa_0^2 a^2 - 2\beta$. Solution of Eq. (P7.9) yields $G_{ij}^L(0,0) = g_i^{-1} \delta_{i,j}$

$$= \frac{2\delta_{i,j}}{m_0 a^2 - 2v_0 \cos k_x^{(i)} a + \sqrt{\left(m_0 a^2 - 2v_0 \cos k_x^{(i)} a\right)^2 - 4v_0^2}}, \quad \text{(P7.10)}$$

The $+$ sign in the denominator is chosen because the square root is imaginary, and we want the retarded Green's function, which must have a negative imaginary part.

The technique of generating Eq. (P7.8) may now be used to calculate $G_{ij}^L(1,1)$. In the operator notation, adding a *disordered* strip can be done in the same way as through the perturbation series, Eq. (P7.6). It follows that

$$G_{ab}^L(1,1) = \left\{\left[\mathbf{G}^0(1,1)\right]^{-1} - v_0^2 \mathbf{G}^L(0,0)\right\}_{ab}^{-1}, \quad \text{(P7.11)}$$

where a, b are the *real space coordinates along the x axis*, and $\mathbf{G}^0(1,1)$ is the Green's function of the first strip of the disordered medium. In real space, it is given by

$$[\mathbf{G}^0(1,1)]^{-1} = \begin{pmatrix} [m_0 - \sigma_1(1)]a^2 & v_0 & 0 & \cdots \\ v_0 & [m_0 - \sigma_1(2)]a^2 & v_0 & \\ 0 & v_0 & [m_0 - \sigma_1(3)]a^2 & \cdots \\ \vdots & \vdots & \vdots & \end{pmatrix},$$
(P7.12)

where $\sigma_1(m)$ means the random perturbation at site ma and strip 1. To be consistent, $\mathbf{G}^L(0,0)$ also has to be expressed in real space; i.e., $G^L_{ab}(0,0)$ may be obtained from $G^L_{ij}(0,0) = \delta_{i,j} g_i^{-1}$ as

$$G^L_{ab}(0,0) = \frac{1}{M} \sum_{i=1}^{M} g_i^{-1} \exp\left[ik_x^{(i)}(x_a - x_b)\right], \qquad (P7.13)$$

where x_a, x_b is an integral multiple of a, ranging from a to Ma. From Eq. (P7.11) $G^L_{ab}(n,n)$ can be obtained recursively as

$$\mathbf{G}^L(n,n) = \left\{ [\mathbf{G}^0(n,n)]^{-1} - v_0^2 \mathbf{G}^L(n-1, n-1) \right\}^{-1}. \qquad (P7.14)$$

Here $\mathbf{G}^0(n,n)$ is the Green's function of the nth strip of the disordered medium. Numerically, the price one has to pay is a matrix inversion at each step. This can continue until $G^L_{ab}(N,N)$ is reached. The strip N is linked to the perfect lead to the right and the strip $N-1$ to the left. That means

$$\mathbf{G}(N,N) = \left\{ [\mathbf{G}^0(N,N)]^{-1} - v_0^2 \mathbf{G}^L(N-1, N-1) \right.$$
$$\left. - v_0^2 \mathbf{G}^R(N+1, N+1) \right\}^{-1}, \qquad (P7.15)$$

where the superscript L is dropped on $\mathbf{G}(N,N)$ because it is now a Green's function of the full system. $\mathbf{G}^R(N+1, N+1)$ is the Green's function of the perfect lead on the right side. In real space, it is given by

$$G^R_{ab}(N+1, N+1) = \frac{1}{M} \sum_{j=1}^{M} g_j^{-1} \exp\left[ik_x^{(j)}(x_a - x_b)\right]. \qquad (P7.16)$$

Having obtained the diagonal terms of the Green's functions, we proceed to calculate $\mathbf{G}(0, N+1)$. To that effect, it is noted that

$$\mathbf{G}^L(0,1) = \mathbf{G}^L(0,0) v_0 \mathbf{G}^0(1,1)$$
$$+ \mathbf{G}^L(0,0) v_0 \mathbf{G}^0(1,1) v_0 \mathbf{G}^L(0,0) v_0 \mathbf{G}^0(1,1) + \cdots$$
$$= \mathbf{G}^L(0,0) v_0 \mathbf{G}^L(1,1), \qquad (P7.17)$$

and $\mathbf{G}^L(0,2) = \mathbf{G}^L(0,1) v_0 \mathbf{G}^L(2,2)$. It follows straightforwardly that

$$\mathbf{G}(0, N+1) = \mathbf{G}^L(0, N) v_0 \mathbf{G}(N+1, N+1) \qquad (P7.18)$$

may be obtained recursively. Once $\mathbf{G}(0, N + 1)$ is evaluated, $|\tau|^2$ is given by

$$|\tau|^2 = a^2 \sum_{i}^{M_0} \sum_{j}^{M_0} \bar{k}_z(i)\bar{k}_z(j)|G_{ij}^+(0, N + 1)|^2, \qquad \text{(P7.19)}$$

where

$$G_{ij}^+(0, N + 1) = \exp\left[ik_z^{(j)}(N + 1)\right] \sum_{x_a}^{Ma} \sum_{x_b}^{Ma}$$

$$\times G_{ab}^+(0, N + 1) \exp\left[i(k_x^{(i)} x_a - k_x^{(j)} x_b)\right]. \quad \text{(P7.20)}$$

The values of x_a, x_b range from a to Ma. By configurationally averaging $\ln|\tau|^2$, ξ_M can be easily obtained as $-2Na/\langle\ln|\tau^2|\rangle_c$.

References

Abrahams, E., Anderson, P. W., Licciardello, D. C., and Ramakrishnan, T. V. (1979). *Phys. Rev. Lett.* **42**, 673.
Edward, J. T., and Thouless, D. J. (1972). *J. Phys.* **C5**, 807.
Landauer, R. (1970). *Philos. Mag.* **21**, 863.
Lee, P., and Fisher, D. (1981). *Phys. Rev. Lett.* **47**, 882.
Li, Q., Soukoulis, C. M., Economou, E. N., and Grest, G. S. (1989). *Phys. Rev.* **B40**, 282.
Licciardello, D. C., and Thouless, D. J. (1975). *J. Phys.* **C8**, 4157.
MacKinnon, A., and Kramer, B. (1981). *Phys. Rev. Lett.* **47**, 1546.
MacKinnon, A., and Kramer, B. (1983). *Z. Phys. B-Condensed Matter* **52**, 1.
Mott, N. F., Pepper, M., Pollitt, S., Wallis, R. H., and Adkins, C. J. (1975). *Proc. Roy. Soc.* **A345** 169.
Thouless, D. J. (1974). *Phys. Rep.* **13C**, 93.
Thouless, D. J. (1979). "Ill-Condensed Matter," (R. Balian, R., Maynard, and G. Toulouse, eds.), p. 43. North-Holland, Amsterdam.
Xue, W., Sheng, P., Chu, Q. J., and Zhang, Z. Q. (1989). *Phys. Rev. Lett.* **63**, 2837.
Zhang, Z. Q., and Sheng, P. (1991a), *Phys. Rev.* **B44**, 3304.
Zhang, Z. Q., and Sheng, P. (1991b), *Phys. Rev. Lett.* **67**, 2541.

8

Localized States and the Approach to Localization

8.1 The Self-Consistent Theory of Localization

The purposes of this chapter are to substantiate the scaling theory results by continuing the Green's function development left off at Chapter 6 and to fill in the details about the localized states and the approach to localization. It may be recalled that aside from the description of 1D localization, the estimates of 2D and 1D localization lengths in Chapter 6 [Eqs. (6.23) and (6.24)] have been based on the weak scattering approximation. Besides being inconsistent, the mathematical framework cannot demonstrate how the localized states come into being. Therefore, further development obviously calls for the extension of the Green's function formalism into the strong scattering regime, where localization is possible.

The starting point of our consideration is Eq. (6.17), without the approximate evaluation of $M_0^{(\mathrm{MC})}$:

$$\frac{1}{D(\omega)} = \frac{1}{D^{(\mathrm{B})}(\omega)} + \frac{\pi \rho_\mathrm{e}(\omega)}{8K_0^2(d\kappa_\mathrm{e}^2/d\kappa_0^2)(-\mathrm{Im}\,\Sigma^+)} M_0^{(\mathrm{MC})}.$$

Multiplying both sides by $D(\omega)D^{(\mathrm{B})}(\omega)$ yields

$$D^{(\mathrm{B})}(\omega) = D(\omega) + D(\omega)D^{(\mathrm{B})}(\omega) \frac{\pi \rho_\mathrm{e}(\omega)}{8K_0^2(-\mathrm{Im}\,\Sigma^+)(d\kappa_\mathrm{e}^2/d\kappa_0^2)} M_0^{(\mathrm{MC})}.$$

After rearrangement, this becomes

$$D(\omega) = D^{(\mathrm{B})}(\omega)\left[1 - \frac{\pi \rho_\mathrm{e}(\omega)}{8K_0^2(-\mathrm{Im}\,\Sigma^+)(d\kappa_\mathrm{e}^2/d\kappa_0^2)} D(\omega)M_0^{(\mathrm{MC})}\right]$$

$$= D^{(\mathrm{B})}(\omega)(1 - C). \qquad (8.1)$$

From Eqs. (6.11) and (6.12), it is straightforward to write the dimensionless correction term C as

$$C = \frac{\pi \rho_e(\omega)}{8K_0^2(-\text{Im }\Sigma^+)(d\kappa_e^2/d\kappa_0^2)} D(\omega) M_0^{(\text{MC})}$$

$$= \frac{i}{4K_0^2} \frac{L^{3d}}{N^3} \iint \frac{d\mathbf{k}_i}{(2\pi)^d} \frac{d\mathbf{k}_f}{(2\pi)^d}$$

$$\times (\Delta G_e)_{\mathbf{k}_i} (\Delta G_e)_{\mathbf{k}_f}^2 \left[\nabla_{k_i} e(\mathbf{k}_i) \cdot \Delta \hat{\mathbf{k}} \right]$$

$$\times \left[\nabla_{k_f} e(\mathbf{k}_f) \cdot \Delta \hat{\mathbf{k}} \right] \frac{-\text{Im }\Sigma^+}{-\frac{i\Delta\omega}{D(\omega)} + \frac{D^{(B)}(\omega)}{D(\omega)} |\mathbf{k}_i + \mathbf{k}_f|^2}. \quad (8.2)$$

A crucial step in extending the Green's function formalism from the weak scattering regime to the strong regime is to replace $D^{(B)}(\omega)$ in the integrand of Eq. (8.2) by the renormalized diffusion constant, $D(\omega)$. That is, instead of using the irreducible vertex $U^{(B)}$ in the evaluation of $U^{(\text{MC})}$ as given by Eq. (6.7), one uses the more accurate form $U^{(B)} + U^{(\text{MC})}$. This has the effect of making the theory self-consistent as first proposed by Vollhardt and Wölfle (1980). Mathematically, this replacement is equivalent to extending the theory to the strong scattering regime. The correction term C after this replacement becomes

$$C = i\left(\frac{-\text{Im }\Sigma^+}{4K_0^2}\right) \frac{L^{3d}}{N^3} \iint \frac{d\mathbf{k}_i}{(2\pi)^d} \frac{d\mathbf{k}_f}{(2\pi)^d}$$

$$\times (\Delta G_e)_{\mathbf{k}_i} \left[\nabla_{k_i} e(\mathbf{k}_i) \cdot \Delta \hat{\mathbf{k}} \right]$$

$$\times \frac{1}{-\frac{i\Delta\omega}{D(\omega)} + |\mathbf{k}_i + \mathbf{k}_f|^2} (\Delta G_e)_{\mathbf{k}_f}^2 \left[\nabla_{k_f} e(\mathbf{k}_f) \cdot \Delta \hat{\mathbf{k}} \right]. \quad (8.3)$$

The term $-i\Delta\omega/D(\omega)$ has two distinct limits when $\Delta\omega \to 0$. Since in the electronic case $D(\omega)$ may be related to the conductivity by the Einstein relation, we can use the conductivity behavior to guide us in getting the limits. In the conducting regime, $D(\omega)$ is always finite so

$$\lim_{\Delta\omega \to 0} \frac{-i\Delta\omega}{D(\omega)} = 0.$$

8.1 The Self-Consistent Theory of Localization

On the other hand, in the insulating (localized) regime the dc diffusivity $D(\omega) = 0$ by definition, so the limiting behavior is not immediately clear. However, in the solution to Problem 8.1 it is shown that at finite modulation frequency $\Delta\omega$ the conductivity may be regarded as complex. Therefore even if the real part of the conductivity is zero, there can still be a 90° out-of-phase (imaginary) component of the conductivity whose physical origin lies in the polarizability of the localized state, in exact analogy to the polarizability of a bound electron in an atom. In that case

$$D(\omega) = -i\,\Delta\omega\,(\text{length scale})^2. \tag{8.4a}$$

Physically, the only length scale in the problem is the size of the localized state as characterized by the localization length ξ. Therefore

$$\frac{-i\,\Delta\omega}{D(\omega)} = \nu^2 \propto \xi^{-2} \tag{8.4b}$$

in the localized regime. The constant of proportionality, $\nu\xi$, evaluated in the 1D classical wave case is on the order of 2 (~ 1.84). For the Anderson model the comparison of numerical results with those calculated from the self-consistent theory gives a constant that is much larger ($\sim 5{-}10$). This difference could be due to the difference in the dispersion relations.

From Eqs. (8.1) and (8.4b), the condition for determining the mobility edge is $C = 1$ with $\nu^2 = 0$. Since $D(\omega)$ cannot be negative, the same $C = 1$ condition gives an equation for ν^2 in the localized regime. In fact, a quick glimpse of the consequences of the self-consistent theory may be obtained by evaluating the multiple integral of Eq. (8.3) in the manner prescribed in Chapter 6. That is, by setting $\mathbf{k}_i = -\mathbf{k}_f$ so that $\nabla_{k_i} e(\mathbf{k}_i) = -\nabla_{k_f} e(\mathbf{k}_f)$, and approximating $(\Delta G_e)^2_{\mathbf{k}} \cong [(\Delta G_e)_{\mathbf{k}} / - i\,\mathrm{Im}\,\Sigma^+](N/L^d)$, Eq. (8.3) becomes

$$C = \frac{1}{K_0} \int \frac{d\mathbf{Q}}{(2\pi)^d} \frac{1}{\nu^2 + Q^2}. \tag{8.5a}$$

It may be argued that there is a physical upper cutoff for the magnitude of Q, defined by the width of the coherent backscattering peak, α/l, where α is some adjustable constant. Alternatively, one can argue that the diffusive behavior exists only for length scale $> l$, and as a result the maximum value of Q is defined by α/l. In any case, by using such a cutoff, we get

$$C = \frac{S_d}{(2\pi)^d K_0} \begin{cases} \dfrac{\alpha}{l} - \nu \tan^{-1} \dfrac{\alpha}{\nu l} & \text{3D} \\ \dfrac{1}{2} \ln\left(1 + \dfrac{\alpha^2}{\nu^2 l^2}\right) & \text{2D}, \\ \dfrac{1}{\nu} \tan^{-1} \dfrac{\alpha}{\nu l} & \text{1D} \end{cases} \qquad (8.5b)$$

where $S_d = 4\pi, 2\pi, 2$ in 3D, 2D, 1D, respectively. It is seen that in 3D, $\nu = 0$ and $C = 1$ give the mobility edge equations

$$l = \frac{\alpha}{2\pi^2 K_0}, \qquad (8.6)$$

which yields a critical randomness for every frequency. In 2D and 1D, ν has to be finite in order for C not to diverge, and the equations for the localization length are

$$\frac{1}{\nu} = \frac{l}{\alpha}[\exp(4\pi K_0) - 1]^{1/2}, \qquad \text{2D} \qquad (8.7a)$$

$$\pi K_0 = \frac{1}{\nu} \tan^{-1} \frac{\alpha}{\nu l}. \qquad \text{1D} \qquad (8.7b)$$

Although the approximate evaluation of C in this manner can give us all the qualitative localization behavior, it is nevertheless desirable to have a look at the consequences of the theory with as little approximation as feasible, so that the true nature of the self-consistent theory may be manifest. Below we treat the Anderson model and the classical scalar wave case separately.

8.2 Localization Behavior of the Anderson Model

The C factor involves a $2d$-fold integral for the MC diagram contributions. To locate the mobility edge and to calculate the localization length in $d = 3$ and 2, this integral is difficult to evaluate due to the nonspherical Brillouin zone. What one can do is to define a wave vector $\bar{k}$ that is the

8.2 Localization Behavior of the Anderson Model

average over the constant energy surface. In accordance with the rules worked out in the solution to Problem 8.2, that means in 3D

$$\bar{ak}(E) = \frac{16\int_0^\pi \int_0^\pi \int_0^\pi dx\,dy\,dz\,\sqrt{x^2 + y^2 + z^2}\,\sqrt{\sin^2 x + \sin^2 y + \sin^2 z}}{S_0(E)},$$

$$\times \delta[E + 2\cos x + 2\cos y + 2\cos z]$$

(8.8a)

where

$$S_0(E) = 16 \int_0^\pi \int_0^\pi \int_0^\pi dx\,dy\,dz\,\sqrt{\sin^2 x + \sin^2 y + \sin^2 z}$$
$$\times \delta[E + 2\cos x + 2\cos y + 2\cos z],$$

(8.8b)

is the constant-E surface area in the **k**-space, in units of $1/a^2$. Figure 8.1 gives a plot of $\bar{ak}(E)$ versus E, for $-6 < E < 6$. Here $E = 2u$, where u is the dimensionless energy/frequency parameter defined in Chapter 2.

Another factor that should be considered in the strong scattering limit is the width of $(\Delta G_e)_\mathbf{k}$. Since $-\text{Im}\,\Sigma^+$ may be appreciable, the delta-function approximation is no longer adequate, and one should use

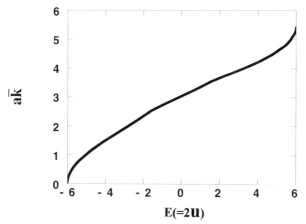

Figure 8.1 Average wave vector $\bar{k}$ over the constant energy surface for the 3D Anderson model plotted as a function of $E = 2u$, where u is the reduced frequency variable as defined in Figure 2.5. Here $\bar{k}$ is defined relative to the lower band edge.

the accurate representation

$$\frac{L^d}{N}(\Delta G_e)_\mathbf{k} = \frac{2i \operatorname{Im} \Sigma^+}{\left[\kappa_0^2 - \operatorname{Re}\Sigma^+ - e(\mathbf{k})\right]^2 + (\operatorname{Im}\Sigma^+)^2}. \quad (8.9)$$

With these considerations in mind, we can approach the evaluation of the maximally crossed (MC) diagrams' contribution by converting the **k** integrals into $dE\, dS$ integrals as shown in Problem 8.2, where dS denotes surface integration over the constant-frequency surface. To see it explicitly, let us denote the double **k**-integral factor in C as (MC); then

$$(\text{MC}) = i\frac{L^{3d}}{N^3}\frac{1}{(2\pi)^6}\int d\mathbf{k}_i \int d\mathbf{k}_f \left[\nabla_{k_i}e(\mathbf{k}_i)\cdot\Delta\hat{\mathbf{k}}\right]$$

$$\times \left[\nabla_{k_f}e(\mathbf{k}_i)\cdot\Delta\hat{\mathbf{k}}\right](\Delta G_e)_{\mathbf{k}_i}(\Delta G_e)_{\mathbf{k}_f}^2 \frac{1}{\nu^2+|\mathbf{k}_i+\mathbf{k}_f|^2}$$

$$\cong -\frac{8}{(2\pi)^6}\int_{E_l}^{E_u}dE_x\int_{E_l}^{E_u}dE_y\int dS_x\int dS_y f(E_x)f^2(E_y)$$

$$\times \frac{\cos\theta_x \cos\theta_y}{\nu^2 a^2 + x^2 + y^2 + 2xy\cos\theta_{xy}}. \quad (8.10)$$

Here the direction of $\nabla_k e(\mathbf{k})$ is approximated as being the same as **k**. This is certainly true at the lower band edge, where $e(\mathbf{k}) = k^2$ so that $\nabla_k e(\mathbf{k}) = 2\mathbf{k}$. Also, in (8.10) the parameters are nondimensionalized, with $y_0 = a^2\Sigma^+$ denoting the dimensionless self-energy, and

$$f(E) = \frac{-\operatorname{Im} y_0}{E^2 + (\operatorname{Im} y_0)^2}$$

denoting the dimensionless form of $(\Delta G_e)_\mathbf{k}$, to be used extensively below. In addition,

$$E_l = 2u - \operatorname{Re} y_0 - 2d,$$
$$E_u = 2u - \operatorname{Re} y_0 + 2d,$$
$$u = \frac{\kappa_0^2 a^2 - 2d}{2}, \quad (\beta = 1) \quad (8.11)$$

and

$$x = a|\mathbf{k}_i|, \quad y = a|\mathbf{k}_f|.$$

8.2 Localization Behavior of the Anderson Model

θ_x, θ_y are the angles between $\Delta \hat{k}$ and $\mathbf{k}_i$, $\mathbf{k}_f$, respectively. θ_{xy} is the angle between $\mathbf{k}_i$ and $\mathbf{k}_f$, and dS_x, dS_y denote the constant $e(\mathbf{k}_i)$, $e(\mathbf{k}_f)$ surface differentials. In (8.10), $x = a|\mathbf{k}_i|$ is related to E_x via the dispersion relation

$$E_x = \kappa_0^2 a^2 - \text{Re } y_0 - a^2 e(\mathbf{k}_i),$$

and similarly for y to E_y. The dS integrals are thus coupled to the dE integrals. To facilitate the evaluation of (MC), let us approximate x, y by

$$x = a\bar{k}_i(E'_x),$$
$$y = a\bar{k}_f(E'_x),$$

where $\bar{k}_i$, $\bar{k}_f$ are defined in accordance with Eq. (8.8a), and

$$E'_{x,y} = 2u - \text{Re } y_0 - E_{x,y}.$$

The approximation has the effect of decoupling the dS integrals from the dE integrations, i.e.,

$$(\text{MC}) \cong -\frac{8}{(2\pi)^6} \int_{E_l}^{E_u} dE_x \int_{E_l}^{E_u} dE_y f(E_x) f^2(E_y) \frac{S_0(E'_x)}{4\pi} \frac{S_0(E'_y)}{4\pi}$$

$$\times \left\{ \int_0^{2\pi} d\phi_x \int_0^{2\pi} d\phi_y \int_{-1}^{1} d\cos\theta_x \int_{-1}^{1} d\cos\theta_y \right.$$

$$\left. \times \frac{\cos\theta_x \cos\theta_y}{\nu^2 a^2 + x^2 + y^2 + 2xy\cos\theta_{xy}} \right\}. \quad (8.12\text{a})$$

The angular integration in the curly brackets can be explicitly evaluated as illustrated in Problem 8.3, resulting in

$$\{\} = \frac{8\pi^2}{3} \frac{1}{xy} \left[1 - \frac{x^2 + y^2 + a^2\nu^2}{4xy} \ln \frac{(x+y)^2 + a^2\nu^2}{(x-y)^2 + a^2\nu^2} \right]. \quad (8.12\text{b})$$

Substitution of (8.12b) into the previous expression yields

$$(\text{MC}) = \frac{4}{3} \frac{1}{(2\pi)^6} A_3[u, y_0(\sigma_w, u), a^2\nu^2], \quad (8.13\text{a})$$

$$A_3[u, y_0(\sigma_w, u), a^2\nu^2]$$

$$= \int_{E_l}^{E_u} dE_x \int_{E_l}^{E_u} dE_y f(E_x) f^2(E_y) S_0(E'_x) S_0(E'_y) (xy)^{-1}$$

$$\times \left[\frac{x^2 + y^2 + a^2\nu^2}{4xy} \ln \frac{(x+y)^2 + a^2\nu^2}{(x-y)^2 + a^2\nu^2} - 1 \right]. \quad (8.13\text{b})$$

The subscript 3 is used to indicate that the function A_3 is for 3D. The logarithmic divergence present in the integrand of (8.13b) is integrable; i.e., it is nondivergent upon integration. This fact is crucial for setting apart the 3D localization behavior from those of 2D, and 1D, as will be seen later.

An important point about the (MC) contribution is that it does not obey the particle–hole symmetry about the center of the band. This is traced to the form of $U^{(MC)} \propto 1/[\nu^2 + |\mathbf{k}_i + \mathbf{k}_f|^2]$, where $\mathbf{k}_i$ and $\mathbf{k}_f$ are *not* the arguments of any trignometric function. That means in our present calculation if the particle–hole symmetry were to be restored, one heuristic way would be to use Eq. (8.8a) only for $-3 \leq u \leq 0 (-6 \leq E \leq 0)$. For $3 \geq u \geq 0$ the magnitude of $\mathbf{k}$ should be defined relative to the upper band edge. It has to be emphasized, however, that the problem of asymmetry with respect to $u = 0$ is not the result of any approximation used to evaluate $M_0^{(MC)}$, but is rather in the form of $U^{(MC)}$, which is only approximate due to the small Q expansion.

For the $2K_0$ factor in C, accurate numerical evaluation may be carried out by expressing it in the following form:

$$2K_0 = \frac{2}{(2\pi)^3} \int d\mathbf{k} \left[\nabla_k e(\mathbf{k}) \cdot \Delta\hat{\mathbf{k}} \right]^2$$

$$\times \left\{ \frac{-\mathrm{Im}\,\Sigma^+}{\left[\kappa_0^2 - \mathrm{Re}\,\Sigma^+ - e(\mathbf{k})\right]^2 + (\mathrm{Im}\,\Sigma^+)^2} \right\}^2$$

$$= \frac{32}{(2\pi)^3 a} B_3[u, y_0(\sigma_w, u)], \qquad (8.14)$$

where

$$B_3[u, y_0(\sigma_w, a)] = \int_{E_l}^{E_u} dE\, W_3(E') f^2(E), \qquad (8.15)$$

and

$$W_3(E') = \int_0^\pi \int_0^\pi \mathrm{Re} \sqrt{1 - \left(\frac{E'}{2} + \cos x + \cos y\right)^2}\, dx\, dy, \qquad (8.16)$$

with $E' = 2u - \mathrm{Re}\,y_0 - E$. The derivation of Eqs. (8.15) and (8.16) is shown in the solutions to Problem 8.2. The mobility edge equation can

8.2 Localization Behavior of the Anderson Model

now be expressed as

$$1 = C = \frac{-\operatorname{Im} \Sigma^+}{(2K_0)^2} (MC)$$

$$= \frac{1}{768} [-\operatorname{Im} y_0(\sigma_w, u)] \frac{A_3[u, y_0(\sigma_w, u), a^2 v^2 = 0]}{\{B_3[u, y_0(\sigma_w, u)]\}^2}. \quad (8.17)$$

Equation (8.17) represents a condition on σ_w for a given u, and vice versa. It can be solved numerically by first tabulating the functions x, y, and $W_3(E)$ and then carrying out the integrations in Eqs. (8.15) and (8.17). The results are plotted in Figure 8.2. At $u = 0$, the mobility edge is found to be at $(\sigma_w a^2)_c = 6$, which compares reasonably with the numerical simulation result of $(\sigma_w a^2)_c = 8\text{--}8.25$ (Zdetsis et al., 1985). The comparison with numerical simulation over the whole mobility-edge curve shows general qualitative agreement. The fact that there is no adjustable parameter in the calculation makes it apparent that there is some overestimation of the localization effect in the self-consistent theory. This could be due to the approximate form of $U^{(MC)}$ (owing to the small $|\mathbf{k}_i + \mathbf{k}_f|$ expansion). Higher-order terms in $|\mathbf{k}_i + \mathbf{k}_f|$ may provide a natural cutoff for the integral and make it smaller.

From Figure 8.2 it is seen that as $\sigma_w a^2$ increases from zero, the first occurrence of localization is near the band edge. Physically, this is due to

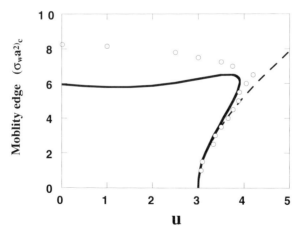

Figure 8.2 Mobility edge trajectory for the 3D Anderson model calculated from the self-consistent theory. The $u < 0$ half is symmetric to the $u > 0$ half. The dashed line is the CPA band edge, defined by the condition $-\operatorname{Im} \Sigma^+ = 0$. Numerical simulation results (Zdetsis et al., 1985) are shown as open circles.

the small group velocity of the band-edge states. As $\sigma_w a^2$ continues to increase, however, the band-edge states are seen to delocalize. The reason for this behavior is that with the increase of randomness the CPA band edge moves toward larger frequencies, thus making the original band-edge states more like those inside the band. These states will eventually localize, however, with further increase in randomness. The rather flat top of the mobility edge curve means that as the randomness is increased toward $(\sigma_w a^2)_c$ and crosses it, the mobility edge closes in very rapidly from the band edges, making all the states in the band localized.

When $C < 1$, Eq. (8.3) may be written alternatively as

$$D(\omega) = D^{(B)}(\omega)\frac{l}{\zeta} = D^{(B)}(\omega)(1 - C), \tag{8.18}$$

(with $\nu^2 = 0$ in the A_3 factor) in accordance with the scaling concept of the correlation length ζ, which determines the scale within which the classical diffusion behavior is significantly modified. It follows that

$$\frac{\zeta}{l} = \frac{1}{1-C}. \tag{8.19}$$

In the localized regime, the equation for the localization length is

$$1 = C = \frac{1}{768}[-\operatorname{Im} y_0(\sigma_w, u)]\frac{A_3[u, y_0(\sigma_w, u), \nu^2 a^2]}{\{B_3[u, y_0(\sigma_w, u)]\}^2}. \tag{8.20}$$

At $u = 0$, the solution of Eq. (8.20) for ξ/a is shown in Figure 8.3 as a function of $\sigma_w a^2$. The value of ζ/l on the other side of the mobility edge is also plotted on the same graph, as well as the value of l/a, expressed as

$$\frac{l}{a} = \sqrt{6}\left\{\frac{B_3[u, y_0(\sigma_w, u)]}{C_3[u, y_0(\sigma_w, u)]}\right\}^{1/2}\frac{1}{(-\operatorname{Im} y_0)^{1/2}}. \tag{8.21}$$

where

$$C_3[u, y_0(\sigma_w, u)] = \int_{E_l}^{E_u} dE\, V_3(2u - \operatorname{Re} y_0 - E)f(E), \tag{8.22}$$

$$V_3(E) = \int_0^\pi \int_0^\pi \operatorname{Re}\left[1 - \left(\frac{E}{2} + \cos x + \cos y\right)^2\right]^{-1/2} dx\, dy. \tag{8.23}$$

Equation (8.21) is the strong-scattering version of the definition for l. In Figure 8.3, the functional form of ξ is found to accurately reflect the relation

$$\xi^{-1} \propto \left[\sigma_w a^2 - (\sigma_w a^2)_c\right],$$

8.2 Localization Behavior of the Anderson Model

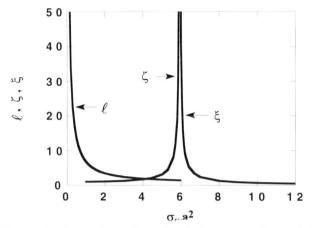

Figure 8.3 Localization length ξ, correlation length ζ, and mean free path l for the 3D Anderson model, plotted as a function of randomness $\sigma_w a^2$. Here $u = 0$ (band center). ξ and ζ are both calculated from the self-consistent theory.

which means that the exponent of divergence is 1 in the self-consistent theory of localization. The value of $\bar{\kappa}_e l$ can also be calculated, given by

$$\bar{\kappa}_e l = \frac{\bar{\kappa}_e^2}{(-\operatorname{Im} \Sigma^+)} = 6 \frac{B_3[u, y_0(\sigma_w, u)]}{C_3[u, y_0(\sigma_w, u)]}. \tag{8.24}$$

The value of $\bar{\kappa}_e l$ along the mobility edge is found to be

$$(\bar{\kappa}_e l)_c = 0.8\text{--}1.$$

The narrow range of variation for this dimensionless parameter may be regarded as support for the Ioffe–Regel criterion in the Anderson model (Zdetsis et al., 1985).

In 2D, a similar approach may be applied to the calculation of C. To make the dispersion relation isotopic, one uses the relation

$$a\bar{k}(E') = \frac{8 \int_0^\pi \int_0^\pi dx\, dy\, \sqrt{x^2 + y^2}\, \sqrt{\sin^2 x + \sin^2 y}\, \delta[E + 2\cos x + 2\cos y]}{L_0(E)}.$$

$$(8.25)$$

where

$$L_0(E) = 8\int_0^\pi\int_0^\pi dx\,dy\,\sqrt{\sin^2 x + \sin^2 y}\,\,\delta[E + 2\cos x + 2\cos y], \quad (8.26)$$

with $-4 \le E \le 4$. Let us first consider the multiple integral factor in C that may be written as

$$(MC) = \frac{i}{(2\pi)^4}\int d\mathbf{k}_i\int d\mathbf{k}_f\left[\nabla_{k_i}e(\mathbf{k}_i)\cdot\Delta\hat{\mathbf{k}}\right]\left[\nabla_{k_f}e(\mathbf{k}_f)\cdot\Delta\hat{\mathbf{k}}\right]$$

$$\times(\Delta G_e)_{\mathbf{k}_i}(\Delta G_e)^2_{\mathbf{k}_f}\frac{1}{\nu^2 + |\mathbf{k}_i + \mathbf{k}_f|^2}\frac{L^{3d}}{N^3}$$

$$\cong -\frac{8a^2}{(2\pi)^4}\int_{E_l}^{E_u}dE_x\int_{E_l}^{E_u}dE_y\,\frac{L_0(E'_x)}{2\pi}\frac{L_0(E'_y)}{2\pi}f(E_x)f^2(E_y)$$

$$\times\left\{\int_0^{2\pi}d\theta_x\int_0^{2\pi}d\theta_y\,\frac{\cos\theta_x\cos\theta_y}{a^2\nu^2 + x^2 + y^2 + 2xy\cos\theta_{xy}}\right\}. \quad (8.27)$$

Just as in the case of 3D, here the direction of $\nabla e(\mathbf{k})$ is approximated as being the same as $\mathbf{k}$, and $x = a\bar{k}_i\,(E'_x)$, $y = a\bar{k}_f(E'_y)$ are evaluated in accordance with Eq. (8.25), with the magnitude of $\mathbf{k}_i$ defined from the lower band edge for $u \le 0$, and from the upper band edge for $u > 0$. $E'_{x,y} = 2u - \text{Re}\,y_0 - E_{x,y}$, $\theta_{x,y}$ is the angle between $\Delta\hat{\mathbf{k}}$ and $\mathbf{k}_{i,f}$, and $\theta_{xy} = \theta_y - \theta_x$, so that $\cos\theta_y = \cos\theta_x\cos\theta_{xy} - \sin\theta_x\sin\theta_{xy}$. The angular integral in the curly brackets may be performed to yield

$$\{\} = \int_0^{2\pi}d\theta_x\cos^2\theta_x\int_0^{2\pi}d\theta_{xy}\,\frac{\cos\theta_{xy}}{a^2\nu^2 + x^2 + y^2 + 2xy\cos\theta_{xy}}$$

$$= \frac{\pi}{2xy}\int_0^{2\pi}d\theta_{xy}\left[1 - \frac{a^2\nu^2 + x^2 + y^2}{a^2\nu^2 + x^2 + y^2 + 2xy\cos\theta_{xy}}\right]$$

$$= \frac{\pi^2}{xy}\left[1 - \frac{a^2\nu^2 + x^2 + y^2}{\sqrt{(a^2\nu^2 + x^2 + y^2)^2 - 4x^2y^2}}\right], \quad (8.28)$$

8.2 Localization Behavior of the Anderson Model

where the $\sin\theta_x \sin\theta_{xy}$ term is noted to integrate to zero. Substitution of (8.28) into (8.27) yields

$$(\text{MC}) = \frac{2a^2}{(2\pi)^4} A_2[u, y_0(\sigma_w, u), a^2v^2], \qquad (8.29a)$$

$$A_2[u, y_0(\sigma_w, u), a^2v^2]$$
$$= \int_{E_l}^{E_u} dE_x \int_{E_l}^{E_u} dE_y f(E_x) f^2(E_y) L_0(E_x') L_0(E_y')(xy)^{-1}$$
$$\times \left[\frac{a^2v^2 + x^2 + y^2}{\sqrt{(a^2v^2 + x^2 + y^2)^2 - 4x^2y^2}} - 1 \right], \qquad (8.29b)$$

$$x = a\bar{k}_i(2u - \operatorname{Re} y_0 - E_x), \qquad (8.30a)$$
$$y = a\bar{k}_f(2u - \operatorname{Re} y_0 - E_y), \qquad (8.30b)$$

and $E_u = 2u - \operatorname{Re} y_0 + 4$, $E_l = 2u - \operatorname{Re} y_0 - 4$. From Eq. (8.29b) it is evident that A_2 diverges when $v \to 0$; i.e., the contribution of the MC diagrams dictates that in order to avoid divergence, $v \neq 0$ in 2D, so *waves are localized for all u and σ_w*. For the factor $2K_0$, we have

$$2K_0 = \frac{2}{(2\pi)^2} \int d\mathbf{k} [\nabla_k e(\mathbf{k}) \cdot \Delta\hat{\mathbf{k}}]^2 \left\{ \frac{-\operatorname{Im}\Sigma^+}{[\kappa_0^2 - \operatorname{Re}\Sigma^+ - e(\mathbf{k})]^2 + (-\operatorname{Im}\Sigma^+)^2} \right\}^2$$

$$= \frac{2}{(2\pi)^2} \int dE' \int d\mathbf{k} [\nabla_k e(\mathbf{k}) \cdot \Delta\hat{\mathbf{k}}]^2 \delta[E' + 4 - a^2 e(\mathbf{k})]$$

$$\times \left\{ \frac{-\operatorname{Im}\Sigma^+}{[\kappa_0^2 - \operatorname{Re}\Sigma^+ - e(\mathbf{k})]^2 + (-\operatorname{Im}\Sigma^+)^2} \right\}^2$$

$$= \frac{16}{(2\pi)^2} B_2[u, y_0(\sigma_w, u)], \qquad (8.31)$$

where $y_0 = a^2\Sigma^+$,

$$B_2[u, y_0(\sigma_w, u)] = \int_{E_l}^{E_u} dE\, W_2(E') f^2(E), \qquad (8.32)$$

$$W_2(E') = \int_0^\pi \operatorname{Re} \sqrt{1 - \left(\frac{E'}{2} + \cos x\right)^2} dx, \qquad (8.33)$$

with $E' = 2u - \text{Re } y_0 - E$. Combining terms yields the 2D localization-length equation as

$$1 = C = \frac{1}{128}[-\text{Im } y_0(\alpha_w, u)] \frac{A_2[u, y_0(\sigma_w, u), v^2 a^2]}{\{B_2[u, y_0(\sigma_w, u)]\}^2}. \quad (8.34)$$

Since A_2 diverges as $\ln v$, one can write

$$A_2 = c_1 \ln v + c_2.$$

Then Eq. (8.34) implies

$$\frac{a}{\xi} \propto \exp\left[\frac{(B_2)^2}{-\text{Im } y_0} \frac{128}{c_1} - \frac{c_2}{c_1}\right].$$

With numerical evaluation of c_1 and c_2, the solution of Eq. (8.34) is plotted as a function of $\sigma_w a^2$ in Figure 8.4a. At small $\sigma_w a^2$, ξ/a is transcendentally large. In the solution to Problem 8.4 it is shown that the 2D localization length should behave as

$$\ln \frac{\xi}{a} \propto \frac{1}{(\sigma_w a^2)^2} \quad (8.35)$$

in the limit of small randomness. Accordingly, the plot is done in the form of $\ln(\xi/a)$ versus $(\sigma_w a^2)^{-2}$. It is seen that at small $\sigma_w a^2$, $\ln(\xi/a)$ is indeed linear in $(\sigma_w a^2)^{-2}$, whereas at large randomness $\ln(\xi/a)$ varies with a smaller power of $1/(\sigma_w a^2)$. The value of ξ/a obtained is noted to differ from those obtained from numerical simulation results by a factor of ~ 8. For a fixed $\sigma_w a^2 = 2.5$, the localization length solved from Eq. (8.34) is plotted in Figure 8.4b after it is multiplied by 8. The circles are the data from numerical simulation. The fact that the localization length is smaller near the center of the band is the result of the large density of states at those frequencies (it is divergent at the band center in the ordered case). As a result, the average $\bar{\kappa}_e$ is small, so that the group velocity $\hbar \bar{\kappa}_e/m$ is small at the band center. It is seen that there is overall general agreement with the simulation results.

8.2 Localization Behavior of the Anderson Model

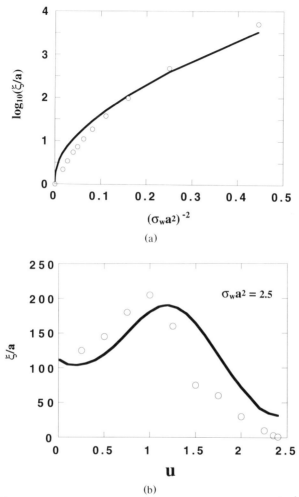

Figure 8.4 (a) Log of localization length ($\times 8$) as a function of $(\sigma_w a^2)^{-2}$ for the 2D Anderson model, calculated from the self-consistent theory at $u = 0$ (band center). (b) 2D localization length ($\times 8$) as a function of u, calculated from the self-consistent theory at $\sigma_w a^2 = 2.5$. Numerical simulation results (Zdetsis et al., 1985) are shown as open circles.

In 1D, no approximation is needed for the evaluation of C. The factor $2K_0$ may be expressed as

$$2K_0 = 2a \int_{-\pi}^{\pi} \frac{d(ka)}{2\pi} 4\sin^2 ka \left[\frac{-\operatorname{Im} y_0}{(2u - \operatorname{Re} y_0 + 2\cos ka)^2 + (\operatorname{Im} y_0)^2} \right]^2$$

$$= \frac{4a}{\pi} \int_{-\pi}^{\pi} dx \sin^2 x \left[\frac{-\operatorname{Im} y_0}{(2u - \operatorname{Re} y_0 + 2\cos x)^2 + (\operatorname{Im} y_0)^2} \right]^2$$

$$= \frac{4a}{2\pi} \int_{-2}^{2} d(2\cos x) \sqrt{1 - \cos^2 x}$$

$$\times \left[\frac{-\operatorname{Im} y_0}{(2u - \operatorname{Re} y_0 + 2\cos x)^2 + (\operatorname{Im} y_0)^2} \right]^2$$

$$= \frac{2a}{\pi} B_1[u, y_0(\sigma_w, u)], \qquad (8.36a)$$

where

$$B_1[u, y_0(\sigma_w, u)] = \int_{E_l}^{E_u} dE \operatorname{Re} \sqrt{1 - \left(\frac{E}{2} - u + \frac{\operatorname{Re} y_0}{2} \right)^2} f^2(E), \quad (8.36b)$$

$E_u = 2u - \operatorname{Re} y_0 + 2$, $E_l = 2u - \operatorname{Re} y_0 - 2$, and $u = (\kappa_0^2 a^2 - 2)/2$. The maximally crossed diagrams' contribution, on the other hand, is given by

$$(MC) = \left[32a^4/(2\pi)^2 \right] A_1[u, y_0(\sigma_{w,u}), a^2\nu^2],$$

where

$$A_1[u, y_0(\sigma_w, u), a^2\nu^2] = \int_{-\pi}^{\pi} dx \int_{-\pi}^{\pi} dy$$

$$\times \left[\frac{-\operatorname{Im} y_0}{(2u - \operatorname{Re} y_0 + 2\cos x)^2 + (\operatorname{Im} y_0)^2} \right].$$

$$\left[\frac{-\operatorname{Im} y_0}{(2u - \operatorname{Re} y_0 + 2\cos y)^2 + (\operatorname{Im} y_0)^2} \right]^2 \cdot \frac{\sin x \sin y}{a^2\nu^2 + \{(x+y)^2\}}. \quad (8.37)$$

8.2 Localization Behavior of the Anderson Model

Here $\{(x+y)^2\}$ means a higher-order evaluation of the term using Eq. (4.96d). In the present case it has the following precise meaning:

$$\{(x+y)^2\} = \frac{1}{2}\frac{\int_{-\pi}^{\pi}\left[g_+(\theta)g_-(\theta) - g_+\left(\theta + \frac{x+y}{2}\right)g_-\left(\theta - \frac{x+y}{2}\right)\right]d\theta}{\int_{-\pi}^{\pi}\sin^2\theta\, g_+^2(\theta)\,d\theta},$$

(8.38)

$$g_\pm(\theta) = \frac{1}{2u - \mathrm{Re}\, y_0 + 2\cos\theta \mp i\,\mathrm{Im}\, y_0}.$$

(8.38b)

The 1D localization length equation has the following form:

$$1 = C = 2[-\mathrm{Im}\, y_0(\sigma_w, u)]\frac{A_1[u, y_0(\sigma_w, u), \nu^2 a^2]}{\{B_1[u, y_0(\sigma_w, u)]\}^2}.$$

(8.39)

The divergence of A_1 as $\nu \to 0$ means there is no solution to Eq. (8.39) unless ν^2 is finite. This fact directly implies the *localization of all* 1D *waves*. The numerically evaluated 1D localization length at $u = 0$ is plotted as a function of $\sigma_w a^2$ in Figure 8.5a after it is multiplied by a factor of 5. In anticipation of ξ/a varying as some inverse power of $\sigma_w a^2$, the plot is done in the log–log scale. It is seen that at small $\sigma_w a^2$, $\alpha\xi^{-1}$ varies as $(\sigma_w a^2)^2$, in complete agreement with what has been calculated in Chapter 6. In fact, comparison with Figure 6.6 shows good agreement between the two calculations performed via very different routes. Of course, this agreement is helped by the fact that the constant (5) between ν and ξ^{-1} [see Eq. (8.7b)] is set by requiring the calculated ξ equal to the exact result at $u = 0$ and $\sigma_w a^2 = 0.5$ ($\xi \cong 105a$) (E. N. Economou, 1983). In Figure 8.5b a plot is made of the localization length as a function of u for $\sigma_w a^2 = 0.5$. Comparison with Figure 6.5 again shows reasonable agreement except near the band edge.

From the foregoing discussion, it is clear that all the localization calculations involve two elements. One element is the $\langle GG \rangle_c$ calculation, which provides the framework for deriving the functional form of the equations. The other equally important element is the CPA calculation of $\langle G \rangle_c$, which gives $\mathrm{Re}\,\Sigma^+$ and $\mathrm{Im}\,\Sigma^+$ as inputs to the equations. It is through Σ^+ that the localization parameters are linked to the randomness of the model. Both elements are essential to the accurate determination of the mobility edge and the localization length. In the next section it will be seen that in the context of the two elements described above, the difficulty with determining the classical wave localization parameters lies mostly in the accurate evaluation of Σ^+ in the intermediate frequency regime, where CPA has no solution or no unique solution.

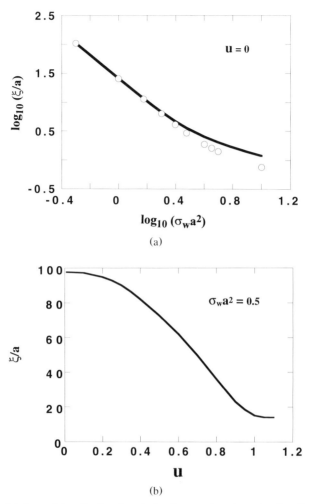

Figure 8.5 (a) Log of localization length (×5) as a function of $\ln(\sigma_w a^2)$ for the 1D Anderson model, calculated from the self-consistent theory at $u = 0$ (band center). Numerical simulation data (Economou *et al.*, 1984) are shown as open circles. (b) Localization length (×5) as a function of u, calculated from the self-consistent theory at $\sigma_w a^2 = 0.5$. This figure should be compared with Figure 6.5.

8.3 Classical Scalar Wave Localization

For classical scalar wave in continuum, $e(\mathbf{k}) = k^2$, and it becomes possible to express the correction term C in terms of just one dimensionless parameter. To proceed, let us define a complex $\kappa_e^* = \kappa + i/2l$ such that

$$G_e^+ = \frac{1}{(\kappa_e^*) - k^2} \frac{N}{L^d} = \frac{1}{\kappa_0^2 - k^2 - \Sigma^+} \frac{N}{L^d}. \qquad (8.40)$$

That means

$$\text{Re}(\kappa_e^*)^2 = \kappa^2 - \frac{1}{4l^2} = \kappa_0^2 - \text{Re}\,\Sigma^+, \qquad (8.41a)$$

$$\text{Im}(\kappa_e^*)^2 = \frac{\kappa}{l} = -\text{Im}\,\Sigma^+. \qquad (8.41b)$$

Note that when $\kappa l \gg 1$, $\kappa^2 \cong \kappa_e^2 = \kappa_0^2 - \text{Re}\,\Sigma^+$ as defined before. However, the possibility that $\kappa l \cong 1$ has now to be taken into account. As a result, $(\Delta G_e)_\mathbf{k}$ should be written as

$$\begin{aligned}(\Delta G_e)_\mathbf{k} &= -2i\frac{-\text{Im}\,\Sigma^+}{(\kappa_0^2 - \text{Re}\,\Sigma^+ - k^2)^2 + (\text{Im}\,\Sigma^+)^2}\frac{N}{L^d} \\ &= -2i\frac{(\kappa/l)\cdot l^4}{\left(\kappa^2 l^2 - \tfrac{1}{4} - k^2 l^2\right)^2 + \kappa^2 l^2}\frac{N}{L^d} \\ &= -2i\,l^2\frac{\kappa l}{\left(\kappa^2 l^2 - \tfrac{1}{4} - k^2 l^2\right)^2 + \kappa^2 l^2}\frac{N}{L^d}. \end{aligned} \qquad (8.42)$$

By treating $\kappa l = \mu$ as a parameter of the problem, $(\Delta G_e)_\mathbf{k}$ may be expressed as

$$(\Delta G_e)_\mathbf{k} = -2i\,l^2\frac{\mu}{\left(\mu^2 - \tfrac{1}{4} - k^2 l^2\right)^2 + \mu^2}\frac{N}{L^d}. \qquad (8.43)$$

From Eq. (8.41a), we have the condition that $\mu > 0.5$, since otherwise the wave would be evanescent.

To determine the 3D mobility edge, it is important to first note that for the classical scalar wave case, the $D(\omega)$ expression as given by Eq. (4.98b) is different from that of the quantum case by a factor $(1 + \delta)^{-1}$, which is due to the noncancellation of $\mathbf{v}^+$ and $\mathbf{v}^-$ in the derivation of the Ward identity when $\Delta\omega \neq 0$. However, since this factor is just a multiplier for the whole $D(\omega)$ expression, it *does not alter the location of the mobility edge*, although it does affect the diffusion constant through the transport velocity v_t. The latter effect will be examined in the next section. For the present case, let us consider the factor $-\operatorname{Im}\Sigma^+/(2K_0)^2$ in C:

$$\frac{-\operatorname{Im}\Sigma^+}{4K_0^2} = \frac{-\operatorname{Im}\Sigma^+}{4}\left\{\frac{4}{(2\pi)^3}\int d\mathbf{k}\,k^2\cos^2\theta\right.$$

$$\left.\times\left[\frac{-\operatorname{Im}\Sigma^+}{(\kappa_0^2-\operatorname{Re}\Sigma^+-k^2)^2+(\operatorname{Im}\Sigma^+)^2}\right]^2\right\}^{-2}. \quad (8.44)$$

The expression in the curly brackets, K_0, may be written as

$$K_0 = \frac{1}{l}\frac{2}{3\pi^2}\int_0^\infty dx\,x^4 f_\mu^2(x), \quad (8.45)$$

where $x = kl$ and

$$f_\mu(x) = \frac{\mu}{\left(\mu^2-\frac{1}{4}-x^2\right)^2+\mu^2}. \quad (8.46)$$

It follows that

$$\frac{-\operatorname{Im}\Sigma^+}{4K_0^2} = \mu\left[\frac{4}{3\pi^2}\int_0^\infty dx\,x^4 f_\mu^2(x)\right]^{-2}. \quad (8.47)$$

For the double-$\mathbf{k}$ integral, it is straightforward to write it as

$$(\mathrm{MC}) = i\iint\frac{d\mathbf{k}_i}{(2\pi)^3}\frac{d\mathbf{k}_f}{(2\pi)^3}\frac{L^d}{N}(\Delta G_e)_{\mathbf{k}_i}\frac{L^{2d}}{N^2}(\Delta G_e)_{\mathbf{k}_f}^2\left[\nabla_{\mathbf{k}_i}e(\mathbf{k}_i)\cdot\Delta\hat{\mathbf{k}}\right]$$

$$\times\frac{1}{\nu^2+|\mathbf{k}_i+\mathbf{k}_f|^2}\left[\nabla_{\mathbf{k}_i}e(\kappa_f)\cdot\Delta\hat{\mathbf{k}}\right]$$

$$= -\frac{32}{(2\pi)^4}\int_0^\infty\int_0^\infty dx\,dy\,x^2y^2 f_\mu(x)f_\mu^2(y)xy\int_{-1}^1\int_{-1}^1 d\cos\theta_i d\cos\theta_f$$

$$\times\frac{\cos\theta_i\cos\theta_f}{l^2\nu^2+x^2+y^2+2xy\cos\theta_{if}}$$

$$= \frac{4}{3\pi^4}\int_0^\infty\int_0^\infty dx\,dy\,x^2y^2 f_\mu(x)f_\mu^2(y)$$

$$\times\left[\frac{l^2\nu^2+x^2+y^2}{4xy}\ln\frac{(x+y)^2+\nu^2l^2}{(x-y)^2+\nu^2l^2}-1\right]. \quad (8.48)$$

8.3 Classical Scalar Wave Localization

Here $x = |\mathbf{k}_i|l$, $y = |\mathbf{k}_f|l$, and the angular integration is done exactly the same way as shown in Problem 8.3.

From Eqs. (8.47) and (8.48), the mobility-edge condition is an equation of only one variable, μ,

$$1 = C = \frac{3}{4}\mu \frac{A_3^{(c)}(\mu, \nu^2 = 0)}{\left[B_3^{(c)}(\mu)\right]^2}, \qquad (8.49)$$

where the superscript (c) and subscript 3 are used to denote classical wave and $d = 3$, respectively, and

$$A_3^{(c)}(\mu, \nu^2 l^2) = \int_0^\infty \int_0^\infty dx\, dy\, x^2 y^2 f_\mu(x) f_\mu^2(y)$$

$$\times \left[\frac{l^2\nu^2 + x^2 + y^2}{4xy} \ln \frac{(x+y)^2 + \nu^2 l^2}{(x-y)^2 + \nu^2 l^2} - 1\right], \quad (8.50)$$

$$B_3^{(c)}(\mu) = \int_0^\infty dx\, x^4 f_\mu^2(x). \qquad (8.51)$$

Both $A_3^{(c)}$ and $B_3^{(c)}$ are convergent integrals regardless of the value of ν^2. Equation (8.49) can be solved numerically, giving

$$\mu_c = (\kappa l)_c = 0.985 \qquad (8.52)$$

as the critical value for the mobility edge. Equation (8.52) may be regarded as a formal expression of the Ioffe–Regel criterion (van Tiggelen, 1992) in the classical scalar wave case. The window for classical wave localization is therefore $0.5 \leq \mu \leq 0.985$. It is indeed remarkable that for this case the two elements of the localization calculation can be so clearly separated—the $\langle GG \rangle_c$ calculation determines the value of μ, and the $\langle G \rangle_c$ calculation links μ to the parameters of the model. In fact, from Eqs. (8.41a) and (8.41b), κ, l may be expressed as

$$l = \left\{\frac{(\kappa_0^2 - \mathrm{Re}\,\Sigma^+) + \sqrt{(\kappa_0^2 - \mathrm{Re}\,\Sigma^+)^2 + (\mathrm{Im}\,\Sigma^+)^2}}{2(\mathrm{Im}\,\Sigma^+)^2}\right\}^{1/2}, \quad (8.53)$$

and $\kappa = (-\mathrm{Im}\,\Sigma^+)l$, so that $\mu = (-\mathrm{Im}\,\Sigma^+)l^2$. Here the question is, what should be the value of κ_0? In the long-wavelength limit where the CPA is valid, κ_0 should be taken to be $\kappa_e = \bar{\epsilon}\omega/v_0$, and Σ^+ is calculated relative to the effective medium according to the rules prescribed in Chapter 3 (e.g., $\mathrm{Re}\,\Sigma^+ = 0$). In this way the parameters of the system, such as the dielectric constant ratio, R/λ, and the amount of disorder, may be linked to the localization criterion. However, localization is very unlikely in the long-wavelength limit because the Rayleigh frequency dependence tells us that the scattering must be weak in the long-wavelength limit, in contrast to the quantum case. At high frequencies geometric optics becomes a good description for classical wave propagation. In that regime the effect of scattering, as expressed by the mean free path, is expected to saturate at a scale comparable to the particle size or interparticle separation, whereas κ, being proportional to ω, diverges. Therefore $\mu \to \infty$ again and localization is unlikely in the high-frequency limit either. It follows that the localization of classical waves is most likely in the intermediate-frequency regime, where the wavelength is comparable to the size of the scatterers.

In the intermediate-frequency regime, the CPA may not have solution (or unique solution), and one needs a way to determine the effective κ_e for the medium to be used for the calculation of l and κ. In Chapter 3 an approach to extending the CPA has been presented. However, the GCPA usually cannot offer a unique κ_e for the medium in the intermediate-frequency regime because of the existence of the multiple peaks of the spectral function. In the next section (8.4), on the transport velocity of classical waves, it is proposed that in the delocalized regime a κ_m may be determined on the basis of wave energy uniformity over the scale of microstructural correlation length. While the basic principle here is very different from that of the CPA, e.g., it does not involve phase coherence at all, at low frequencies it turns out $\kappa_m = \kappa_e$, the CPA value, because the two requirements are mutually compatible. The advantage of κ_m is that it has unique solution at every frequency, and it plays a natural role in determining the wave transport velocity, v_t. Therefore, κ_m is the natural "effective medium" wave vector to be used in the calculation of κ and l. Accordingly,

$$l = \frac{1}{-\text{Im } \Sigma_m^+} \frac{1}{\sqrt{2}} \left\{ \kappa_m^2 - \text{Re } \Sigma_m^+ + \sqrt{(\kappa_m^2 - \text{Re } \Sigma_m^+)^2 + (\text{Im } \Sigma_m^+)^2} \right\}^{1/2},$$

(8.54a)

$$\kappa = (-\text{Im } \Sigma_m^+) l,$$

(8.54b)

where

$$\Sigma_m^+ \equiv \sum_i n_i \langle \kappa_m | \bar{t}^+ | \kappa_m \rangle,$$

(8.55)

8.3 Classical Scalar Wave Localization

is calculated by solving a scattering problem where the scattering structural unit is embedded in a medium characterized by $v_m = \omega/\kappa_m$. Σ_m^+ is approximated by the sum of the forward scattering components of the t matrix for the various scattering units, weighted by their respective number densities.

For a system of randomly dispersed spherical scatterers of dielectric constant ϵ_1 and volume fraction p in a matrix of dielectric constant ϵ_2, the dispersion microstructure applies, and one can write, from the solution to problem 3.11 [Eq. (P3.95)],

$$\Sigma_m^+ \cong n\langle \kappa_m|\bar{t}^+|\kappa_m\rangle = -\frac{4\pi i}{\kappa_m} n \sum_{j=0}^{\infty} (2j+1)D_j,$$

where D_j is given by Eq. (P3.100). Here n is the number density of the coated sphere structural unit. With this Σ_m^+ and the value of κ_m determined in the manner specified in Section 8.4, the value of $\mu = \kappa l$ may be calculated. In Figure 8.6 the results for $\phi = 0.1$ are plotted as a function of the reduced frequency variable $\kappa_0 R(\epsilon_2 = 1$ and $\kappa_0 = \omega/v_0)$. It is seen that for $\sqrt{\epsilon_1/\epsilon_2} = 5$, there is no frequency regime which satisfies the Ioffe–Regel criterion. However, the value of κl is indeed lowest at the intermediate-frequency range. Although the results shown in Figure 8.6 do

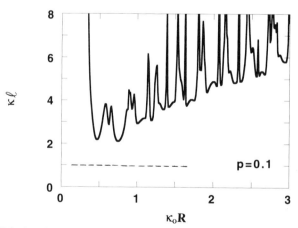

Figure 8.6 Calculated value of $\mu = \kappa l$ as a function of reduced frequency $\kappa_0 R$ for the classical scalar wave in a strong-scattering medium, as determined by the approach described in the text. The random medium is a dispersion of dielectric spheres with $\sqrt{\epsilon_1} = 5$ and $p = 1$. The Ioffe–Regel criterion, $\mu_c = 0.985$, is indicated by the dashed line. No localized state is seen in this case, despite the fact that the value $p = 0.1$ is optimized to yield the lowest μ values in the intermediate-frequency range.

not constitute a proof that localization is impossible for classical waves (the fact that the medium surrounding the scatterer is homogenized in our calculation certainly underestimates the actual scattering involved), they do confirm the difficulty of classical wave localization in 3D.

In Figure 8.6, every dip in the κl value is noted to be associated with a sphere resonance. Physically, it is a well-known fact that for almost every kind of resonance, the real part of the response function goes through a sign change at the resonance frequency, with the higher-frequency side being negative. In analogy with optics, a negative dielectric constant means the index of refraction is purely imaginary, thus implying a nonpropagating, exponentially decaying wave. Not only is the wave being excluded from the inside of the scatterers in this frequency region, but the fact that the scattering cross section can be larger than the physical cross section of each scatterer means that the wave could be attenuated in the interscatterer space as well. However, one can *not* carry this line of reason to the conclusion that wave must localize at a high volume fraction of spheres. This is because if the sphere concentration is too high, the effective medium in which the coated sphere is embedded can approach the property of the scatterer itself, thus making the effective scattering weaker rather than stronger. Another way of saying the same thing is that the multiple scattering between the spheres can renormalize the resonance property of the single scatterer, and at a high enough scatterer concentration the whole medium is inverted, with the scatterer material resembling the matrix and the interstitial spaces resembling the scatterers. Therefore, if classical scalar wave localization were to occur, it must happen not only within narrow frequency ranges in the intermediate-frequency regime but also at intermediate concentrations of the scatterers.

An obvious reason for the difficulty of classical wave localization in 3D lies in the absence of a minimum length scale for classical waves in continuum. The existence of such a scale in the Anderson model is responsible for the existence of an upper band edge. Near the band edge the slow group velocity implies that the lattice wave is literally "waiting" to be localized by the addition of randomness, as seen from the fact that as randomness increases from zero, the mobility edge of the Anderson model always nucleates from the CPA band edges.

One can deduce from such reasoning that one way to facilitate the localization of 3D classical waves is to start with a periodic system and then add randomness to it (John, 1987; Yablonovitch, 1987). Since periodicity would induce band edges, in the simple way of thinking one might expect 3D localization must occur, in direct analogy to the lattice models. However, complications arise because the scatterers in continuum are not point particles; therefore as frequency increases there can be additional

8.3 Classical Scalar Wave Localization

frequency bands associated with the various scatterer resonances. The presence of additional bands means while a gap always exists for a wave of a fixed propagation direction, such a gap need not be isotropic with respect to the propagation directions. In particular, the upper band edge of the lower-frequency band in *one direction* can overlap with the lower band edge of the higher-frequency band in *another direction*. If randomness is introduced into such a periodic system, randomization of the wave propagation direction would mean that there is always some escape direction(s) for the random wave. Therefore, for the periodic scheme to work, it is not enough just to have periodicity. Instead, it is necessary that there be a "true" frequency gap in the sense that what described above does not happen. This proves to be possible only for certain periodic structures in which $\sqrt{\epsilon_1/\epsilon_2}$ has to exceed a certain minimum threshold. For electromagnetic waves, it has been found that whereas it is possible for the diamond structure to have a true gap, it is not possible for many other periodic structures (Ho *et al.*, 1990; Leung and Liu, 1990). It turns out that the scalar wave is a poor approximation to the vector wave in the selection of periodic structures possessing true frequency gaps. In this particular case, the vector character of the electromagnetic wave plays an important role.

Suppose one finds a localized classical-wave state near the band edge of a "true gap" system. It may be argued that such a localized state is different from a localized state of a purely random system. This difference can be made clear from their respective sensitivities to the environment beyond their localization length. For the "true gap + randomness" system, the localization length can be extremely small, e.g., comparable to the interparticle separation for the states well inside the gap region. Suppose the material around such a localized state, beyond a few ξ, is removed. From the Thouless argument presented in Chapter 7, there should be very little effect on the state because a localized state "feels" its environment only through its exponential tail. However, in the present case the existence of the localized state is dependent on the existence of a frequency gap, which in turn is dependent on the existence of long-range order. As material is removed so that only a small cube remains, the frequence gap would be drastically affected. As true gap becomes not so much of a gap any more, the "localized" state is also expected to be not so localized any more, much more so than is implied by the Thouless argument. In other words, a "localized" state in the "true gap + randomness' system has sensitivity to its environment *beyond* its localization length. As a result, a more accurate description of such states should be the evanescent states rather than the localized states, in direct analogy with the impurity state in a periodic system.

Classical wave localization has been an active research area in recent years. So far, no experiment has given a "smoking-gun" demonstration of classical wave localizations in 3D, whether in random or in "true gap + randomness" systems. However, strong indications of localization have been found (Drake and Genack, 1989).

In 2D, similar calculations as those in 3D yield

$$\frac{-\operatorname{Im} \Sigma^+}{4K_0^2} = \frac{\mu}{l^2} \left[\frac{2}{\pi} \int_0^\infty dx\, x^3 f_\mu^2(x) \right]^{-2}, \quad (8.56)$$

whereas the double-**k** integral may be expressed as $2A_2^{(c)}(\mu, v^2 l^2)/\pi^2$, with

$$A_2^{(c)}(\mu, v^2 l^2) = \int_0^\infty \int_0^\infty dx\, dy\, xy f_\mu(x) f_\mu^2(y)$$

$$\times \left[\frac{l^2 v^2 + x^2 + y^2}{\sqrt{(l^2 v^2 + x^2 + y^2)^2 - 4x^2 y^2}} - 1 \right]. \quad (8.57)$$

The 2D localization length equation for classical scalar wave is therefore

$$1 = C = \frac{\mu}{2} \frac{A_2^{(c)}(\mu, v^2 l^2)}{\left[B_2^{(c)}(\mu) \right]^2}, \quad (8.58)$$

where

$$B_2^{(c)}(\mu) = \int_0^\infty dx\, x^3 f_\mu^2(x). \quad (8.59)$$

To obtain ξ/R as a function of reduced frequency $\kappa_0 R$, it is noted that the argument of wave energy uniformity no longer applies here because the states are localized. It is therefore necessary to use the CPA solution, where it is well defined. In the same spirit as in the 3D case, Σ^+ may be approximated by

$$\Sigma^+ \cong 4in \sum_{j=-\infty}^{\infty} D_j, \quad (8.60)$$

where n is the real density of the scatterers, D_j is given by Eq. (P3.100) with the spherical Bessel functions replaced by the regular Bessel functions, and the value of κ_e is taken to be $k_{\max}$ where $-\operatorname{Im} G_e(\omega, \mathbf{k})$ has its maximum. For a given μ, the solution of Eq. (8.58) gives vl, which is multiplied by $0.46\, R/l$ to get R/ξ. The factor 0.46 is obtained from the 1D calculation, as shown below.

8.3 Classical Scalar Wave Localization

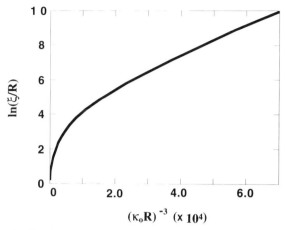

Figure 8.7 2D localization length as a function of reduced frequency $\kappa_0 R$ for a classical scalar wave in a medium consisting of randomly dispersed disks with $p = 0.35$ and $\sqrt{\epsilon_1} = 5$, calculated in the framework of self-consistent theory with the self-energy input from the single-site CPA. At low frequencies it is seen that $\ln \xi \propto \omega^{-3}$.

Figure 8.7 shows the results of the 2D localization length calculations for a system of randomly dispersed disks. The value of p is chosen to be 0.35, and $\sqrt{\epsilon_1} = 5$, $\epsilon_2 = 1$ (matrix). In accordance with the solution to problem 8.4, it is expected that at small ω,

$$\ln \frac{\xi}{R} \propto l \propto \frac{1}{\omega^3}, \qquad (8.61)$$

[see Eq. (3.77)]. Therefore, $\ln(\xi/R)$ is plotted as a function of $(\kappa_0 R)^{-3}$. Indeed, at small ω a good straight-line correlation is obtained.

In 1D, the equation for the localization length is

$$1 = C = \frac{1}{4} \mu \frac{A_1^{(c)}(\mu, \nu^2 l^2)}{\left[B_1^{(c)}(\mu)\right]^2}, \qquad (8.62)$$

where

$$A_1^{(c)}(\mu, \nu^2 l^2) = \int_0^\infty \int_0^\infty dx\, dy\, xy f_\mu(x) f_\mu^2(y)$$

$$\times \left[\frac{1}{\nu^2 l^2 + (x-y)^2} - \frac{1}{\nu^2 l^2 + (x+y)^2} \right], \qquad (8.63)$$

and

$$B_1^{(c)}(\mu) = \int_0^\infty dx\, x^2 f_\mu^2(x). \qquad (8.64)$$

Equation (8.62) can be solved for νl in terms of μ. The results show $\nu l \cong 0.46$ for all values of μ. That implies ξ is directly proportional to l. Since in the $\omega \to 0$ limit $l/\xi = 0.25$ as shown in Chapter 6 [Eq. (6.41)], it follows that $1/\nu \cong 0.5\xi$ in the scalar wave case. This may be interpreted as an indication that $1/\nu$ gives the decay length for wave intensity rather than for wave amplitude.

The 1D localization length calculation can be carried out in exactly the same manner as in the 2D case. For a binary system in which the combined thickness of two components is fixed but the relative fraction fluctuates over a flat distribution, one has

$$\Sigma^+ \cong 2ik_{\max} n \int D(p')[\tau(p',k_{\max}) - 1]\, dp', \tag{8.65}$$

where n is the lineal density of the ϵ_1 component. Here τ is given by Eq. (P3.110). The localization length has been calculated for the case of $p = 0.35$, $D(p') = 5/3$ for $0 \le p' \le 0.6$ and zero otherwise, and $\sqrt{\epsilon_1} = 5$, $\epsilon_2 = 1$. The results are plotted in Figure 8.8 as a function of $\kappa_0 R$, where $2R$ is the mean width of the ϵ_1 component, and κ_0 is defined in the ϵ_2 component. The results of numerical simulations are also shown in the figure. Good agreement is obtained at low frequencies. At high frequencies the jitter in the calculated results reflects the instability of the CPA solution in that frequency regime. Surprisingly, however, the numerical value of ξ never departs far from the simulated results.

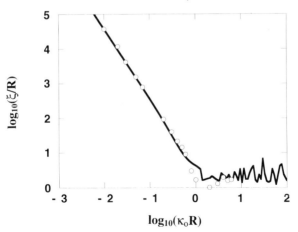

Figure 8.8 1D localization length as a function of reduced frequency $\kappa_0 R$ for classical scalar wave in a binary random medium defined in the text, with $\sqrt{\epsilon_1} = 5$, $\epsilon_2 = 1$, and $p = 0.35$, calculated in the framework of self-consistent theory with the self-energy input from the single-site CPA. At low frequencies $\xi \propto \omega^{-2}$, and at high frequencies $\xi \sim$ constant. Numerical simulation results are shown as open circles.

8.4 Transport Velocity of Classical Scalar Waves

In the beginning of the previous section it was mentioned that the factor δ, which is particular to the classical waves, has no effect on the mobility edge and localization length calculations. However, δ does have an effect on the transport velocity and the overall magnitude of the diffusion constant.

From statistical mechanics, the diffusion constant of a randomly scattered particle can be written as

$$D = \tfrac{1}{3} v l$$

in 3D, where v is the ballistic particle velocity between collisions and l is the mean free path. For wave diffusion, the relevant velocity is v_t, which is shown in Chapter 4 to be [Eq. (4.144)]

$$v_t = \frac{v_0^2}{(\omega/\kappa_e)(1 + \delta)}$$

for the scalar wave. It is clear that v_t is not the phase velocity of the effective medium, which is given by ω/κ_e, nor is it the group velocity due to the $1 + \delta$ factor. In fact, here δ may be viewed as a delay additional to the group velocity, and which can be especially large near a scatterer resonance. To get an interpretation of v_t, we recall that in Chapter 4, it has been shown that the first moment of the Bethe–Salpeter equation may be interpreted as a continuity equation, i.e.,

$$\frac{1}{2}\left(\frac{d\kappa_0^2}{d\omega}\right)\frac{\partial S}{\partial t} + \nabla \cdot \mathbf{J} = \text{source term}.$$

Since from Eq. (4.44) one can write

$$\Delta \hat{\mathbf{k}} \cdot \mathbf{J} = -\frac{i}{2} D(\omega) |\Delta \mathbf{k}|^2 S, \tag{8.66}$$

it follows that v_t, which is part of $D(\omega)$, is proportional to the ratio $|\mathbf{J}|/S$, i.e., as the ratio of the local energy flux to the local energy density. In this context, the resonances of a continuum scatterer are expected to have a significant effect on v_t because S is large near a resonance (since the density of states peaks at the resonant frequency), and v_t should therefore exhibit dips near resonant frequencies. However, in the regime of strong resonant scattering the interparticle scattering is also strong. As a result, the "effective medium" surrounding a given scatterer may approach the property of the scatterer itself, and the resonances will thus become "leaky" and broadened in frequency (resonances would disappear if the medium surrounding a scatterer had the same property of the scatterer). From this intuitive point of view, v_t is expected to exhibit pronounced dips as a function of frequency only at low scatterer concentrations. At high scatterer concentrations the variation with frequency should be reduced.

To capture the effect of resonances and interparticle interaction, it is proposed that the choice of the medium property be based on the principle that the wave energy density should be uniform on the scale of the microstructural correlation length or larger. That is, consider the case of the dispersed microstructure where the structural unit may be approximated by a coated sphere. Let $\kappa_m = \omega/v_m$ characterize the embedding medium. The expectation value of the energy may be written as

$$\text{wave energy in a structural unit} \propto \langle \psi | \nabla^2 + \kappa_m^2 | \psi \rangle$$
$$+ \langle \psi | \kappa^2(\mathbf{r}) - \kappa_m^2 | \psi \rangle$$
$$= \langle \psi | \nabla^2 + \kappa^2(\mathbf{r}) | \psi \rangle, \qquad (8.68)$$

where ψ is the scattering wave function for a plane wave incident on a coated sphere. Its solution is given in Chapter 3 (Problem 3.11). The condition for the *determination of κ_m* is that it should homogenize $\kappa(\mathbf{r})$, i.e.,

$$\langle \psi | \nabla^2 + \kappa_m^2 | \psi \rangle = \langle \psi | \nabla^2 + \kappa(\mathbf{r}) | \psi \rangle,$$

or

$$\langle \psi | \kappa^2(\mathbf{r}) - \kappa_m^2 | \psi \rangle = 0. \qquad (8.69)$$

That immediately implies

$$\delta_m = \frac{3p}{4\pi R_0^3} \int d\mathbf{r} |\psi(\mathbf{r})|^2 [\epsilon(\mathbf{r}) - \epsilon_m] = 0, \qquad (8.70\text{a})$$

where $\epsilon_m = (\kappa_m/\kappa_0)^2$, and δ_m is the δ parameter defined relative to the medium κ_m. Physically, Eq. (8.70a) means that relative to the homogenized medium, there is no additional delay. In Eq. (8.70a), ϵ_m is the parameter to be determined. It should be noted that $\psi(\mathbf{r})$, being a scattering wave function, implicitly depends on ϵ_m. When Eq. (8.70a) is satisfied we have

$$v_t = \frac{v_m^2}{\omega} \sqrt{\kappa_m^2 - \text{Re}\,\Sigma_m^+}, \qquad (8.70\text{b})$$

where Σ_m^+ is calculated with the embedding medium characterized by κ_m. Equations (8.70a) and (8.70b) together define a mean field approach to the calculation of the transport velocity. The interparticle interaction is implicitly taken into account by the requirement of energy density homogeneity. This condition is reasonable because over the scale of the structural correlation length or larger, the medium is geometrically homogeneous. For a delocalized wave, its energy density should be homogeneous over the same spatial scale. On the other hand, this condition should not hold when the states are localized because each localized state has a center, and the

8.5 The Scaling Function $\beta(\ln \gamma)$

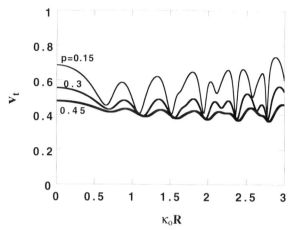

Figure 8.9 Calculated transport velocity v_t plotted as a function of the reduced frequency $\kappa_0 R$ for classical scalar wave in a 3D medium, consisting of a random dispersion of identical spheres with $\sqrt{\epsilon_1} = 5$ and $p = 0.15$, 0.3, and 0.45.

decay of the localized states can exhibit large fluctuations even when the scale of observation is larger than the structural correlations length. As a result, Eqs. (8.70a) and (8.70b) make sense only in the 3D delocalized regime.

The calculation of v_t for $\sqrt{\epsilon_1} = 2.91$ (the index of refraction for alumina at the microwave frequency) has been carried out for three values of $p = 0.15$, 0.3, and 0.45. The results are plotted as a function of $\kappa_0 R$ in Figure 8.9. It is seen that whereas at the low volume fractions there are pronounced dips in v_t, as p increases the dips become smeared out, as expected.

8.5 The Scaling Function $\beta(\ln \gamma)$

Up to now all the localization calculations in this chapter have been for infinite-sized samples. However, Chapter 7 has shown that an examination of wave localization in finite-sized samples can afford some insight into the crossover behavior from an extended state to a localized state. The key quantity to be examined in this regard is the parameter $\gamma = \delta\omega/\Delta\omega$. As shown in chapter 7, γ may be expressed as

$$\gamma = 2\pi\rho_e(\omega)D(\omega)L^{d-2} \tag{8.71}$$

for delocalized states. In order to use the Green's function approach for the calculation of a $\gamma(L)$ that can interpolate between the extended and the localized regimes, two aspects of Eq. (8.71) need to be modified.

The first aspect concerns the expression for $D(\omega)$,
$$D(\omega) = D^{(B)}(\omega)(1 - C).$$
It is important to recognize that C depends on the sample size L, which was already demonstrated in Chapter 6. In order to make that dependence more precise, it is preferred to have an evaluation of C in which one focuses not on its numerical accuracy but rather on its correct variation with L. Since the end result, the function β, depends only on the functional form of $\gamma(L)$, this approach is well suited for our purpose. To this end, let us rewrite Eq. (8.4) in the form of (8.5a):

$$C = \frac{1}{K_0} \int \frac{d\mathbf{Q}}{(2\pi)^d} \frac{1}{\nu^2 + Q^2}. \tag{8.72}$$

Since $D^{(B)}(\omega) = K_0(d\kappa_e^2/d\kappa_0^2)/\pi\rho_e(\omega)$, an equivalent form of (8.72) is

$$C = \frac{(d\kappa_e^2/d\kappa_0^2)}{\pi\rho_e(\omega)D^{(B)}(\omega)} \frac{1}{L^d} \sum_{\mathbf{Q}} \frac{1}{\nu^2 + Q^2}. \tag{8.73}$$

The sample size dependence of C comes in as the lower limit to the magnitude of Q, $1/L$, to be summed over. This is physically reasonable from the discussion in Chapter 5 because the sample size restricts the scattering path length, which may be expressed as a constraint on the diffusion-mode wave vector Q be larger than $1/L$. If in addition we use the fact that in the localized regime the equation for the determination of ν is $C = 1$, with lower limit $Q = 0$, then $(1 - C)$ is essentially the difference between two integrals which differ only in the lower integration limits, i.e.,

$$\sum_{Q=0}^{1/l} - \sum_{Q=1/L}^{1/l} = \sum_{Q=0}^{1/L},$$

so

$$D(L) = D^{(B)}(\omega)(1 - C) = \frac{d\kappa_e^2/d\kappa_0^2}{\pi\rho_e(\omega)} \frac{1}{L^d} \sum_{\mathbf{Q}}^{1/L} \frac{1}{\nu^2 + Q^2}$$

$$= \frac{d\kappa_e^2/d\kappa_0^2}{\pi\rho_e(\omega)} \begin{cases} \dfrac{1}{2\pi^2} \dfrac{1}{L}\left[1 - \nu L \tan^{-1}\left(\dfrac{1}{L\nu}\right)\right] & \text{3D} \\[1ex] \dfrac{1}{4\pi} \ln\left[1 + \dfrac{1}{(\nu L)^2}\right] & \text{2D} \\[1ex] \dfrac{1}{\pi\nu} \tan^{-1}\left(\dfrac{1}{L\nu}\right) & \text{1D.} \end{cases} \tag{8.74}$$

8.5 The Scaling Function $\beta(\ln \gamma)$

The second point concerns the amount of mobile flux for a finite sample. For delocalized states, a pulse injected at origin would result in a measurable diffusive flux that is proportional to the density of states $\rho_e(\omega)$. However, for localized states the measured total flux is expected to decrease as L increases. To account for this phenomena quantitatively, it is necessary to recall Eq. (4.35), which is the first moment of the Bethe–Salpeter equation. As interpreted in Chapter 4, the $\nabla \cdot \mathbf{J}$ term in that equation represents the outgoing flux due to a source term (the right-hand side of the equation), i.e.,

$$\nabla \cdot \mathbf{J} = \frac{d\omega}{d\kappa_e^2} \pi \rho_e(\omega) \frac{N}{L^d} + \frac{1}{2} \frac{d\kappa_0^2}{d\omega} i\Delta\omega S,$$

or

$$i\Delta\mathbf{k} \cdot \mathbf{J} = \frac{d\omega}{d\kappa_e^2} \pi \rho_e(\omega) \frac{N}{L^d} \left[1 + \frac{i\Delta\omega}{-i\Delta\omega + D(\omega)|\Delta\mathbf{k}|^2}\right], \quad (8.75)$$

where the expression for S, Eq. (4.45), has been used. By expressing $\nu^2 = -i\Delta\omega/D(\omega)$ in the $\Delta\omega \to 0$ limit, it is seen that

$$i\Delta\mathbf{k} \cdot \mathbf{J} = \frac{d\omega}{d\kappa_e^2} \frac{N}{L^d} \pi \rho_e(\omega) \frac{|\Delta\mathbf{k}|^2}{\nu^2 + |\Delta\mathbf{k}|^2}. \quad (8.76)$$

Equation (8.76) tells us that for a unit source, the amount of disturbance away from the source is given by

$$\chi(\Delta\mathbf{k}) = \frac{|\Delta\mathbf{k}|^2}{\nu^2 + |\Delta\mathbf{k}|^2} \frac{N}{L^d}. \quad (8.77)$$

Not surprisingly, for $\nu = 0$ (delocalized states), this factor is 1 (up to the normalization factors N/L^d).

Let us now consider a finite-sized electronic sample on which a potential $U(z)$ is applied along one direction, given by

$$U(z) = L - z. \quad (8.78)$$

The applied potential will cause a change in the electronic density $\delta\rho$ such that the net diffusion current would exactly cancel the electric current. To calculate $\delta\rho$, one first Fourier transforms (8.77) to get

$$\chi(z) = \frac{L^d}{N} \int_{-\infty}^{\infty} \exp(i\Delta kz) \chi(\Delta\mathbf{k}) \frac{d(\Delta k)}{2\pi} = \delta(z) - \frac{\nu}{2} \exp(-\nu|z|). \quad (8.79)$$

The first term on the right-hand side is the local term of the response function. It would be the only term present if the states are delocalized. The second term is nonlocal in character. It is particular to the localized states. The presence of the nonlocal term means that

$$\delta\rho = \rho_e(\omega) \int_0^L \chi(z - z')U(z') \, dz'. \tag{8.80}$$

For delocalized states, $\rho_e(\omega)$ in (8.71) gives the density of states participating in electrical transport. In the more general case this may be measured by the quantity

$$-\left[\frac{d(\delta\rho)}{dz}\right]_{z=L},$$

since it gives the number of states participating in the diffusive counter current. When $\chi(z) = \delta(z)$, we get

$$-\frac{d(\delta\rho)}{dz}\bigg|_{z=L}$$

$$= -\rho_e(\omega) \lim_{\Delta z \to 0} \frac{\int_0^L \delta(L + \Delta z - z')U(z') \, dz' - \int_0^L \delta(L - \Delta z - z')U(z') \, dz'}{2\Delta z}$$

$$= \tfrac{1}{2}\rho_e(\omega). \tag{8.81}$$

The factor $1/2$ comes from the fact that for the current, there is an abrupt drop to zero outside the sample, so at the boundary what is measured is the mean between zero and one. For delocalized states, therefore, $-2[d(\delta\rho)/dz]$ at $z = L$ is equivalent to $\rho_e(\omega)$. In the case where $\chi(z)$ is given by Eq. (8.79), direct integration plus some care at $z = L$ gives

$$-2\frac{d(\delta\rho)}{dz}\bigg|_{z=L} = (\nu L + 1)\exp(-\nu L)\rho_e(\omega). \tag{8.82}$$

In the general case, therefore, $\rho_e(\omega)$ should be represented by $(\nu L + 1)\exp(-\nu L)\rho_e(\omega)$ (Vollhardt and Wölfle, 1982).

8.5 The Scaling Function $\beta(\ln \gamma)$

By substituting Eq. (8.82) and Eq. (8.74) into Eq. (8.71), a generalized expression for γ is obtained:

$$\gamma = 2(\nu L + 1)\exp(-\nu L)\pi \rho_e(\omega)D(L)L^{d-2}$$

$$= (\nu L + 1)\exp(-\nu L)\begin{cases} \dfrac{1}{\pi^2}\left(1 - \nu L \tan^{-1}\dfrac{1}{L\nu}\right) & \text{3D} \\ \dfrac{1}{2\pi}\ln\left[1 + \dfrac{1}{\nu^2 L^2}\right] & \text{2D} \\ \dfrac{2}{\pi\nu L}\tan^{-1}\dfrac{1}{L\nu} & \text{1D,} \end{cases} \quad (8.83)$$

where $d\kappa_c^2/d\kappa_0^2$ is set equal to 1; i.e., the starting medium is taken to be the effective medium. Since $1/\nu$ is proportional to ξ, it is seen that γ is a function of ξ/L only, thus guaranteeing the scaling behavior as pointed out in the last chapter. By writing $x = \nu L$, the scaling function in the localized regime is given by

$$\beta = \frac{d\ln\gamma}{d\ln x} = x\frac{d\ln\gamma}{dx}$$

$$= \begin{cases} -\dfrac{x^2}{1+x} + \dfrac{x^2}{1+x\tan^{-1}x^{-1}}\left[\dfrac{1}{1+x^2} - \dfrac{1}{x}\tan^{-1}\dfrac{1}{x}\right] & \text{3D} \\ -\dfrac{x^2}{1+x} - \dfrac{2}{1+x^2}\left[\ln\left(1 + \dfrac{1}{x^2}\right)\right]^{-1} & \text{2D} \\ -\dfrac{1+x+x^2}{1+x} - \dfrac{x}{\tan^{-1}x^{-1}}\dfrac{1}{1+x^2} & \text{1D.} \end{cases} \quad (8.84)$$

Note that in order to write β in terms of $\ln \gamma$, x has to be solved in terms of γ (or $\ln \gamma$) through Eq. (8.83).

The limiting behaviors of β may be obtained by letting $x \to 0$ and $x \to \infty$. For 2D and 1D, the $x \to 0$ limit is given by

$$\lim_{x\to 0}\beta = \begin{cases} \dfrac{1}{\ln x} = -\dfrac{1}{\pi\gamma} & \text{2D} \\ -1 - \dfrac{2}{\pi}x = -1 - \dfrac{2}{\pi\gamma} & \text{1D} \end{cases}, \quad (8.85)$$

and the $x \to \infty$ limit is given by

$$\lim_{x\to\infty}\beta = -x = \ln\gamma + \text{constant} \quad (8.86)$$

to the leading order for $d = 1$, 2, and 3. These are precisely the desired asymptotic behaviors as spelled out in Chapter 7. In 3D, the $x \to 0$ limit yields

$$\lim_{x \to 0} \beta = -\frac{\pi}{2} x = (\pi^2 \gamma - 1). \tag{8.87}$$

That means the fixed point at which $\beta = 0$ occurs at

$$\gamma_c = \frac{1}{\pi^2}. \tag{8.88}$$

This value is very close to the γ_c value of $1/3\pi$ estimated in Chapter 7. There is another branch of the scaling function in 3D associated with the delocalized states. For that branch, $\nu = 0$, so that

$$D(L) = D^{(B)} \left[1 - \frac{1}{\pi \rho_e(\omega)} \frac{1}{D^{(B)}} \frac{1}{2\pi^2} \left(\frac{\alpha}{l} - \frac{1}{L} \right) \right], \tag{8.89}$$

where α/l is the upper cutoff of the Q integral. It follows that

$$\gamma = 2\pi \rho_e(\omega) D(L) L = \frac{1}{\pi^2} + 2\pi \rho_e D^{(B)} l \frac{L}{\zeta}. \tag{8.90}$$

By labeling $x = L/\zeta$, we have $x(d\gamma/dx) = \gamma - 1/\pi^2$ or

$$\beta = \frac{x}{\gamma} \frac{d\gamma}{dx} = 1 - \frac{1}{\pi^2 \gamma} \tag{8.91}$$

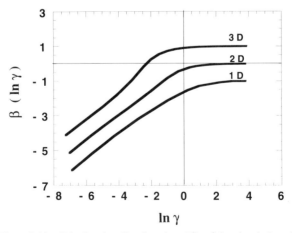

Figure 8.10 Calculated scaling function $\beta(\ln \gamma)$ for $d = 1$, 2, and 3.

as the delocalized branch of β in 3D. When $\gamma \to \infty$, $\beta \to 1$ as expected. The scaling function $\beta(\ln \gamma)$ is plotted in Figure 8.10 for $d = 1, 2, 3$. In the case of 1D, the present scaling function is not identical to that expressed by Eq. (7.50). However, since the asymptotic behaviors are similar, the difference is not that significant.

The derivation of the scaling function $\beta(\ln \gamma)$ indicates that the mechanism of coherent backscattering, with the added self-consistency assumption, is compatible with the scaling hypothesis as enunciated in the last chapter.

Problems and Solutions

8.1. Derive from Maxwell's equations an expression for the complex conductivity.

Only one of Maxwell's equations is needed for this purpose:

$$\nabla \times \mathbf{H} = \frac{4\pi}{c} \mathbf{j} + \frac{1}{c} \frac{\partial \mathbf{D}}{\partial t}, \qquad (P8.1)$$

where

$$\mathbf{D} = \epsilon \mathbf{E},$$
$$\mathbf{j} = \sigma_E \mathbf{E},$$

c is the speed of light, $\mathbf{E}$ is the electric field, $\mathbf{j}$ is the current density, and $\mathbf{H}$ is the magnetic field. For a harmonic $[\exp(-i \Delta \omega t)]$ time variation, Eq. (P8.1) becomes

$$\nabla \times \mathbf{H} = \frac{4\pi}{c} \left(\sigma_E - \frac{i \Delta \omega}{4\pi} \epsilon \right) \mathbf{E}$$
$$= \frac{4\pi}{c} \sigma_E^* \mathbf{E}. \qquad (P8.2)$$

Consistent with the notations used in this book, here $\Delta \omega$ is used to denote the modulation frequency. $\sigma_E^* \mathbf{E}$ is a complex current density whose imaginary part (the component that is 90° out of phase with the electric field) represents the dielectric polarization current, which arises from bound (localized) electrons.

8.2. Express the $\mathbf{k}$ integral of K_0 in terms of a frequency/energy integration.

The essential point in the conversion of the $\mathbf{k}$ integration to a frequency/energy integration is the division of $d\mathbf{k}$ into $dS\, dk_\perp$, where S is the equal-energy surface and $k_\perp$ is the normal to that surface. Since

$$\frac{dk_\perp}{de} = \frac{1}{\nabla_k e(\mathbf{k})}, \tag{P8.3}$$

it follows that

$$d\mathbf{k} = dS \frac{dk_\perp}{de} de = dS\, de\, \frac{1}{|\nabla_k e(\mathbf{k})|}, \tag{P8.4}$$

and

$$S_0(e = \kappa_0^2) = \int d\mathbf{k} |\nabla_k e(\mathbf{k})| \delta[\kappa_0^2 - e(\mathbf{k})]$$

$$= \int dS\, de\, \delta[\kappa_0^2 - e(\mathbf{k})]$$

$$= \int_{e = \kappa_0^2} dS \tag{P8.5a}$$

is the constant energy surface area for the ordered lattice. In particular, for the 3D simple cubic lattice the **k**-space surface area (in units of a^{-2}) is

$$S_0(E) = 16 \int_0^\pi \int_0^\pi \int_0^\pi dx\, dy\, dz \sqrt{\sin^2 x + \sin^2 y + \sin^2 z}$$

$$\times \delta[E + 2\cos x + 2\cos y + 2\cos z], \tag{P8.5b}$$

where $-6 \le E \le 6$.

With this preliminary, it is easy to write

$$2K_0 = \frac{2}{(2\pi)^3} \int d\mathbf{k} \left[\nabla_k e(\mathbf{k}) \cdot \Delta\hat{\mathbf{k}}\right]^2 \left\{ \frac{-\operatorname{Im} \Sigma^+}{\left[\kappa_0^2 - \operatorname{Re}\Sigma^+ - e(\mathbf{k})\right]^2 + (\operatorname{Im}\Sigma^+)^2} \right\}^2$$

$$= \frac{1}{(2\pi)^3} \frac{2}{3} \int dE' \int d\mathbf{k} |\nabla_k e(\mathbf{k})|^2 \delta[E' - e(\mathbf{k})]$$

$$\times \left\{ \frac{-\operatorname{Im}\Sigma^+}{\left[\kappa_0^2 - \operatorname{Re}\Sigma^+ - e(\mathbf{k})\right]^2 + (\operatorname{Im}\Sigma^+)^2} \right\}^2$$

$$= \frac{1}{(2\pi)^3} \frac{2}{3} \int_0^{12/a^2} dE' \int_{E'} dS |\nabla_k e(\mathbf{k})|$$

$$\times \left\{ \frac{-\operatorname{Im}\Sigma^+}{\left[\kappa_0^2 - \operatorname{Re}\Sigma^+ - E'\right]^2 + (\operatorname{Im}\Sigma^+)^2} \right\}^2. \tag{P8.6}$$

In (P8.6) the integral of $(\nabla_k e(\mathbf{k}) \cdot \Delta \hat{\mathbf{k}})^2$ is written as the integral of $|\nabla_k e(\mathbf{k})|^2/3$ because $\Delta \hat{\mathbf{k}}$ is an independent unit vector ($\Delta \mathbf{k}$ is recalled to be the vector that is the difference between the + channels and the − channels in the $\langle GG \rangle_c$ evaluation. See Chapter IV.) and therefore may be taken as the z axis. Since the integral is symmetrical with respect to k_x, k_y, and k_z, it follows that the integration of $|\nabla_k e(\mathbf{k})|^2 = 4(\sin^2 k_x a + \sin^2 k_y a + \sin^2 k_z a)$ would give three times $(\nabla_k e(\mathbf{k}) \cdot \Delta \hat{\mathbf{k}})^2 = 4(\sin^2 k_z)$.

To evaluate the dS integral in (P8.6), it is noted that from (P8.5),

$$\int_{E'} dS |\nabla_k e(\mathbf{k})| = \int d\mathbf{k} |\nabla_k e(\mathbf{k})|^2 \delta[E' - e(\mathbf{k})]$$

$$= \frac{1}{a^3} \int_{-\pi}^{\pi} \int_{-\pi}^{\pi} \int_{-\pi}^{\pi} dx\, dy\, dz$$

$$\times 4(\sin^2 x + \sin^2 y + \sin^2 z)\, \delta[a^2 E' - a^2 e]$$

$$= \frac{96}{a^3} \int_0^{\pi} \int_0^{\pi} \int_0^{\pi} dx\, dy\, dz \sin^2 z$$

$$\times \delta[E - 6 + 2\cos x + 2\cos y + 2\cos z]$$

$$= \frac{96}{a^3} \frac{1}{2} \int_0^{\pi} \int_0^{\pi} \int_0^{\pi} dx\, dy \frac{d(2\cos z)}{|-\sin z|}$$

$$\times \sin^2 z\, \delta[E - 6 + 2\cos x + 2\cos y + 2\cos z]$$

$$= \frac{48}{a^3} \int_0^{\pi} \int_0^{\pi} \text{Re} \sqrt{1 - \left(\frac{E-6}{2} + \cos x + \cos y\right)^2}\, dx\, dy.$$

(P8.7)

Combining (P8.6) and (P8.7) yields

$$2K_0 = \frac{32}{(2\pi)^3 a} \int_{E_l}^{E_u} dE\, W_3(2u - \text{Re}\, y_0 - E) \left\{ \frac{-\text{Im}\, \Sigma^+}{E^2 + (\text{Im}\, y_0)^2} \right\}^2, \quad \text{(P8.8)}$$

where $(\kappa_0^2 - \text{Re}\, \Sigma^+ - E/a^2)a^2$ is relabeled as E, $y_0 = a^2 \Sigma^+$, and $u = (\kappa_0^2 a^2 - 6)/2\beta$. $E_l = 2u - \text{Re}\, y_0 - 6$, $E_u = 2u - \text{Re}\, y_0 + 6$, and

$$W_3(E') = \int_0^{\pi} \int_0^{\pi} \text{Re} \sqrt{1 - \left(\frac{E'}{2} + \cos x + \cos y\right)^2}\, dx\, dy. \quad \text{(P8.9)}$$

8.3. Carry out the angular integration in Eq. (8.12).

There are three unit vectors in the integration: $\Delta\hat{\mathbf{k}}$, $\hat{\mathbf{k}}_i$, and $\hat{\mathbf{k}}_f$. Since it does not matter which one of the three vectors is held fixed in the angular integration, let us choose $\hat{\mathbf{k}}_i$ as being the z axis. Then, from the addition theorem of spherical harmonics,

$$\cos\theta_f = \cos\theta_i \cos\theta_{if} + \sin\theta_i \sin\theta_{if} \cos\phi_{if}, \quad (P8.10)$$

where $\cos\theta_f = \hat{\mathbf{k}}_f \cdot \Delta\hat{\mathbf{k}}$, $\cos\theta_i = \hat{\mathbf{k}}_i \cdot \Delta\hat{\mathbf{k}}$, and $\cos\theta_{if} = \hat{\mathbf{k}}_i \cdot \hat{\mathbf{k}}_f$. It follows that the expression in the curly brackets of the Eq. (8.12a) can be written as

$$\int_0^{2\pi} d\phi_i \int_0^{2\pi} d\phi_f \int_{-1}^1 d\cos\theta_i \int_{-1}^1 d\cos\theta_f \frac{\cos\theta_i \cos\theta_f}{a^2\nu^2 + x^2 + y^2 + 2xy\cos\theta_{if}}$$

$$= \int d\Omega_{\Delta\hat{\mathbf{k}}} \int d\Omega_{\hat{\mathbf{k}}_f} \frac{(\Delta\hat{\mathbf{k}} \cdot \hat{\mathbf{k}}_i)(\Delta\hat{\mathbf{k}} \cdot \hat{\mathbf{k}}_f)}{a^2\nu^2 + x^2 + y^2 + 2xy(\hat{\mathbf{k}}_i \cdot \hat{\mathbf{k}}_f)}$$

$$= \int d\Omega_{\Delta\hat{\mathbf{k}}} \int_0^{2\pi} d\phi_{if} \int_{-1}^1 d\cos\theta_{if}$$

$$\times \frac{\cos\theta_i (\cos\theta_i \cos\theta_{if} + \sin\theta_i \sin\theta_{if} \cos\phi_{if})}{a^2\nu^2 + x^2 + y^2 + 2xy\cos\theta_{if}}$$

$$= 2\pi \int_{-1}^1 d(\cos\theta_i)\cos^2\theta_i \left[2\pi \int_{-1}^1 du \frac{u}{(a^2\nu^2 + x^2 + y^2) + 2xyu}\right]$$

$$= 4\pi^2 \frac{2}{3}\left[2 - \frac{a^2\nu^2 + x^2 + y^2}{2xy}\ln\frac{(x+y)^2 + a^2\nu^2}{(x-y)^2 + a^2\nu^2}\right]\frac{1}{2xy}$$

$$= 4\pi^2 \frac{2}{3}\left[1 - \frac{a^2\nu^2 + x^2 + y^2}{4xy}\ln\frac{(x+y)^2 + a^2\nu^2}{(x-y)^2 + a^2\nu^2}\right]\frac{1}{xy}. \quad (P8.11)$$

8.4. Determine the functional dependence of the 2D localization length in the weak scattering limit of the Anderson model.

To determine the functional dependence of ξ in 2D, it is simpler to use the approximate evaluation of the maximally crossed diagrams' contribution, and the resulting equation

$$\frac{1}{\nu} = \frac{l}{\alpha}[\exp(4\pi K_0) - 1]^{1/2}, \quad (P8.12)$$

where $\xi \propto \nu^{-1}$, α/l is the cutoff for the Q wave vector, and $K_0 \propto D^{(B)}\rho_e \propto l$. We know $l \propto (\sigma_w a^2)^{-2}$ in the weak scattering limit because the mean free path is inversely proportional to the scattering cross section, which is the angular integral of |scattering amplitude|^2. The scattering amplitude must be proportional to $\sigma_w a^2$ to the first order because the scattering amplitudes $\to 0$ as $\sigma_w \to 0$, and, moreover, the real part of the scattering amplitude (the t matrix element) changes sign as σ_w changes sign. Together they imply linear dependence. It follows that $\ln \xi \propto 1/(\sigma_w a^2)^2$ in the weak scattering limit.

References

Drake, J. M., and Genack, A. Z. (1989). *Phys. Rev. Lett.* **63**, 259.
Economou, E. N. (1983). "Green's Functions in Quantum Physics," 2nd Ed. p. 173. Springer-Verlag, Berlin.
Economou, E. N., Soukoulis, C. M., and Zdetsis, A. D. (1984). *Phys. Rev.* **B30**, 1686.
Ho, K. M., Chan, C. T., and Soukoulis, C. M. (1990). *Phys. Rev. Lett.* **65**, 3152.
John, S. (1987). *Phys. Rev. Lett.* **58**, 2486.
Leung, K. M., and Liu, Y. F. (1990). *Phys. Rev. Lett.* **65**, 2646.
Sheng, P. and Zhang, Z. Q. (1986). *Phys. Rev. Lett.* **57**, 1879.
van Tiggelen, B. A. (1992). "Multiple Scattering and Localization of Light," p. 77. Ph.D. Thesis, University of Amsterdam.
Vollhardt, D., and Wölfle, P. (1980). *Phys. Rev.* **B22**, 4666.
Vollhardt, D., and Wölfle, P. (1982). *Phys. Rev. Lett.* **48**, 699.
Yablonovitch, E. (1987). *Phys. Rev. Lett.* **58**, 2059.
Zdetsis, A. D., Soukoulis, C. M., Economou, E. N., and Grest, G. S. (1985). *Phys. Rev.* **B32**, 7811.

9

Localization Phenomena in Electronic Systems

9.1 Finite Temperatures and the Effect of Inelastic Scattering

In previous chapters the term "scattering" has been meant to denote elastic scattering in which the wave frequency is not altered. However, the absence of inelastic scattering is at best an idealization. To be more realistic means to account for inelastic scattering in the interpretation of observed localization phenomenon.

The effect of inelastic scattering is generally expressed through a length scale, the inelastic scattering length l_{in}. The value of l_{in} is usually very different for the quantum case and the classical wave case. For classical waves, l_{in} may be macroscopic in size and is generally insensitive to temperature. Just as in the observation of the coherent backscattering effect, classical wave localization effects may thus be observed with macroscopic samples, relatively free from the hindrance of inelastic scattering. This is not the case, however, with the electronic case. The electronic l_{in} generally varies with temperature as $1/T^p$, where p is some constant on the order of 1–2, and at finite temperatures l_{in} is invariably microscopic in scale. Therefore, the manifestation of electronic localization effects in macroscopic samples is almost always accompanied by inelastic scattering.

For classical waves, it is interesting to point out that inelastic scattering does not necessarily diminish localization effects. In Chapter 5, it has been argued that the effect of inelastic scattering is analogous to that of finite sample size, i.e., cutting off the long scattering paths. However, an important distinction is that inelastic scattering is inherently a probabilistic process. Therefore, even in a semi-infinite dissipative sample there can still be very long backscattering paths. Since for classical waves inelastic scattering is just energy dissipation (frequency down-conversion with no attendant up-conversion), those long paths can be easily picked up by measurement at the incident wave frequency, and they are expected to display the full coherent backscattering effect because dissipation acts *equally* on both the scattering path and its time-reversed counterpart, as pointed out by Weaver (1993). However, the amplitudes of such long paths are necessarily diminished, and consequently when they are considered *together* with the shorter paths (which have higher amplitudes) as in a static backscattering experiment, where the consideration is for a single-frequency wave (which may be regarded as a Fourier addition of paths of all different travel times) the result is similar to that of finite sample size, as shown in Chapter 5. But if these long scattering paths were separated out from the short scattering paths, e.g., by time-of-flight experiments instead of experiments in the frequency domain, then the localization effect should be observable even in a dissipative random medium.

The same is not true for the electronic systems. Here the effect of inelastic scattering is to establish a *thermal equilibrium* in which energy down-conversion is balanced with energy up-conversion. Therefore, even if the long scattering paths could be separated out from the short scattering paths, these long paths would very likely contain inelastic scattering contributions from other energies and hence have random phase relations. For the electronic case the analogy between inelastic scattering and finite sample size is thus fairly accurate. The intention of this chapter is to describe a few electronic phenomena that result from the localization effect.

It should be noted that there can be two types of electronic inelastic scattering. Besides the electron–phonon scattering, which is caused by electrons losing (or gaining) energy to (from) phonons, there is also the electron–electron scattering. The electron–electron interaction is an extensive subject in itself and will not be treated here. However, the effect of the electron–electron interaction in 2D disordered films is noted to be very similar to that of coherent backscattering. Below we will note the effect of electron–electron interactions only when it is relevant.

9.2 Temperature Dependence of the Resistance in 2D Disordered Films

From Eq. (5.18) a quantitative correspondence between sample size L and l_{in} is given by

$$L_{in} = \sqrt{\frac{l^* l_{in}}{d}}, \qquad (9.1)$$

9.2 Temperature Dependence of the Resistance in 2D Disordered Films

where the subscript "in" on L is intended to emphasize its inelastic scattering origin, and l^* is the elastic transport mean free path. From the temperature dependence of l_{in}, it follows that the effect of inelastic scattering may be translated into a temperature-dependent L_{in}:

$$L_{in}(T) \propto T^{-p/2}. \tag{9.2}$$

The value of p is ~ 2 if the inelastic scattering is caused by electron–phonon interaction. Below 5 K, however, there is evidence that the electron–electron interaction is the dominant inelastic scattering mechanism, for which p has a lower value. From the sample-size dependence of renormalized σ_E, which has been addressed in both Chapter 6 and Chapter 8, Eq. (9.2) directly implies a temperature dependence for σ_E which could be observed.

The rationale underlying the substitution of $L_{in}(T)$ for the sample size L is that, in a macroscopic sample, L_{in}^d is the basic volume beyond which all phase coherence effects are lost. That means the macroscopically measured value of σ_E may be obtained from $\{\sigma_E(L_{in})\}$, defined as the values of σ_E measured on the scale of L_{in}, through a simple resistor-network calculation. If elastic scattering is weak so that the values of $\sigma_E(L_{in})$ are narrowly distributed, then

$$\sigma_E \sim \text{mean of } \{\sigma_E(L_{in})\}.$$

Therefore, although the actual sample size L is macroscopic, nevertheless σ_E corresponds to the mean of $\{\sigma_E(L_{in})\}$. In other words, L_{in} defines the minimum scale beyond which the intensive nature of conductivity is preserved. However, this picture is valid only when the sample is fairly conducting. When the states are strongly localized, an alternative picture emerges. This is discussed in Section 9.4.

Now consider a film made by sputtering or evaporation of metal atoms onto an insulating substrate. The film is generally disordered due to the presence of either impurities or structural inhomogeneities. In regard to the localization effect, the film may be regarded as 2D if the thickness L is much less than L_{in} because the coherent backscattering effect is cut off in the third spatial dimension. For most metallic films, $L < 100$ Å may be regarded as a 2D film at $T < 20$ K. From Eq. (8.74) and the Einstein relation, one can write

$$\sigma_E = \frac{2e^2}{\hbar} \frac{1}{4\pi^2} \ln\left[1 + \frac{1}{(\nu L_{in})^2}\right]$$

$$\cong -\frac{e^2}{\hbar} \frac{1}{\pi^2} \ln(L_{in}\nu), \tag{9.3}$$

where a factor of 2 has been included to account for the electronic spin degeneracy, and the factor 1 in the argument of $\ln[1 + (\nu L_{\text{in}})^{-2}]$ is neglected because it is assumed that the localization length is much larger than L_{in}; i.e., the system is weakly localized at those temperatures considered. Combining Eqs. (9.3) with (9.2) yields

$$\sigma_{\text{E}} \cong \frac{e^2}{\hbar} \frac{p}{2\pi^2} \ln T + \text{constant}. \tag{9.4}$$

As $T \to 0$, σ_{E} decreases as $\ln T$. Since the constant term is usually larger than the $\ln T$ term, $\rho_{\text{E}} = 1/\sigma_{\text{E}} \propto -\ln T$. Therefore, at $T < 10$ K, the resistance of a thin disordered film is expected to increase logarithmically. In fact, experimentally this is a fairly general phenomenon, as shown in Figure 9.1. As T increases, however, a transition to (delocalized) 3D behavior occurs, and the role of the coherent backscattering becomes that of a correction to the classical conductivity expression. In that regime the temperature dependence of σ_{E} is dominated by the temperature dependence of the usual metallic behavior, in which σ_{E} decreases as T increases due to increased electron–phonon scattering. Thus, disordered metallic films generally display a shallow resistance minimum at $T \approx 10$ K.

From Eq. (9.4), the slope of the $\ln T$ behavior is seen to be a constant. However, that is based on a pure 2D calculation. In actual experiments the slope of the $\ln T$ behavior can deviate from that predicted by Eq. (9.4). This could be an indication that the finite thickness of the film plays a role

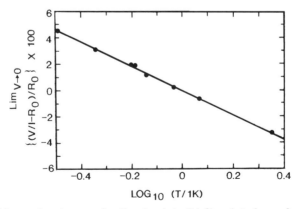

Figure 9.1 Measured resistance of a disordered Au-Pd film plotted as a function of $\ln T$. The data are from Dolan and Osheroff (1979). Here V/I is the measured resistance and R_0 is the wire resistance in series with the sample and therefore has to be subtracted from the measurements.

9.3 Magnetoresistance of Disordered Metallic Films

In a metallic film, the normal magnetoresistance is positive because when a magnetic field is applied perpendicular to the film (with an electric field applied in the plane of the film), the electrons tend to move normal to the electric field direction (in the plane of the film). Since the current is deflected away from the electric field direction, the measured resistance increases. Magnetoresistance is generally expressed as $\Delta\rho_E(H)/\rho_E(0)$, where $\Delta\rho_E(H)$ is the change in resistance as a function of applied magnetic field H, and $\rho_E(0) = [\sigma_E(0)]^{-1}$ is the resistivity at zero magnetic field. Magnetoresistance in metallic films is typically on the order of 1% or less for magnetic field up to 5–10 tesla.

Just as inelastic scattering tends to weaken the localization effect, the presence of a magnetic field also tends to lessen the coherent backscattering effect by destroying the time-reversal symmetry. Since the localization effect increases the resistance, the renormalized resistance of a disordered film must be *higher* than that without the coherent backscattering correction. By using the magnetic field to quench the coherent backscattering correction, one should therefore see a decrease in the measured resistance, i.e., negative magnetoresistance, just opposite to what is expected for normal metallic behavior.

To calculate quantitatively the effect of a magnetic field, it is necessary to observe that, in quantum mechanics, the wave vector (momentum) of an electron in a magnetic field is given by

$$\mathbf{k} - \frac{e\mathbf{A}}{\hbar c},$$

where c is the speed of light and $\mathbf{A}$ is the vector potential, related to $\mathbf{H}$ by

$$\mathbf{H} = \nabla \times \mathbf{A}. \tag{9.5}$$

If $\mathbf{H} = H\hat{\mathbf{k}}$, then one choice of $\mathbf{A}$ would be

$$\mathbf{A} = (-Hy, 0, 0). \tag{9.6}$$

In the present case, where we are interested in monitoring the transport behavior of wave intensity, it is noted that the momenta of both the +

and − channels are modified:

$$\mathbf{k}_+ \to \mathbf{k}_+ - \frac{e\mathbf{A}}{\hbar c}$$

$$\mathbf{k}_- \to \mathbf{k}_- + \frac{e\mathbf{A}}{\hbar c},$$

where the sign of the vector potential in the − channel is reversed in accordance with the time arrow. The net result for $\Delta \mathbf{k}$ is

$$\Delta \mathbf{k} = \mathbf{k}_+ - \mathbf{k}_- \to \Delta \mathbf{k} - \frac{2e\mathbf{A}}{\hbar c}, \tag{9.7}$$

i.e., there is an additional factor of 2 associated with $\mathbf{A}$ for $\Delta \mathbf{k}$. In Chapter IV, it has been shown that $\Delta \mathbf{k}$ is the conjugate variable to the propagation distance $\mathbf{r} - \mathbf{r}'$, so that in real space it may be represented by the operator $-i\nabla$. It follows that whereas $|\Delta \mathbf{k}|^2$ may be regarded as the eigenvalue of the operator $-\nabla^2$, which takes on continuous values, now the corresponding operator is

$$-\nabla^2 \to \left(i\nabla + \frac{2e\mathbf{A}}{\hbar c}\right)^2, \tag{9.8}$$

which has a discrete set of eigenvalues for the momenta perpendicular to the magnetic field as shown in the solution to Problem 9.1. These discrete levels correspond to the well-known Landau levels.

In Eq. (6.17), the coherent backscattering correction to the diffusion constant is expressed as an integral over $\mathbf{Q}$, where $\mathbf{Q}$ plays exactly the role of $\Delta \mathbf{k}$ (see Chapter 6). In view of the quantized Landau levels, the integration over the Q^2 values should be replaced by a sum over the eigenvalues of the new operator, given by

$$Q^2 \to \left(n + \frac{1}{2}\right)\frac{4}{l_H^2} + k_z^2, \quad n = 0, 1, 2, \ldots. \tag{9.9}$$

where k_z is the wave vector in the direction normal to the film, and

$$l_H = \sqrt{\frac{\hbar c}{eH}}$$

is a new magnetic length scale which measures the size of the quantized orbits in the magnetic field. It is on the order of 400 Å for $H = 1$ tesla.

9.3 Magnetoresistance of Disordered Metallic Films

We will treat the magnetoresistance in both the thick- and the thin-film limits. From Eq. (6.17), it is easy to deduce that when the film is thick, the $\delta\sigma_H$ due to the coherent backscattering is given by

$$\delta\sigma_H = -\frac{e^2/\hbar}{\pi} \frac{2}{(2\pi)^3} \sum_{n=0}^{N_0} \int_{-1/l}^{1/l} dQ_z \frac{\pi(4/l_H^2)}{-\frac{i\Delta\omega}{D(\omega)} + \frac{4}{l_H^2}(n+\frac{1}{2}) + Q_z^2}.$$

(9.10)

Here $N_0 = (l_H/l)^2$, l being the mean free path, $dQ_x\, dQ_y$ is converted into $\pi\, dQ_\parallel^2$, where $Q_\parallel^2$ means the component of Q^2 perpendicular to $\overline{H}$ (parallel to the film), and the eigenvalues of $Q_\parallel^2$ are quantized in units of $4/l_H^2$. A factor of 2 has been included to account for the spin degeneracy, and the integral and summation are both cut off at $|Q_z|, |Q_\parallel| = 1/l$.

If the disordered material is in the form of a thin film with thickness L, then the dQ_z integral would have to be split into two parts: one from $-1/l$ to $-1/L$ and another one from $1/L$ to $1/l$. As L approaches l, however, wave quantization in the z direction is expected where each quantized state corresponds to a standing wave. In the limit where the film is so thin that the z variation of the wave may be regarded as uniform, one can replace Q_z^2 in the denominator of the integrand by zero and do away with the dQ_z integration. The resulting expression involves only a simple summation, as shown by Eq. (9.11b).

An important point to be considered now is the factor $-i\Delta\omega/D(\omega)$ [following the self-consistent approximation, $D^{(B)}(\omega)$ is replaced by $D(\omega)$], which was argued to be proportional to $1/\xi^2$ in the last chapter. In the presence of inelastic scattering, if $L_{\text{in}} \ll \xi$, then L_{in} plays a much more dominant role. That is, in addition to the $-i\Delta\omega$ term, there is a term, $1/\tau_{\text{in}}$, whose origin lies in the presence of the additional relaxation term in the continuity equation for S [see Eq. (4.35)]:

$$\frac{1}{2}\frac{d\kappa_0^2}{d\omega}\left(\frac{\partial S}{\partial t} + \frac{S}{\tau_{\text{in}}}\right) + \nabla \cdot \vec{J} = \text{source term}.$$

Here τ_{in} denotes the inelastic scattering time. Therefore,

$$-\frac{i\Delta\omega}{D(\omega)} \rightarrow \nu^2 + \frac{1}{\tau_{\text{in}} D(\omega)}.$$

By substituting $l^*v_t/3$ for $D(\omega)$ and defining $l_{in} = \tau_{in}v_t$, we get

$$\tau_{in} D(\omega) = l_{in}l/3 = L_{in}^2$$

exactly. Since $L_{in} \ll \xi$, the $\nu^2 \propto 1/\xi^2$ term can be neglected, and the thick-film expression for $\delta\sigma_H$ is given by

$$\delta\sigma_H = -\frac{e^2}{\hbar} \frac{1}{\pi^3} \frac{1}{l_H^2} \sum_{n=0}^{N_0} \int_{-1/l}^{1/l} dQ_z \frac{1}{\frac{1}{L_{in}^2} + \frac{4}{l_H^2}(n + \frac{1}{2}) + Q_z^2}$$

$$= -\frac{e^2}{\hbar} \frac{1}{l_H\pi^3} \sum_{n=0}^{N_0} \int_{-l_H/l}^{l_H/l} dp \frac{1}{(l_H/L_{in})^2 + 4(n + \frac{1}{2}) + p^2}$$

$$= -\frac{e^2}{\hbar} \frac{2}{l\pi^3} \sum_{n=0}^{N_0} \frac{l/l_H}{\sqrt{(l_H/L_{in})^2 + 4(n + \frac{1}{2})}}$$

$$\times \tan^{-1} \frac{l_H/l}{\sqrt{(l_H/L_{in})^2 + 4(n + \frac{1}{2})}}. \quad (9.11a)$$

Equation (9.11a) gives the amount of conductivity correction arising from coherent backscattering, in the presence of a magnetic field. The thin-film, 2D version of $\delta\sigma_H$ is given by

$$\delta\sigma_H = -\frac{2e^2}{\hbar} \frac{1}{\pi^3} \sum_{n=0}^{N_0} \frac{1}{(l_H/L_{in})^2 + 4(n + \frac{1}{2})}. \quad (9.11b)$$

The magnetoresistance is defined relative to $\delta\sigma_H$ at $H = 0$. Therefore, let

$$\Delta\sigma_E(H) = \delta\sigma_H - \delta\sigma_0,$$

where $\delta\sigma_0$ may be obtained from (9.11a) by converting it into an integral in the limit of $H = 0$, as shown in the solution to Problem 9.2:

$$\delta\sigma_0 = -\frac{e^2}{\hbar} \frac{1}{l\pi^3} \left(\sqrt{4 + \left(\frac{l}{L_{in}}\right)^2} \tan^{-1} \left[4 + \left(\frac{l}{L_{in}}\right)^2\right]^{-1/2} \right.$$

$$\left. - \left(\frac{l}{L_{in}}\right) \tan^{-1}\left(\frac{L_{in}}{l}\right) + \frac{1}{2} \ln \frac{5L_{in}^2 + l^2}{L_{in}^2 + l^2} \right). \quad (9.12)$$

9.3 Magnetoresistance of Disordered Metallic Films

It follows that the magnetoresistance $\Delta\rho_E(H)/\rho_E \cong -\Delta\sigma_E(H)/\sigma_E$ is given by

$$\frac{\Delta\rho_E(H)}{\rho_E} = -\frac{\Delta\sigma_E(H)}{\sigma_E}$$

$$= -\frac{3}{\pi(\kappa_e l)^2}$$

$$\times \left\{ \sqrt{4+a^{-2}} \tan^{-1}\frac{1}{\sqrt{4+a^{-2}}} + \frac{1}{2}\ln\frac{5a^2+1}{a^2+1} - \frac{1}{a}\tan^{-1}a \right.$$

$$\left. - \sum_{n=0}^{1/\overline{H}} \frac{2\sqrt{\overline{H}}}{\sqrt{(\overline{H}a^2)^{-1}+4(n+\frac{1}{2})}} \tan^{-1}\frac{1/\sqrt{\overline{H}}}{\sqrt{(\overline{H}a^2)^{-1}+4(n+\frac{1}{2})}} \right\}.$$

(9.13a)

Here

$$a = \frac{L_{in}}{l}, \qquad \overline{H} = \left(\frac{l}{l_H}\right)^2,$$

and we have used the relation

$$\sigma_E = 2\frac{e^2}{\hbar}\rho_e(\omega)\frac{1}{3}vl = \frac{e^2}{\hbar}\frac{1}{3\pi^2}(\kappa_e l)^2\frac{1}{l},$$

where the spherical Brillouin zone approximation is used to get $\rho_e(\omega) = 4\pi\kappa_e^2(d\kappa_e/d\omega)/(2\pi)^3 = \kappa_e^2/2\pi^2 v$. When $1/\overline{H}$ is not an integer, the last term in the summation of Eq. (9.13a) is understood to be multiplied by the fraction of $1/\overline{H}$ that is over the largest integer smaller than $1/\overline{H}$. What is clear from Eq. (9.13a) is that the maximum amount of magnetoresistance is $\sim -(\kappa_e l)^{-2}$. Since in 3D the value of $\kappa_e l$ is a measure of the proximity to localization, for a disordered conducting film with $\kappa_e l \gtrsim 10$ the maximum magnetoresistance is on the order of -1% or less. A plot of $\Delta\rho_E(H)/\rho_E$ is shown in Figure 9.2 for $\kappa_e l = 10$ and $a = 2, 10, 50, \infty$. The magnetoresistance is seen to vanish as the constant a becomes smaller, i.e., as T increases.

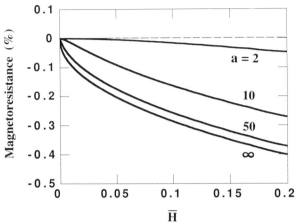

Figure 9.2 Calculated magnetoresistance for $\kappa_e l = 10$, plotted as a function of the dimensionless magnetic field $\overline{H}$. Four cases of $a = 2, 10, 50, \infty$ are considered, where $a = \infty$ corresponds to the $T = 0$ case.

Similar considerations for the 2D film yield

$$\frac{\Delta \rho_E(H)}{\rho_E} = -\frac{1}{\pi^2(\kappa_e l)}\left[\ln(1 + 4a^2) - \sum_{n=0}^{1/\overline{H}} \frac{1}{(4a^2\overline{H})^{-1} + (n + \tfrac{1}{2})}\right]. \tag{9.13b}$$

Figure 9.3 shows the measured variation of resistance versus the magnetic field for a thin Mg film (top) and the resistance variation of the same film after a partial layer of Au is superposed (bottom) (Bergmann, 1984). The top set of curves has been shown to be well fitted by Eq. (9.11b), whereas the bottom set of curves illustrates another effect.

The additional magnetoresistance effect is due to the coupling of the electronic spin to the wave vector **k**, called the spin–orbit coupling. When the spin–orbit coupling is strong, the sign of the spin switches as **k** → −**k** so that the overall sign of the coupling energy is not altered. That means for a scattering path and its time-reversed counterpart the original constructive interference in the backward direction now becomes destructive (due to the reversal of the sign of the time-reversed path), thus implying a positive magnetoresistance since the effect of a magnetic field becomes just the opposite—reducing the coherent backscattering means increasing the resistance. As the magnetic field increases in strength, however, the spins eventually have to be aligned with the magnetic field (the spin–orbit

9.4 Transport of Localized States at Finite Temperatures—Hopping Conduction

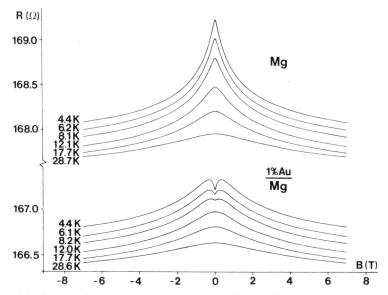

Figure 9.3 Measured magnetoresistance of an Mg film at different temperatures, from Bergmann (1984). Both the magnitude of the effect and its variations with the temperature and the magnetic field are similar to those shown in Figure 2. The top part shows the magnetoresistance to be negative, as predicted. The curves have been quantitatively fitted by Eq. (9.11b). The bottom part shows that by adding only a partial layer of Au on top of the Mg film, the magnetoresistance changes sign at small magnetic field. Here the magnetic film is denoted by B, with T denoting tesla.

effect no longer dominates), and the magnetoresistance turns negative again. The magnitude of the spin–orbit coupling generally increases with the atomic charge. That is why the presence of a heavy metal, e.g., Au, alters the sign of the magnetoresistance at small magnetic field.

9.4 Transport of Localized States at Finite Temperatures—Hopping Conduction

Suppose for whatever reason an electron becomes strongly localized; then at $T = 0$ it is by definition nonmobile and therefore cannot contribute to current transport beyond its localization length. This situation is altered at finite temperatures, because although the localized states are separated in both their energies and spatial locations, an electron can nevertheless "hop" from one localized state to another. The term "hopping" is coined to denote the combination of two mechanisms—thermal activation (up in energy) and tunneling (across in space)—which makes current transport possible.

Although thermal activation and tunneling are two nominally independent processes, Mott (1968) was the first one to show that these two processes can actually couple to result in a nonactivated temperature dependence that is characteristic of hopping conduction. This coupling may be elucidated through the so-called critical-path method first proposed by Ambegaokar, Halperin, and Langer (1971).

Consider the conductance Γ_{ij} between two localized sites i and j. Γ_{ij} may be expressed as the product of the probabilities for thermal activation and tunneling:

$$\Gamma_{ij} = \Gamma_0 \exp(-2\chi S_{ij} - E_{ij}/k_B T), \tag{9.14}$$

where $\chi - \xi^{-1}$ is the tunneling constant, S_{ij} is the tunneling distance between the two sites, k_B is Boltzmann's constant, and

$$E_{ij} = \max[E_i, E_j] = \tfrac{1}{2}(|E_i - E_j| + |E_i| + |E_j|) \tag{9.15}$$

is the activation energy between the two sites. In Eq. (9.14), the indices i and j are not required to denote only the nearest-neighbor sites. Rather, they span all pairs of sites. The expression for E_{ij}, Eq. (9.15), requires some explanation. Suppose $E_i > E_j$, where E_{ij} is measured from the Fermi level; then for a charge to go from j to i it has to overcome an energy barrier $E_i - E_j$ through thermal activation. However, for the charge to hop from i to j, there is no such barrier. In other words, the barrier is asymmetric. But this reasoning ignores the probability that a charge is on site i or j in the first place, before the hopping occurred. If that probability is taken into account, which in thermal equilibrium is proportional to $\exp(-E_i/k_B T)$, then the two conductances become symmetrical because going from j to i now means

$$\exp(-E_j/k_B T) \exp[-(E_i - E_j)/k_B T] = \exp(-E_i/k_B T),$$

which is the same as going from i to j.

The conductances Γ_{ij} as expressed by Eq. (9.14) are noted to be correlated and not independent of each other. Consider Γ_{ij} and Γ_{ik}, which share the same site i. If E_i is varied, then Γ_{ij} and Γ_{ik} would both vary in a correlated way. This correlation is an important character of the hopping conduction.

To apply the critical-path method, let us consider the following rule: Pick a Γ and consider the pair ij connected if $\Gamma_{ij} \geq \Gamma$ and disconnected otherwise. If the value of Γ is sufficiently large, then only a few Γ_{ij}'s can

9.4 Transport of Localized States at Finite Temperatures—Hopping Conduction

satisfy the connection criterion, and the connected sites form only disjoint clusters. Now lower the value of Γ by a small amount and apply the rule again. More links will be added to the existing clusters. By lowering Γ continuously, the connected clusters are expected to increase in size until at $\Gamma = \Gamma_c$, the *percolation* conductance, an infinite network of connected sites is formed. The fact that there must be such a Γ_c is based on the phenomenon of percolation, in which an infinite conduction network must form at some critical concentration of bonds connecting the sites.

The value of Γ_c may be regarded as the hopping conductance of the macroscopic sample for the following reasons. First, since at $\Gamma > \Gamma_c$ the network is not percolating, the last added resistance with the value of $(\Gamma_c)^{-1}$ must be in series with the resistance of the rest of the network. Since Eq. (9.14) gives an exponential form for Γ_{ij}, even a relatively narrow distribution in the values of S_{ij} and E_{ij} can translate into an exponentially broad distribution for Γ_{ij}. As a result, most of the resistances in the percolating network may be regarded as having zero resistance compared with $(\Gamma_c)^{-1}$, and only an exponentially small number of resistances are comparable with $(\Gamma_c)^{-1}$. The overall network resistance should, therefore, be on the order of $(\Gamma_c)^{-1}$. Second, those conductances $\Gamma_{ij} < \Gamma_c$ can add additional *parallel* conduction channels; however, here again the exponentially broad distribution of Γ_{ij} means the additional parallel channels will mostly have conductances exponentially smaller than Γ_c, and only very few are comparable to Γ_c. The net result of these two considerations is that Γ_c characterizes the overall hopping conductance, if not accurately in absolute value, certainly in qualitative features such as the temperature dependence.

To examine the temperature dependence of Γ_c, let us start with the inequality $\Gamma_{ij} \geq \Gamma_c$, or

$$\Gamma_0 \exp(-2\chi S_{ij} - E_{ij}/k_B T) \geq \Gamma_c. \qquad (9.16)$$

By taking the logarithms of both sides of the inequality and substituting Eq. (9.15) for E_{ij}, one gets

$$\frac{S_{ij}}{S_m} + \frac{|E_i| + |E_j| + |E_i - E_j|}{E_m} \leq 1, \qquad (9.17)$$

where

$$S_m = \frac{1}{2\chi} \ln \frac{\Gamma_0}{\Gamma_c}, \qquad (9.18a)$$

$$E_m = 2k_B T \ln \frac{\Gamma_0}{\Gamma_c}. \qquad (9.18b)$$

Equation (9.17) immediately tells us that there is indeed a correlation between thermal activation and tunneling. This correlation arises from the global optimization of the hopping conduction paths based on the concept of percolation. Besides relating S_{ij} to E_{ij}, Eq. (9.17) tells us that in order for the inequality to hold, S_{ij} must be less than S_m, and E_i or E_j must be less than E_m. Therefore, S_m and E_m serve as the respective upper bounds to S_{ij} and E_i (or E_j).

The requirement for the formation of a critical percolation network may be cast in the form that the average number of bonds emanating from a given site must exceed a critical number (Shante, 1973). Since the actual value of this number is not important, let us denote it as b_c. This condition must be an equality at the percolation threshold, and may be written in the following form for the site i:

$$b_c = \rho_0 \int_{-E_m}^{E_m} dE_j \int_0^{S_m} dS_{ij} S_d S_{ij}^{d-1}. \tag{9.19}$$

Here the right-hand side expresses the average number of bonds emanating from site i that satisfy the condition of (9.17), under the assumption of homogeneous distribution of the sites and a constant density of states ρ_0. $S_d = 4\pi, 2\pi, 2$ denotes the area of a d-dimensional sphere of unit radius. By performing the integral, it is seen that

$$b_c = \left(\frac{2\rho_0}{d} S_d\right) S_m^d E_m. \tag{9.20}$$

By substituting Eqs. (9.18a) and (9.18b) into Eq. (9.20), it is seen that

$$\left[\ln \frac{\Gamma_0}{\Gamma_c}\right]^{d+1} \frac{k_B T}{(2\chi)^d} \frac{4 S_d}{d} \rho_0 = b_c, \tag{9.21}$$

so that

$$\Gamma_c = \Gamma_0 \exp\left[-(T_0/T)^{1/(d+1)}\right], \tag{9.22}$$

where

$$T_0 = \frac{(2\chi)^d d}{4 S_d \rho_0 k_B}. \tag{9.23}$$

Equation (9.22) is called Mott's law, which is widely observed in amorphous materials (Mott, 1968).

9.4 Transport of Localized States at Finite Temperatures—Hopping Conduction

Physically, the fractional temperature dependence comes from the continuous shifting of the globally optimal conduction paths as the temperature is varied. This can best be described in terms of Mott's argument. At high temperatures, activation is easy, so the limiting factor is the tunneling distance. The optimal path is thus the one with the smallest tunneling distances between the successive sites, and the temperature dependence is activated, with the activation energy given by the largest E_i of the conduction path sites. As temperature is lowered, activation becomes more of a limiting factor, and the trade-off between a lower activation energy and a larger tunneling distance becomes favorable to the latter. Another way of looking at the trade-off is that as T is lowered, the number of thermally accessible states becomes smaller, and it forces the tunneling distances to be larger so that there can still be a percolatng network for conduction. Since the tunneling distance can be continuously increased as T decreases, this type of conduction is referred to as "variable-range hopping." A heuristic way of estimating the optimal hopping distance is to minimize the exponent,

$$\frac{E_{ij}}{k_B T} + 2\chi S_{ij},$$

by letting

$$E_{ij} \approx \left(C_d S_{ij}^d \rho_0\right)^{-1},$$

where $C_d = 4\pi/3$, π, 2 for $d = 3, 2, 1$, respectively. That is, E_{ij} is approximated by the inverse of the average level separation within a sphere of radius S_{ij}. By differentiating the exponent with respect to S_{ij} and setting the result to zero, one gets the optimal tunneling distance $(S_{ij})_o$ as

$$(S_{ij})_o = \left(\frac{d}{2\chi C_d \rho_0 k_B T}\right)^{1/(d+1)}. \tag{9.24}$$

Substitution of Eq. (9.24) back into the exponent yields the same fractional temperature dependence of the hopping conduction as that given by (9.22). From $(S_{ij})_o$ it is seen that the tunneling distance varies as $T^{-1/4}$.

The spatial dimensionality dependence of the temperature variation is an inherent characteristic of the hopping conduction. However, in 1D this behavior has to be qualified. This is so because if there is a continuous distribution of S_{ij}, then as the change is transported along the chain it will inevitably encounter a particularly large value of S_{ij}. In 1D, there is no chance that a barrier can be circumvented as in 2D or 3D. Therefore, if the distribution of S_{ij} is continuous, the tunneling part of the hopping mechanism will always be stopped by a barrier with $S_{ij} \to \infty$. In that case, simple activation is the only allowed mechanism possible, and the temperature dependence must therefore be activated. However, if the distribution of S_{ij} has an upper bound then the hopping behavior would be recovered when the temperature is low enough so that $(S_{ij})_o$ exceeds the upper bound.

Problems and Solutions

9.1. Calculate the eigenvalue spectrum of the operator $(i\nabla + 2e\mathbf{A}/\hbar c)^2$, where $\mathbf{A} = (-Hy, 0, 0)$.

By writing the operator in its component form, we have

$$\left[\left(i\frac{\partial}{\partial x} - \frac{2eHy}{\hbar c}\right)^2 - \frac{\partial^2}{\partial y^2} - \frac{\partial^2}{\partial z^2}\right]\phi(x,y,z) = K^2\phi(x,y,z). \quad (P9.1)$$

Since the operator $2e\mathbf{A}/\hbar c$ does not have a x or z coordinate, one can use the following form for ϕ:

$$\phi(x, y, z) = \exp(ik_x x + ik_z z)\psi(y). \quad (P9.2)$$

Substitution of (P9.2) into (P9.1) yields

$$-\frac{d^2\psi(y)}{dy^2} + \left[k_z^2 + \left(k_x + \frac{2eHy}{\hbar c}\right)^2\right]\psi(y) = K^2\psi(y). \quad (P9.3)$$

By defining

$$l_H = \sqrt{\frac{\hbar c}{eH}},$$

one gets

$$-\frac{d^2\psi(y)}{dy^2} + \frac{4}{l_H^4}\left(y + \frac{1}{2}k_x l_H^2\right)^2\psi(y) + k_z^2\psi(y) = K^2\psi(y). \quad (P9.4)$$

The first two terms correspond exactly to the linear oscillator, so the eigenvalue K^2 is given by

$$K^2 = \left(n + \frac{1}{2}\right)\frac{4eH}{\hbar c} + k_z^2, \quad n = 0, 1, 2, \ldots. \tag{P9.5}$$

9.2. Derive an analytical expression for $\delta\sigma_H$ in the limit of $H \to 0$. From Eq. (9.11), $\delta\sigma_H$ may be written as

$$\delta\sigma_H = -\frac{e^2}{\hbar}\frac{1}{l\pi^3}\sum_{n=0}^{1/\overline{H}} \frac{2\sqrt{\overline{H}}}{\sqrt{(\overline{H}a^2)^{-1} + 4(n+\frac{1}{2})}} \tan^{-1} \frac{1}{\sqrt{a^{-2} + 4\overline{H}(n+\frac{1}{2})}}$$

$$= -\frac{e^2}{\hbar}\frac{1}{l\pi^3}\sum_{n=0}^{1/\overline{H}} \frac{2\overline{H}}{\sqrt{a^{-2} + 4\overline{H}(n+\frac{1}{2})}} \tan^{-1} \frac{1}{\sqrt{a^{-2} + 4\overline{H}(n+\frac{1}{2})}}. \tag{P9.6}$$

Let $x = \overline{H}n$. As $\overline{H} \to 0$ and $n \to \infty$, x becomes a continuous variable. We thus obtain

$$\delta\sigma_0 = \lim_{H \to 0} \delta\sigma_H = -\frac{e^2}{\hbar}\frac{2}{l\pi^3}\int_0^1 \frac{dx}{\sqrt{a^{-2} + 4x}} \tan^{-1} \frac{1}{\sqrt{a^{-2} + 4x}}$$

$$= -\frac{e^2}{\hbar}\frac{1}{l\pi^3}\frac{1}{2}\int_{a^{-2}}^{4+a^{-2}} \frac{dy}{\sqrt{y}} \tan^{-1} \frac{1}{\sqrt{y}}. \tag{P9.7}$$

By using integration by parts, this integral may be explicitly evaluated, giving

$$\delta\sigma_0 = -\frac{e^2}{\hbar}\frac{1}{l\pi^3}\left[\sqrt{4+a^{-2}} \tan^{-1}\frac{1}{\sqrt{4+a^{-2}}} - \frac{1}{a}\tan^{-1} a + \frac{1}{2}\ln\frac{5a^2+1}{a^2+1}\right]. \tag{P9.8}$$

When $T \to 0$ so that $a \to \infty$, we have

$$\delta\sigma_0 = -\frac{e^2}{\hbar}\frac{1}{l\pi^3}\left[2\tan^{-1}\tfrac{1}{2} + \tfrac{1}{2}\ln 5\right]. \tag{P9.9}$$

References

Ambegaokar, V., Halperin, B. I., and Langer, J. S. (1971). *Phys. Rev.* **B4**, 2612.
Bergmann, G. (1984). *Phys. Rep.* **107**, 1.
Dolan, G. J., and Osheroff, D. D. (1979). *Phys. Rev. Lett.* **43**, 721.
Kawabata, A. (1980). *J. Phys. Soc. Japan* **49**, 628.
Mott, N. F. (1968). *J. Non-Cryst. Solids* **1**, 1.
Shante, V. K. (1973). *Phys. Lett.* 249.
Weaver, R. (1993). *Phys. Rev.* **B47**, 1077.
Zhang, Z. Q., and Sheng, P. (1991). *Phys. Rev.* **44**, 3304.

10

Mesoscopic Phenomena

10.1 What Is "Mesoscopic"?

In the literature, the term "mesoscopic" generally denotes a sample size regime intermediate between the molecular and the bulk. However, in this chapter we wish to define the term more precisely as the regime where $L < L_{in}$. For electronic systems, this more precise definition coincides with the usual definition at low temperature ($T \lesssim 10$ K). But for classical waves L_{in} is generally macroscopic in size, and a mesoscopic sample for classical waves is thus equivalent to a bulk sample.

What is special about a mesoscopic sample is that the wave transport behavior may be observed without being filtered by inelastic scattering. Localization is therefore a mesoscopic phenomenon observable when $\xi < L < L_{in}$. However, even when waves are delocalized, or localized but with $L < L_{in} < \xi$, the behavior of mesoscopic samples can still differ significantly from that of bulk samples, where inelastic scattering plays a role. In particular, mesoscopic phenomena are related in one way or the other to the preservation of wave characteristics, such as interference and phase memory, even after strong multiple scattering. A striking example in this respect is the speckle pattern formed by scattered light in transmission through a random medium, alluded to in Chapter 6 (speckle pattern also exists in mesoscopic electronic systems, but it is much more difficult to observe). The spatial intensity fluctuation in this case is a direct result of wave interference. These fluctuations have a component which remains constant over a range of wave propagation distances, a characteristic which is called "universal conductance fluctuations" (Lee *et al.*, 1987).

An intriguing question for mesoscopic electronic samples is the definition of electrical conductivity. The traditional definition of σ_E is associ-

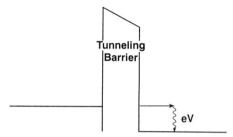

Figure 10.1 A schematic picture of the two-step conduction process across a tunnel junction. In the first step, an electron tunnels elastically through the barrier. In the second step, the electron gives up its excess energy. Dissipation occurs in the second step.

ated with dissipation: $\sigma_E E^2$ is the power dissipated per unit volume. Since in a mesoscopic sample the absence of inelastic scattering implies the absence of dissipation, a new definition of electrical conductivity must be devised. Based on the analogy with a tunnel junction, Landauer (1957) proposed that the electrical conduction in mesoscopic samples be viewed as a two-step process. Consider a voltage difference V applied across a tunnel junction shown schematically in Figure 10.1. An electron tunneling elastically through the barrier will end up on the other side with energy eV above the Fermi level. The subsequent decay to the Fermi level completes the conduction process across the junction. In this picture the electrical conduction consists of two separate steps: tunneling and energy dissipation. If the energy barrier is replaced by a mesoscopic sample, it is clear that one can identify the wave transmission probability across the sample as something that is *proportional* to the electrical conductance. In this definition of σ_E the element of dissipation still exists. However, instead of occurring inside the sample, dissipation now occurs in the leads attached to the sample; i.e., it is nonlocal. Landauer's formula of electrical conductance for mesoscopic samples is the result of these considerations.

Detailed mathematical descriptions of mesoscopic phenomena are given below.

10.2 Intensity Distribution of the Speckle Pattern

A speckle pattern results from the interference of randomly scattered waves. For a given configuration of random scatterers, the intensity at a given observation point is the absolute square of the sum of amplitudes for

10.2 Intensity Distribution of the Speckle Pattern

all the different wave paths emanating from the source:

$$I(\mathbf{r}) = |A(\mathbf{r})|^2 = \frac{1}{M}\left|\sum_{k=1}^{M} a_k \exp(i\phi_k)\right|^2, \qquad (10.1)$$

where for definitiveness the total number of path is given by the number M. The factor $1/M$ is present so that the defined amplitude a_k remains finite as $M \to \infty$. Let us make the plausible assumption that a_k and ϕ_k are statistically independent, with the mean and second moment of a_k given by a and a^2, respectively, and that ϕ_k is distributed uniformly between 0 and 2π. Equation (10.1) can be rewritten as

$$I(\mathbf{r}) = r^2 + i^2, \qquad (10.2)$$

with

$$r = \frac{1}{\sqrt{M}} \sum_{k=1}^{M} a_k \cos \phi_k, \qquad (10.3a)$$

$$i = \frac{1}{\sqrt{M}} \sum_{k=1}^{M} a_k \sin \phi_k. \qquad (10.3b)$$

It is clear that $\langle r \rangle_c = \langle i \rangle_c = 0$. Moreover,

$$\langle r^2 \rangle_c = \frac{1}{M} \sum_{k=1}^{M} \sum_{n=1}^{M} \langle a_k a_n \rangle_c \langle \cos \phi_k \cos \phi_n \rangle_c$$

$$= \frac{1}{M} \sum_{k=1}^{M} \sum_{n=1}^{M} \langle a_k a_n \rangle_c \frac{1}{2} \delta_{k,n}$$

$$= \frac{a^2}{2}, \qquad (10.4)$$

and $\langle i^2 \rangle_c = \langle r^2 \rangle_c$. It is also easy to see that $\langle ri \rangle_c = 0$, so the variables r and i are uncorrelated. From Eqs. (10.3a) and (10.3b), it is seen that r and i are sums of independent random variables. The central limit theorem states that the distributions of r and i must each be Gaussian (Feller, 1957), so that the joint distribution is given by

$$D(r, i) = \frac{1}{2\pi(a^2/2)} \exp\left(-\frac{r^2 + i^2}{2(a^2/2)}\right)$$

$$= \frac{1}{\pi a^2} \exp\left(-\frac{r^2 + i^2}{a^2}\right). \qquad (10.5)$$

By changing the variable from r, i to I, it is immediately clear that

$$D(I) = \frac{1}{\langle I \rangle_c} \exp\left(-\frac{I}{\langle I \rangle_c}\right), \tag{10.6}$$

where $I = r^1 + i^2$, $\langle I \rangle_c = a^2$, and $dr\, di = 2\pi\sqrt{I}\, d\sqrt{I} = \pi dI$. Equation (10.6) is the main result of this section. By assuming that the intensity distribution of the speckle pattern for a given configuration is the same as that for a given observation point varied over different configurations, Eq. (10.6) states that the intensity has an exponential distribution. A special feature of the exponential distribution is that its mean, $\langle I \rangle_c$, is equal to its standard deviation $[\langle I^2 \rangle_c - \langle I \rangle_c^2]^{1/2}$. The intensity fluctuations are therefore not small. Below we calculate the frequency and spatial correlations in a speckle pattern.

10.3 Correlations in the Speckle Pattern

Let us first calculate the average intensity at $\mathbf{r}$ resulting from a point source at $\mathbf{r}'$. This is given by $\langle |G^+(\omega, \mathbf{r}, \mathbf{r}')|^2 \rangle_c$. From Chapter 4, Equation (4.8), this is expressible as

$$\langle I(\omega, |\mathbf{r} - \mathbf{r}'|) \rangle_c = S(\Delta\omega = 0, \mathbf{r} - \mathbf{r}' \mid \omega)$$

$$= \frac{L^d}{N} \int \frac{d\Delta\mathbf{k}}{(2\pi)^d} S(\Delta\omega = 0, \Delta\mathbf{k} \mid \omega) \exp[i\Delta\mathbf{k} \cdot (\mathbf{r} - \mathbf{r}')]. \tag{10.7}$$

Since from Eq. (4.45) it is known that

$$S(\Delta\omega = 0, \Delta\mathbf{k} \mid \omega) = \frac{2(d\omega/d\kappa_0^2)(d\omega/d\kappa_e^2)\rho_e(\omega)\pi}{D(\omega)|\Delta\mathbf{k}|^2} \frac{N}{L^d}, \tag{10.8}$$

it follows that

$$\langle I(\omega, |\mathbf{r} - \mathbf{r}'|) \rangle_c = \frac{2\pi\rho_e(\omega)(d\omega/d\kappa_0^2)(d\omega/d\kappa_e^2)}{(v_t l/3)(2\pi)^d}$$

$$\times \int \frac{d\Delta\mathbf{k}}{|\Delta\mathbf{k}|^2} \exp[i\Delta\mathbf{k} \cdot (\mathbf{r} - \mathbf{r}')]$$

$$= \frac{3(d\omega/d\kappa_0^2)(d\omega/d\kappa_e^2)\rho_e(\omega)}{2v_t l |\mathbf{r} - \mathbf{r}'|} \tag{10.9}$$

10.3 Correlations in the Speckle Pattern

for $d = 3$. The integral diverges for $d \leq 2$, which is an indication that the diffusive wave intensity in 1D or 2D must depend on the boundary conditions, however large the sample. The dependence of the intensity on $|\mathbf{r} - \mathbf{r}'|^{-1}$ is noted to differ from the wave behavior (which should give a $|\mathbf{r} - \mathbf{r}'|^{-2}$ dependence) but to be characteristic of diffusion. For classical scalar waves with $d\omega/d\kappa_0^2 = v_0^2/2\omega$, $d\omega/d\kappa_e^2 = v_e^2/2\omega$, and $\rho_e(\omega) = 4\pi\omega^2/v_e^3$, we get

$$\langle I(\omega, |\mathbf{r} - \mathbf{r}'|)\rangle_c = \frac{3\pi v_0^2}{2v_t v_e l |\mathbf{r} - \mathbf{r}'|}$$

$$\cong \frac{3\pi \bar{\epsilon}}{2l|\mathbf{r} - \mathbf{r}'|} \qquad (10.10)$$

in the low-frequency limit, where $v_e \cong v_t$, and $\bar{\epsilon} = v_0^2/v_e^2$.

From the calculation of the average diffusive intensity at a point, it is only a small step to calculate the frequency correlation of the intensity at a given observation point. Let us define the frequency correlation function as

$$C(\Delta\omega, |\mathbf{r} - \mathbf{r}'|) = \langle I(\omega, |\mathbf{r} - \mathbf{r}'|)I(\omega + \Delta\omega, |\mathbf{r} - \mathbf{r}'|)\rangle_c$$
$$- \langle I(\omega, |\mathbf{r} - \mathbf{r}'|)\rangle_c \langle I(\omega + \Delta\omega, |\mathbf{r} - \mathbf{r}'|)\rangle_c. \quad (10.11)$$

The first term on the right-hand side can be written alternatively as

$$\langle G^+(\omega, \mathbf{r}, \mathbf{r}') G^-(\omega, \mathbf{r}', \mathbf{r}) G^+(\omega + \Delta\omega, \mathbf{r}, \mathbf{r}') G^-(\omega + \Delta\omega, \mathbf{r}', \mathbf{r})\rangle_c.$$

We look for pairings involving G^+ and G^-, which would give the largest contribution. The configurational averaging can be done in subsets of $\langle G^+ G^- \rangle_c$ in two ways. One is given by the second term on the right-hand side of Eq. (10.11), which is subtracted off. The other remaining one is given by

$$\langle G^+(\omega, \mathbf{r}, \mathbf{r}') G^-(\omega + \Delta\omega, \mathbf{r}', \mathbf{r})\rangle_c$$
$$\times \langle G^+(\omega + \Delta\omega, \mathbf{r}, \mathbf{r}') G^-(\omega, \mathbf{r}', \mathbf{r})\rangle_c = |S(\Delta\omega, \mathbf{r} - \mathbf{r}'|\omega)|^2. \quad (10.11)$$

It follows that

$$C(\Delta\omega, |\mathbf{r} - \mathbf{r}'|) \cong |S(\Delta\omega, \mathbf{r} - \mathbf{r}'|\omega)|^2. \quad (10.12)$$

Since in the $\Delta\omega$, $\Delta\mathbf{k}$ representation S has a diffusive pole, it is seen that

$$S(\Delta\omega, \mathbf{r} - \mathbf{r}'|\omega) = \frac{2\pi\rho_e(\omega)(d\omega/d\kappa_0^2)(d\omega/d\kappa_e^2)}{(2\pi)^d} \int \frac{d(\Delta\mathbf{k})\exp[i\,\Delta\mathbf{k}\cdot(\mathbf{r}-\mathbf{r}')]}{-i\,\Delta\omega + v_t l|\Delta\mathbf{k}|^2/3}.$$

We will consider only the case of $d = 3$, where there is a finite limit for $\Delta\omega \to 0$. The integral is done in the solution to Problem 10.1. The result is

$$S(\Delta\omega, \mathbf{r} - \mathbf{r}'|\omega) \frac{3\rho_e(\omega)(d\omega/d\kappa_0^2)(d\omega/d\kappa_e^2)}{2v_t l|\mathbf{r} - \mathbf{r}'|} \exp\left[(i-1)\sqrt{\frac{3\Delta\omega}{2v_t l}}|\mathbf{r} - \mathbf{r}'|\right]. \tag{10.13}$$

From Eqs. (10.9) and (10.12), we have

$$C(\Delta\omega, |\mathbf{r} - \mathbf{r}'|) \cong \langle I(\omega, |\mathbf{r} - \mathbf{r}'|)\rangle_c^2 \exp\left[-\sqrt{\frac{6\Delta\omega}{v_t l}}|\mathbf{r} - \mathbf{r}'|\right]. \tag{10.14}$$

At $\Delta\omega = 0$, $C(\Delta\omega = 0, |\mathbf{r} - \mathbf{r}'|)$ is exactly the variance of the speckle pattern. This is consistent with the result of our earlier calculation, which predicts the standard deviation to be the same as the mean. Equation (10.14) states that at a given observation point, a variation in the frequency of the wave will cause a corresponding variation in the intensity, with a correlation which decays exponentially as the square root of the frequency variation (Shapiro, 1986).

Suppose the frequency is held fixed but the observation point is varied; what would be the *spatial* correlation? That is, what is the spatial correlation in the speckle pattern? To calculate that quantity, let us define

$$\mathbf{r}_1 - \mathbf{r}' = \mathbf{R}$$

to be the separation vector between the source and the observation point $\mathbf{r}_1$, and $\Delta\mathbf{R} = \mathbf{r}_2 - \mathbf{r}_1$ to be the vector defining the separations between the two observation points $\mathbf{r}_2$ and $\mathbf{r}_1$. The spatial correlation $C^{(1)}(\omega, \mathbf{R}, \Delta\mathbf{R})$ is defined by

$$\langle I(\omega, \mathbf{R})I(\omega, \mathbf{R} + \Delta\mathbf{R})\rangle_c$$
$$= \langle G^+(\omega, \mathbf{R})G^-(\omega, \mathbf{R})G^+(\omega, \mathbf{R} + \Delta\mathbf{R})G^-(\omega, \mathbf{R} + \Delta\mathbf{R})\rangle_c$$
$$= \langle I(\omega, \mathbf{R})\rangle_c \langle I(\omega, \mathbf{R} + \Delta\mathbf{R})\rangle_c + C^{(1)}(\omega, \mathbf{R}, \Delta\mathbf{R}). \tag{10.15}$$

To calculate $C^{(1)}(\omega, \mathbf{R}, \Delta\mathbf{R})$, it is noted that in Eq. (10.15), besides the $\langle G^+ G^- \rangle_c$ pairing that gives the $\langle I(\omega, \mathbf{R})\rangle_c \langle I(\omega, \mathbf{R} + \Delta\mathbf{R})\rangle_c$ term, there can also be the pairing choice

$$\langle G^+(\omega, \mathbf{R})G^-(\omega, \mathbf{R} + \Delta\mathbf{R})\rangle_c \langle G^+(\omega, \mathbf{R} + \Delta\mathbf{R})G^-(\omega, \mathbf{R})\rangle_c.$$

10.3 Correlations in the Speckle Pattern

That means

$$C^{(1)}(\omega, \mathbf{R}, \Delta\mathbf{R}) \equiv \langle G^+(\omega, \mathbf{R})G^-(\omega, \mathbf{R} + \Delta\mathbf{R})\rangle_c$$
$$\times \langle G^+(\omega, \mathbf{R} + \Delta\mathbf{R})G^-(\omega, \mathbf{R})\rangle_c. \quad (10.16)$$

Let us recall from Chapter 4 that

$$\langle \mathbf{G}^+ \otimes \mathbf{G}^- \rangle_c = \mathbf{G}_e^+ \otimes \mathbf{G}_e^- + (\mathbf{G}_e^+ \otimes \mathbf{G}_e^-):\mathbf{\Gamma}:\langle \mathbf{G}_e^+ \otimes \mathbf{G}_e^- \rangle_c. \quad (10.17)$$

In the present case, the output channels in real space are noted to be different. In Eq. (10.17), the first term on the right-hand side is negligible, because both $\mathbf{G}_e^+$ and $\mathbf{G}_e^-$ represent the coherent component of the wave, which decays exponentially with a short decay length l. The second term, however, can be separated into two parts. The first part, $(\mathbf{G}_e^+ \otimes \mathbf{G}_e^-):\mathbf{\Gamma}$, is diffusive in character as can be seen as follows. Since from Eq. (4.18)

$$\mathbf{\Gamma} = \mathbf{U} + \mathbf{U}:(\mathbf{G}_e^+ \otimes \mathbf{G}_e^-):\mathbf{\Gamma},$$

and in the Boltzmann limit (ladder diagrams with isotropic scattering approximation) the inner product of $(\mathbf{G}_e^+ \otimes \mathbf{G}_e^-)$ with $\mathbf{\Gamma}^{(B)}$ is represented by the operations of (1) setting the observation points of $\mathbf{G}_e^+$ and $\mathbf{G}_e^-$ the same (same scatterer) and (2) summing over all scatterer positions [see Eq. (4.61)], it follows that in the k-representation ($U^{(B)}$ is recalled to be $\mathbf{k}$ independent), we have

$$\Gamma^{(B)} = \frac{U^{(B)}}{1 - (U^{(B)}/N)\Sigma_\mathbf{k}(G_e^+ G_e^-)_\mathbf{k}}, \quad (10.18)$$

where the summation over $\mathbf{k}$ follows from the summation over all scatterers (since in the effective medium picture scatterers fill all space, this summation is equivalent to summation over all space). Comparison with Eq. (4.96d) shows that

$$(\mathbf{G}_e^+ \otimes \mathbf{G}_e^-):\mathbf{\Gamma} \to \frac{(1/N)\Sigma_\mathbf{k}(G_e^+ G_e^-)_\mathbf{k}}{1 - (U^{(B)}/N)\Sigma_\mathbf{k}(G_e^+ G_e^-)_\mathbf{k}} U^{(B)},$$
$$= S^{(B)}U^{(B)}. \quad (10.19)$$

Since $S^{(B)}$ is diffusive in character, the incoherent component of the wave thus decays algebraically from the source point $\mathbf{r}'$ to the observation point $\mathbf{r}_1 = \mathbf{r}' + \mathbf{R}$.

The second part of the term $(\mathbf{G}_e^+ \otimes \mathbf{G}_e^-):\mathbf{\Gamma}:\langle \mathbf{G}_e^+ \otimes \mathbf{G}_e^- \rangle_c$ is given by the term $\langle \mathbf{G}_e^+ \otimes \mathbf{G}_e^- \rangle_c$, where now the input point is at $\mathbf{r}_1$, and the output points are $\mathbf{r}_1$ and $\mathbf{r}_2$, separated by $\Delta\mathbf{R}$. Since the inner product with Γ means summation over all input channels $\mathbf{r}_1$, $\langle \mathbf{G}^+ \otimes \mathbf{G}^- \rangle_c$ can be written as

$$\sum_{\mathbf{r}_1} G_e^+(\omega, \mathbf{r}_1 - \mathbf{r}_1) G_e^-(\omega, \mathbf{r}_2 - \mathbf{r}_1)$$

$$= \frac{L^d}{N(2\pi)^d} \int d\mathbf{k}\, (G_e^+ G_e^-)_\mathbf{k} \exp(i\mathbf{k} \cdot \Delta\mathbf{R}).$$

By noting that

$$(G_e^+ G_e^-)_\mathbf{k} \cong (G_e^+)_\mathbf{k} (G_e^-)_\mathbf{k} \cong \frac{(\Delta G_e)_\mathbf{k}}{2i\,\text{Im}\,\Sigma^+} \frac{N}{L^d},$$

it follows from the approximation, that Σ^+ is $\mathbf{k}$-independent, that

$$\frac{L^d}{N(2\pi)^d} \int d\mathbf{k}(G_e^+ G_e^-)_\mathbf{k} \exp(i\mathbf{k} \cdot \Delta\mathbf{R})$$

$$\cong \frac{1}{2i\,\text{Im}\,\Sigma^+} \int \frac{d\mathbf{k}}{(2\pi)^d} (\Delta G_e)_\mathbf{k} \exp(i\mathbf{k} \cdot \Delta\mathbf{R})$$

$$= \frac{\Delta G_e(\Delta\mathbf{R})}{2i\,\text{Im}\,\Sigma^+} \frac{N}{L^d}. \quad (10.20)$$

Combining Eq. (10.20) with Eq. (10.19) results in

$$\langle G^+(\omega, \mathbf{R}) G^-(\omega, \mathbf{R} + \Delta\mathbf{R}) \rangle_c \cong S^{(B)}(\Delta\omega = 0, \mathbf{R}|\omega) \frac{\Delta G_e(\Delta\mathbf{R})}{\Delta G_e(\Delta R = 0)}, \quad (10.21)$$

where we have used the identity $U^{(B)} = [2i\,\text{Im}\,\Sigma^+ / \Delta G_e(\Delta R = 0)] L^d / N$.

For the case of classical waves in 3D, one can go a step further and write

$$G_e^\pm(\Delta\mathbf{R}) = -\frac{\exp(\pm i\kappa \Delta R) \exp(-\Delta R/2l)}{4\pi \Delta R}.$$

10.3 Correlations in the Speckle Pattern

from which it follows that

$$\Delta G_e(\Delta \mathbf{R}) = -\frac{i}{2\pi \Delta R} \sin(\kappa \Delta R) \exp(-\Delta R/2l). \quad (10.22)$$

Since Im $\Sigma^+ = -\kappa/l$ [Eq. (8.41b)], one gets

$$\sum_{\mathbf{r}_1} \langle G^+(\omega, \mathbf{r}_1, \mathbf{r}_1) G^-(\omega, \mathbf{r}_1, \mathbf{r}_2) \rangle_c$$

$$\cong \frac{l}{4\pi\kappa \Delta R} \sin(\kappa \Delta R) \exp(-\Delta R/2l) \frac{N}{L^d}. \quad (10.23)$$

Therefore,

$$\langle G^+(\omega, \mathbf{R}) G^-(\omega, \mathbf{R} + \Delta \mathbf{R}) \rangle_c$$

$$\cong S^{(B)}(\Delta\omega = 0, \mathbf{R}|\omega) U^B \frac{l \sin(\kappa \Delta R)}{4\pi\kappa \Delta R} \exp(-\Delta R/2l) \frac{N}{L^d}. \quad (10.24)$$

By using the Ward identity,

$$U^{(B)} \frac{N}{L^d} = \frac{2i \operatorname{Im} \Sigma^+}{\Delta G_e(\Delta R = 0)} = \frac{2i \operatorname{Im} \Sigma^+}{-i\kappa/2\pi} = \frac{4\pi}{l}, \quad (10.25)$$

Eq. (10.24) becomes, for classical scalar waves,

$$\langle G^+(\omega, \mathbf{R}) G^-(\omega, \mathbf{R} + \Delta \mathbf{R}) \rangle_c$$

$$\cong \langle I(\omega, \mathbf{R}) \rangle_c \frac{\sin(\kappa \Delta R)}{\kappa \Delta R} \exp(-\Delta R/2l). \quad (10.26)$$

Here we have identified $S^{(B)}(\Delta\omega = 0, \mathbf{R}|\omega)$ as $\langle I(\omega, \mathbf{R}) \rangle_c$. Since the other term, $\langle G^+(\omega, \mathbf{R} + \Delta \mathbf{R}) G^-(\omega, \mathbf{R}) \rangle_c$, can be obtained from Eq. (10.26) by exchanging $\mathbf{r}_1$ and $\mathbf{r}_2$, it follows from Eqs. (10.16) and (10.21) that

$$C^{(1)}(\omega, \mathbf{R}, \Delta \mathbf{R}) \cong \langle I(\omega, \mathbf{R}) \rangle_c \langle I(\omega, \mathbf{R} + \Delta \mathbf{R}) \rangle_c \frac{\Delta G_e(\Delta \mathbf{R})}{\Delta G_e(\Delta R = 0)} \quad (10.27a)$$

in the general case and

$$C^{(1)}(\omega, \mathbf{R}, \Delta \mathbf{R}) \cong \langle I(\omega, |\mathbf{R}|) \rangle_c \langle I(\omega, |\mathbf{R} + \Delta \mathbf{R}|) \rangle_c$$

$$\times \frac{\sin^2(\kappa \Delta R)}{(\kappa \Delta R)^2} \exp\left(-\frac{\Delta R}{l}\right) \quad (10.27b)$$

for the case of 3D classical waves. Again, when $\Delta R = 0$, $C^{(1)}(\omega, \mathbf{R}, \Delta \mathbf{R} = 0)$ is noted to give the variance of the intensity fluctuations as expected.

Equation (10.27b) tells us that the spots in the speckle pattern are correlated over a distance l (Shapiro, 1986), and such correlation exists no matter how far the wave travels. The result can be viewed physically as the consequence of diffusion plus interference. Since the effect of interference is to produce rapid spatial fluctuations in intensity, its effect on diffusion may be approximated by a random spatial source term in the diffusive process. Such a viewpoint was first put forth by Spivak and Zyuzin (1988). Let us follow this approach and examine the *long-range* implications of the short-range interference effect.

10.4 Long-Range Correlation in Intensity Fluctuations

Consider the division of $I(\mathbf{r})$ into

$$I(\mathbf{r}) = \langle I(\mathbf{r}) \rangle_c + \delta I(\mathbf{r}) + \Delta I(\mathbf{r}), \tag{10.28}$$

where the frequency dependence is suppressed. The separation of the fluctuation part of $I(\mathbf{r})$ into $\delta I(\mathbf{r})$ and $\Delta I(\mathbf{r})$ is based on the difference in spatial scales: Whereas $\delta I(\mathbf{r})$ denotes the diffusive relaxation of the intensity over distances $\gg l$, $\Delta I(\mathbf{r})$ represents the short-range fluctuations manifest as the speckle pattern. From the discussion above, it is also clear that while $\delta I(\mathbf{r})$ is part of the diffusive flux, $\Delta I(\mathbf{r})$ is not, because it is caused by coherent interference. The approach of Spivak and Zyuzin is to treat $\Delta I(\mathbf{r})$ as a random source, where $\langle \Delta I(\mathbf{r}_1) \Delta I(\mathbf{r}_2) \rangle$ is given by the correlation function of Eq. (10.27). Below we treat the long-range correlation only in the classical scalar wave case. In the calculation of long-range correlation, a reasonable approximation is to treat the short-range correlation as delta-function-like, i.e., with $|\mathbf{r}_2 - \mathbf{r}_1| = \Delta R$,

$$\langle \Delta I(\mathbf{r}_1) \Delta I(\mathbf{r}_2) \rangle_c \cong \delta(\mathbf{r}_2 - \mathbf{r}_1) \langle I(\mathbf{r}_1) \rangle_c^2 \, 4\pi$$
$$\times \int_0^\infty (\Delta R)^2 \, d(\Delta R) \frac{\sin^2(\kappa \Delta R)}{(\kappa \Delta R)^2} \exp\left(-\frac{\Delta R}{l}\right)$$
$$\cong \delta(\mathbf{r}_2 - \mathbf{r}_1) \langle I(\mathbf{r}_1) \rangle_c^2 \frac{2\pi l}{\kappa^2}. \tag{10.29}$$

In terms of wave flux, we have

$$\mathbf{J}(\mathbf{r}) = -\frac{v_t l}{3} \nabla [\langle I(\mathbf{r}) \rangle_c + \delta I(\mathbf{r})] + \mathbf{v}_t \, \Delta I(\mathbf{r}), \tag{10.30}$$

10.4 Long-Range Correlation in Intensity Fluctuations

where the first term on the right-hand side is the diffusive flux, whereas the second term is the random flux source

$$\mathbf{j}_s = \mathbf{v}_t \, \Delta I(\mathbf{r}), \tag{10.31}$$

with

$$\langle j_s^{(i)}(\mathbf{r}_1) j_s^{(k)}(\mathbf{r}_2) \rangle_c = \frac{1}{3} \delta_{ik} \frac{2\pi v_t^2 l}{\kappa^2} \langle I(\mathbf{r}_1) \rangle_c^2 \, \delta(\mathbf{r}_2 - \mathbf{r}_1). \tag{10.32}$$

Here i, k denote the vectorial components of $\mathbf{j}_s$. In steady state, the continuity condition requires

$$\nabla \cdot \mathbf{J}(\mathbf{r}) = 0. \tag{10.33}$$

Since $\nabla^2 \langle I(\mathbf{r}) \rangle_c = 0$ in the steady state, it follows from Eq. (10.30) that

$$\frac{v_t l}{3} \nabla^2 \delta I(\mathbf{r}) = \nabla \cdot \mathbf{j}_s(\mathbf{r}). \tag{10.34}$$

To calculate the correlation of $\delta I(\mathbf{r})$, let us specialize to the slab geometry where z is the sample thickness direction, L is the sample thickness, and the desired quantity is the correlation

$$C^{(2)}(\boldsymbol{\rho}_1, \boldsymbol{\rho}_2) = \langle \delta I(\boldsymbol{\rho}_1, L-l) \delta I(\boldsymbol{\rho}_2, L-l) \rangle_c, \tag{10.35}$$

where $\boldsymbol{\rho}_1, \boldsymbol{\rho}_2$ are the coordinates on the (x, y) plane. Here we want $|\boldsymbol{\rho}_1 - \boldsymbol{\rho}_2| \gg l$ so as to distinguish $C^{(2)}$ from $C^{(1)}$. Through Fourier transform in the (x, y) plane, we obtain the equation for $\delta I(\mathbf{k}, z)$, where $\mathbf{k}$ is the 2D wave vector, as

$$\frac{d}{dz^2} \delta I(\mathbf{k}, z) - k^2 \, \delta I(\mathbf{k}, z)$$
$$= \frac{3}{v_t l} \left[i \left(k_x j_s^{(x)}(\mathbf{k}, z) + k_y j_s^{(y)}(\mathbf{k}, z) \right) + \frac{d}{dz} j_s^{(z)}(\mathbf{k}, z) \right]. \tag{10.36}$$

The boundary conditions are $\delta I(\mathbf{k}, z) = 0$ at $z = 0, L$. That is, $I(z = 0) = \langle I \rangle_c$, and at $z = L$ the condition $I = 0$ is imposed. One can be more general by specifying an extrapolation length (see Section 4.8); however, this would not alter the qualitative conclusions to be reached. The Green's function for the above equation is given by

$$G(\mathbf{k}, z, z') = -\frac{1}{k} \frac{\sinh(kz_<) \sinh[k(L - z_>)]}{\sinh(kL)}, \tag{10.37}$$

where $z_< = \min[z, z']$, and $z_> = \max[z, z']$. The fact that $G(\mathbf{k}, z, z')$ indeed satisfies

$$\frac{d}{dz^2} G(\mathbf{k}, z, z') - k^2 G(\mathbf{k}, z, z') = \delta(z, z')$$

is shown in the solution to Problem 10.2. With the knowledge of the Green's function, $\delta I(\mathbf{k}, z)$ is given by

$$\delta I(\mathbf{k}, z) = \frac{3}{vl} \left\{ \int_0^L dz' G(\mathbf{k}, z, z') i \left[k_x j_s^{(x)}(\mathbf{k}, z') + k_y j_s^{(y)}(\mathbf{k}, z') \right] \right.$$
$$\left. - \int_0^L dz' \left[\frac{d}{dz'} G(\mathbf{k}, z, z') \right] j_s^{(z)}(\mathbf{k}, z') \right\}, \quad (10.38)$$

where the second term is obtained through the integration by parts and the fact that $G(\mathbf{k}, z, z')$ vanishes at $z_< = 0$ ad $z_> = L$. Now the quantity $\langle \delta I(\mathbf{k}_1, z_1) \delta I^*(\mathbf{k}_2, z_2) \rangle_c$ may be written as

$$\langle \delta I(\mathbf{k}_1, z_1) \delta I^*(\mathbf{k}_2, z_2) \rangle_c = \left(\frac{3}{v_t l} \right)^2 \int_0^L \int_0^L dz_1' dz_2' \langle j_s^{(i)}(\mathbf{k}_1, z_1') j_s^{(i)*}(\mathbf{k}_2, z_2') \rangle_c$$

$$\times \left[(\mathbf{k}_1 \cdot \mathbf{k}_2) G(\mathbf{k}_1, z_1, z_1') G(\mathbf{k}_2, z_2, z_2') \right.$$
$$\left. + \frac{dG(\mathbf{k}_1, z_1, z_1')}{dz_1'} \frac{dG(\mathbf{k}_2, z_2, z_2')}{dz_2'} \right], \quad (10.39)$$

where the averages of the cross terms $\langle j_s^{(i)} j_s^{(j)*} \rangle_c$, $i \neq j$, are noted to vanish. From Eq. (10.32) one can easily deduce

$$\langle j_s^{(i)}(\mathbf{k}_1, z_1') j_s^{(i)*}(\mathbf{k}_2, z_2') \rangle_c$$
$$= \frac{N^2}{W^4} \frac{(2\pi)^3 v_t^2 l}{3\kappa^2} \delta(z_2' - z_1') \delta(\mathbf{k}_1 - \mathbf{k}_2) \langle I(z_1') \rangle_c^2, \quad (10.40)$$

where W, the lateral dimension of the sample, arises from the 2D Fourier transform of the correlation relation, Eq. (10.29). Here it is also assumed that the incident wave is a plane wave, so that $\langle I(\mathbf{r}_1) \rangle_c$ depends only on the z coordinate. In that case the solution to the 1D diffusion equation,

$$\nabla^2 \langle I(z) \rangle_c = 0,$$

10.4 Long-Range Correlation in Intensity Fluctuations

with the boundary conditions that $\langle I(z)\rangle_c = 1$ at $z = 0$ and 0 at $z = L$, gives the solution

$$\langle I(z)\rangle_c = 1 - \frac{z}{L}. \tag{10.41}$$

By substituting Eqs. (10.40) and (10.41) into Eq. (10.39), one gets zero for $z_1 \neq z_2$, and for $z_1 = z_2$ the in-plane correlation in the $\mathbf{k}$ representation is given by (with $z' = z_<$)

$$\langle \delta I(\mathbf{k}_1, z_1) \delta I^*(\mathbf{k}_2, z_2)\rangle_c = \frac{N^2}{W^4} \frac{3(2\pi)^3}{\kappa^2 l} \delta(\mathbf{k}_1 - \mathbf{k}_2)$$

$$\times \frac{\sinh^2 k_1(L - z_1)}{\sinh^2 k_1 L} \int_0^L \cosh 2k_1 z' \left(1 - \frac{z'}{L}\right)^2 dz'$$

$$= \frac{N^2}{W^4} \frac{3(2\pi)^3}{\kappa^2 l} \delta(\mathbf{k}_1 - \mathbf{k}_2) \frac{\sinh^2 k_1(L - z_1)}{\sinh^2 k_1 L}$$

$$\times \frac{\sinh 2k_1 L - 2k_1 L}{4k_1^3 L^2}. \tag{10.42}$$

By choosing $z_1 = L - l$ instead of $z_1 = L$ (at which $I = 0$ by boundary condition), one gets the long-range correlation in real space as

$$C^{(2)}(\boldsymbol{\rho}_1, \boldsymbol{\rho}_2) = C^{(2)}(\Delta\rho)$$

$$= \frac{3}{2\pi\kappa^2 l} \int_0^\infty dk\, k \int_0^{2\pi} d\theta \exp(ik\Delta\rho \cos\theta) \frac{\sinh^2 kl}{\sinh^2 kL}$$

$$\times \frac{\sinh 2kL - 2kL}{4k^3 L^2}$$

$$\cong \frac{3l^3}{4\kappa^2 l^2 L^3} \int_0^\infty J_0\left(x \frac{\Delta\rho}{L}\right) \frac{\sinh 2x - 2x}{\sinh^2 x} dx, \tag{10.43}$$

where we have made the approximation $\sinh kl \cong kl$, which is consistent with the assumption expressed by Eq. (10.29). It is seen that the correlation $C^{(2)}$ is measured in terms of sample thickness L, which shows its long-range nature (Stephen and Cwilich, 1987). The function

$$F(\Delta\rho/L) = \int_0^\infty J_0\left(x \frac{\Delta\rho}{L}\right) \frac{\sinh 2x - 2x}{\sinh^2 x} dx$$

is plotted in Figure 10.2.

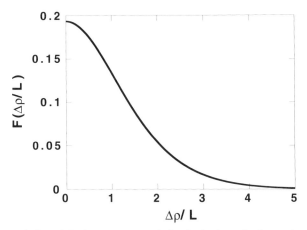

Figure 10.2 Variation of the long-range correlation in the intensity fluctuation as a function of $\Delta\rho/L$.

There can also be fluctuation in the total transmitted intensity. This may be obtained from Eq. (10.42) by integrating over $\mathbf{k}_2$ and then setting $k_1 \to 0$. The result is

$$\langle \delta I^2(L, k = 0) \rangle_c = \frac{1}{W^2} \frac{2\pi}{\kappa^2} \frac{l}{L}. \tag{10.44}$$

Here a factor of N has been divided out by normalizing the result by $\langle \mathbf{k}_1 | \mathbf{k}_1 \rangle$. Equation (10.44) is interesting because it tells us that the decay of the long-range fluctuation is much slower as a function of L than that predicted by the simple-minded picture of $\langle \delta I^2 \rangle_c = \langle I \rangle_c^2$, which follows from the exponential distribution of $I(\mathbf{r})$ and give a much faster $1/L^2$ decay. The reason for the larger-than-expected fluctuation is easy to understand. The effect of interference causes random intensity variation everywhere in space, which counters the smoothing effect of the diffusion process. Therefore, intensity fluctuations contain both a fast varying component, given by $C^{(1)}$, and a slow varying component, given by $C^{(2)}$.

The calculations so far follow the pattern of diffusion–interference giving rise to the speckle pattern, and diffusion–interference–diffusion giving rise to the long-range fluctuation correlation. This process can be continued by calculating the diffusion on a background of long-range correlated fluctuations. Due to the interaction that interference introduces into the propagation mechanism, one can expect the existence of even slower-decay fluctuations that dominate at long distances. In fact, as we shall show below through numerical simulation results, there can be a range of propagation distances in which the intensity fluctuation is roughly a *constant*, i.e., independent of L. This phenomenon is called the universal conductance fluctuations. A heuristic explanation of this unusual mesoscopic behavior is given at the end of section 10.6. However, the universal conductance fluctuation phenomenon is not expected to extend to infinite propagation distance, because the effect of coherent backscattering, i.e., the maximally crossed diagrams, has been neglected so far. In order to take all these effects into consideration, it is necessary to have a formula for the accurate evaluation of the wave diffusion constant, or the conductance, of a mesoscopic sample. This is provided by Landauer's formula, derived next in the context of electronic systems at zero temperature.

10.5 Landauer's Formula and Quantized Conductances

To derive Landauer's formula for the calculation of mesoscopic conductance, let us refer to Figure 10.1, where a tunneling barrier (the mesoscopic sample) is shown together with the two "perfect" leads connected to its right and left. The leads are in turn connected to the electron reservoirs with Fermi levels μ_1 and μ_2. A voltage is applied across the sample, so $\mu_1 > \mu_2$. The electron current in this case is given by

$$j = \frac{ev_e}{2\hbar} \rho_e(\omega)(\mu_1 - \mu_2)|\tau|^2, \qquad (10.45)$$

where $|\tau|^2$ is the intensity transmission coefficient, and the factor $1/2$ arises because we are considering only the right-traveling electrons. If one writes for the quantum wave case $v_e = \sqrt{2\hbar\omega/m}$, then $\rho_e(\omega) = (\pi v_e)^{-1}$ for both quantum and classical waves [see Eqs. (2.26), (2.27)]. Therefore,

$$j = \frac{e^2}{h}\left(\frac{\mu_1 - \mu_2}{e}\right)|\tau|^2.$$

Since $(\mu_1 - \mu_2)/e = V$, we have

$$\Gamma = \frac{e^2}{h}|\tau|^2. \qquad (10.46)$$

If the 1D sample is broadened into a finite-width strip, the only change in the above formula is in the calculation of $|\tau|^2$. As seen in Problem 7.1, for the case of a strip with width M there can be a number $M_0 < M$ of (non-evanescent) incident channels and the same number of exit channels. By defining

$$T_i = \sum_j |\tau_{ij}|^2$$

as the transmission coefficient for the ith output channel, we get

$$\Gamma = \frac{e^2}{h} \sum_{i=1}^{M_0} T_i. \tag{10.47}$$

If the mesoscopic sample is sufficiently small and ordered, then it is possible that $T_i = 1$ for all $i = 1, \ldots, M_0$ (total transmission, no reflection). In that case the conductance is given by

$$\Gamma = \frac{e^2}{h} M_0, \tag{10.48}$$

i.e., Γ *is quantized in units of* e^2/h. Quantized conductances have indeed been observed experimentally in mesoscopic structures (van Wees *et al.*, 1988). Due to spin degeneracy, the observed unit of quantization is $2e^2/h$ since each channel can accomodate two electrons.

The above situation applies in the case of two-terminal measurement, i.e., the current leads coincide with the voltage leads. However, this is not necessarily always the case experimentally. One can equally well have four-terminal measurement where the voltage measurement is conducted via two separate leads, apart from the two current leads. For the same sample, the four-terminal measurement of mesoscopic conductance can differ from the two-terminal measurement. To describe the four-terminal measurement, let us refer to the configuration depicted in Figure 10.3. Besides the two current reservoirs at μ_1 and μ_2, just as in the two-terminal case, we wish to measure the voltage difference across the sample by

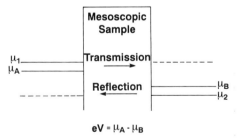

Figure 10.3 A schematic depiction of a mesoscopic sample with four-leads connected to its left and right. The leads connected to electron reservoirs with Fermi levels μ_1 and μ_2 supply the current. Two additional leads are for the purpose of measuring the voltage across the sample. Due to the reflected and transmitted currents, the chemical potentials of the voltage leads are at $\mu_A \neq \mu_1$ and $\mu_B \neq \mu_2$. The voltage difference across the sample is given by $(\mu_A - \mu_B)/e$.

10.5 Landauer's Formula and Quantized Conductances

attaching two additional leads just outside the sample, to its left and right. The chemical potentials of the two voltage leads are denoted by μ_A and μ_B. Due to the current flow, the chemical potentials μ_A and μ_B in the left and right perfect leads, respectively, are not the same as μ_1 and μ_2. μ_A and μ_B are determined by the condition that the number of occupied states above them is the same as the number of unoccupied states below them. By applying this principle to the right lead, we get the number of occupied states above μ_B to be

$$\frac{1}{2\hbar} \rho_e(\omega)|\tau|^2(\mu_1 - \mu_B).$$

Since the total number of states between μ_B and μ_2 is

$$\frac{1}{\hbar} \rho_e(\omega)(\mu_B - \mu_2),$$

it follows that the number of unoccupied states below μ_B is

$$\frac{1}{\hbar}\left(1 - \tfrac{1}{2}|\tau|^2\right)\rho_e(\omega)(\mu_B - \mu_2).$$

μ_B is thus determined by the condition

$$\frac{1}{\hbar}\rho_e(\omega)\tfrac{1}{2}|\tau|^2(\mu_1 - \mu_B) = \frac{1}{\hbar}\left(1 - \tfrac{1}{2}|\tau|^2\right)\rho_e(\omega)(\mu_B - \mu_2), \quad (10.49)$$

or

$$\mu_B = \tfrac{1}{2}|\tau|^2(\mu_1 - \mu_2) + \mu_2. \quad (10.50)$$

To the left, the number of occupied states above μ_A is given by the addition of incident and reflected electrons, i.e.,

$$\frac{1}{2\hbar}\left(1 + |\rho|^2\right)\rho_e(\omega)(\mu_1 - \mu_A),$$

where $|\rho|^2$ denotes the intensity reflection coefficient. Similar consideration leads to the number of unoccupied states as

$$\frac{1}{\hbar}\left[1 - \tfrac{1}{2}(1 + |\rho|^2)\right]\rho_e(\omega)(\mu_A - \mu_2).$$

Equating the two gives

$$\mu_A = \mu_1 - \tfrac{1}{2}(1 - |\rho|^2)(\mu_1 - \mu_2). \quad (10.51)$$

By writing $|\tau|^2 = 1 - |\rho|^2$ in Eq. (10.50), it follows that

$$\mu_A - \mu_B = |\rho|^2(\mu_1 - \mu_2). \tag{10.52}$$

Since now the voltage $V = (\mu_A - \mu_B)/e$, it follows from Eq. (10.45) that

$$\Gamma = \frac{e^2 v_e}{2\hbar} \rho_e(\omega) \frac{|\tau|^2}{|\rho|^2} = \frac{e^2}{h} \frac{|\tau|^2}{|\rho|^2}, \tag{10.53}$$

where we have again used the expression $\rho_e(\omega) = 2(d\kappa_e/d\omega)/2\pi = 1/\pi v_e$. The dimensionless conductance of a mesoscopic sample is thus given by

$$\frac{\Gamma}{(e^2/h)} = \gamma = \frac{|\tau|^2}{|\rho|^2}. \tag{10.54}$$

Equation (10.54) should be contrasted with Eq. (10.46) for the two-terminal case. To generalize this (four-terminal) formula to 2D and 3D samples, one simply notes that there are many more incident and transmission "channels," as defined in the solution to Problem 7.1. Here we will define the channels in the momentum representation. Let k be the channel index for the left lead and i be the index for the right lead. If $|\tau_{ik}|^2$ is the intensity transmission coefficient between the kth input channel and the ith output channel, the current in the ith channel is now given by

$$j_i = \frac{e^2}{h}(\mu_1 - \mu_2) \sum_{k=1}^{M_0} |\tau_{ik}|^2 = \frac{e^2}{h}(\mu_1 - \mu_2)T_i, \tag{10.55}$$

where M_0 is the number of propagating channels (see the solution to Problem 7.1), and T_i is the sum of $|\tau_{ik}|^2$ over all k. The total current is thus

$$j = \sum_i j_i = \frac{e^2}{h}(\mu_1 - \mu_2) \sum_{i=1}^{M_0} T_i. \tag{10.56}$$

The current in channel k of left lead is given by the flux of right-going electrons in channel k, minus the reflected flux:

$$j_k = \frac{e^2}{h}(\mu_1 - \mu_2)\left[1 - \sum_{i=1}^{M_0} |\rho_{ik}|^2\right] = \frac{e^2}{h}(\mu_1 - \mu_2)(1 - R_k),$$

10.5 Landauer's Formula and Quantized Conductances

where R_k is defined as the sum of $|\rho_{ik}|^2$ over all i. The total current is therefore

$$j = \frac{e^2}{h}(\mu_1 - \mu_2)\sum_{k=1}^{M_0}(1 - R_k). \tag{10.57}$$

Since the total current must be conserved, we have the identity

$$\sum_{i=1}^{M_0} T_i = \sum_{k=1}^{M_0}(1 - R_k). \tag{10.58}$$

By following the same argument as the 1D case, μ_B is determined by the condition [see Eq. (10.49)]

$$\sum_{i=1}^{M_0} v_i^{-1} T_i(\mu_1 - \mu_B) = \sum_{i=1}^{M_0} v_i^{-1}(2 - T_i)(\mu_B - \mu_2), \tag{10.59}$$

where the density of states for the ith channel is given by $(\pi v_i)^{-1}$. The only complication here, as compared with the 1D case, is that the density of states can differ from channel to channel. From (10.59) it follows that

$$\mu_B = \frac{\sum_{i=1}^{M_0} v_i^{-1} T_i}{2\sum_{i=1}^{M_0} v_i^{-1}}(\mu_1 - \mu_2) + \mu_2. \tag{10.60a}$$

Similar reasoning leads to the formula

$$\mu_A = \mu_1 - \frac{\sum_{i=1}^{M_0} v_i^{-1}(1 - R_i)}{2\sum_{i=1}^{M_0} v_i^{-1}}(\mu_1 - \mu_2). \tag{10.60b}$$

From the fact $V = (\mu_A - \mu_B)/e$ and $\Gamma = j/V$, we have

$$\Gamma = \frac{e^2}{h}\frac{(\sum_{i=1}^{M_0} v_i^{-1})\sum_{i=1}^{M_0} T_i}{\frac{1}{2}\sum_{i=1}^{M_0}(1 + R_i - T_i)v_i^{-1}}. \tag{10.61}$$

This is the general form of the Landauer's, or the Landauer–Buttiker, formula (Buttiker et al., 1985) for four-terminal measurement of mesoscopic conductances. It should be noted that spin degeneracy of the electrons is not taken into account in Eq. (10.61).

With Landauer's formula and the recursive Green's function method as presented in the solution to Problem 7.1, the conductance of mesoscopic samples can be evaluated numerically. Some of their special characteristics are given below.

10.6 Characteristics of Mesoscopic Conductance

Small electronic samples at low temperatures can realize the condition of $L_{in} > L$. When this happens a number of interesting, nonclassical behaviors occur. First, the dimensionless γ can vary significantly from one sample to the next, even if the samples are made in the same batch. The reason is described below. Second, γ can be sample size dependent. This is due to the coherent backscattering effect as discussed in earlier chapters. Moreover, $\gamma \to 0$ in 1D and 2D as $L \to \infty$, since all states are localized in 1D and 2D.

To be more specific, it is instructive to look at a particular system, such as a thin disordered metallic film. In practice, a thin film may be regarded as 2D if its thickness is less than a few hundred angstroms. We will thus model this system by a 2D square lattice which has a certain fraction p of the sites randomly occupied by metal particles and the rest by insulating particles. Just as in the Anderson model [see Eq. (2.53)], the wave equation may be written as

$$\mathbf{M}|\phi\rangle = \frac{2m}{\hbar^2}(\hbar\omega - \epsilon_0)|\phi\rangle, \tag{10.62}$$

where

$$\mathbf{M} = \sum_{\mathbf{l}} \sigma(\mathbf{l})|\mathbf{l}\rangle\langle\mathbf{l}| + t \sum_{\mathbf{l},\mathbf{n}} |\mathbf{l}\rangle\langle\mathbf{l} + \mathbf{n}|, \tag{10.63}$$

t is a constant, and $\mathbf{n}$ is the vector that points from $\mathbf{l}$ to one of its nearest neighbors. However, unlike the Anderson model, here $\sigma(\mathbf{l})$ will take only two values:

$$\sigma(\mathbf{l}) = \begin{cases} 0 & \text{if site } \mathbf{l} \text{ is occupied by a metal particle} \\ \infty & \text{if site } \mathbf{l} \text{ is occupied by insulator}. \end{cases} \tag{10.64}$$

Since by setting $\sigma(\mathbf{l}) = \infty$ means the electron will be prevented from entering the site, the randomness in this model is noted to be in the

10.6 Characteristics of Mesoscopic Conductance

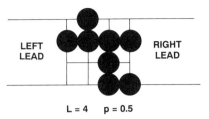

Figure 10.4 A schematic picture of a 2D, $L = 4$, $p = 0.5$ quantum percolation model configuration. Filled circles denote metallic particles.

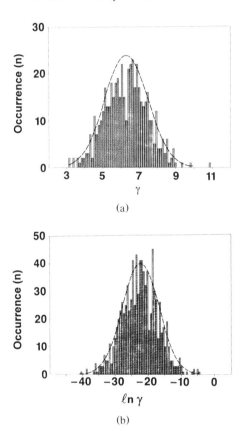

Figure 10.5 Histogram of the conductance distribution. In (a), 500 random configurations are calculated with the parameters $L = 60$ (in units of lattice constant), $p = 0.95$, and $2m(\hbar\omega - \epsilon_0)/\hbar^2 t = 0.01$ (near the bottom of the band). In (b), 900 configurations are calculated with the parameters $L = 60$, $p = 0.75$, and $2m(\hbar\omega - \epsilon_0)/\hbar^2 t = 0.01$. The dashed lines denote the Gaussian lineshape to guide the eye. It is seen that in (a), the distribution is Gaussian in terms of γ; whereas in (b), the distribution is Gaussian in terms of $\ln\gamma$. The figures are from Sheng and Zhang (1991).

position of the metal particles, in contrast to the Anderson model, where the randomness is in the value of $\sigma(l)$. In the literature the present model is usually called the *quantum percolation model*. A finite-sized sample in this model may be depicted schematically as shown in Figure 10.4.

By using the method of recursive Green's function and Eq. (10.61), the conductance of the model may be evaluated numerically. For a fixed p, there can be many configurations of the sample as the arrangement of metal particles can vary randomly. As s result, the conductance will also vary randomly. However, the *distribution* of the conductances is nevertheless well defined. Figures 10.5a and b show the forms of the distribution for two samples with $L = 60$ (in units of the lattice constant) but different values of p. In Figure 10.5a the value of $p = 0.95$, the sample's total transmission coefficient is close to 1, and the distribution is Gaussian in character. In Figure 10.5b the value of $p = 0.75$, the sample's total transmission is rather low, and the distribution is well described by a lognormal distribution, i.e., Gaussian in terms of $\ln \gamma$. Between $p = 0.95$ and 0.75, the distribution can be difficult to describe analytically.

The distribution of γ and its variation with p and L contain all the essential information about mesoscopic conductances. For example, by looking at the logarithmic mean of γ, i.e., $\langle \ln \gamma \rangle_c$, as a function of L it is easy to deduce the localization length. This is shown in Figure 10.6. Also, the standard deviation of the distribution, defined as $\Delta \gamma = \sqrt{\langle \gamma^2 \rangle - \langle \gamma \rangle^2}$, displays interesting variation with L as shown in Figure 10.7. At small L,

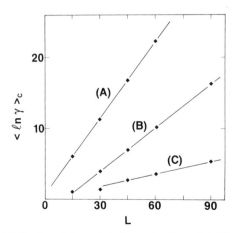

Figure 10.6 The average value of $\ln \gamma$ plotted as a function of L. The data points show a linear behavior. The inverse of the slope gives the localization length. (a) is for $p = 0.75$, (b) is for $p = 0.8$, and (c) is for $p = 0.85$. The value of $2m(\hbar \omega - \epsilon_0)/\hbar^2 t = 0.01$. The figure is from Sheng and Zhang (1991).

10.6 Characteristics of Mesoscopic Conductance

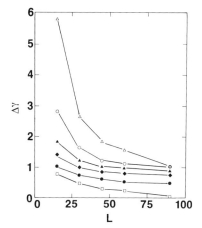

Figure 10.7 The conductance fluctuation, expressed in terms of the standard deviation $\Delta\gamma$ of the distribution, plotted as a function of L for various values of p: $\triangle$ is for $p = 0.9$, $\bigcirc$ is for $p = 0.85$, $\blacktriangle$ is for $p = 0.8$, $\blacklozenge$ is for $p = 0.75$, $\bullet$ is for $p = 0.7$, and $\square$ is for $p = 0.65$. The value of $2m(\hbar\omega - \epsilon_0)/\hbar^2 t$ is 0.5. The figure is from Sheng and Zhang (1991).

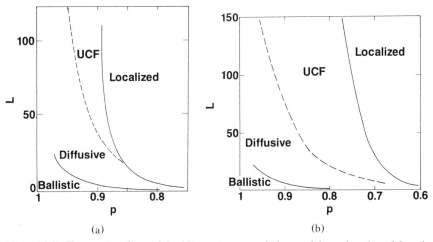

Figure 10.8 Transport regimes of the 2D quantum percolation model as a function of L and p. In both cases the upper solid line denotes the localization length ($\xi = L$), the lower solid line denotes the mean free path ($l = L$). Dashed line denotes ten times the mean free path ($10l = L$), and $2m(\hbar\omega - \epsilon_0)/\hbar^2 t = 0.01$ in (a), 0.5 in (b). The figures are from Sheng and Zhang (1991).

$\Delta\gamma$ displays fast decay. But at a certain point $\Delta\gamma$ becomes roughly a constant $\equiv 1$, independent of L. This is precisely the phenomenon of universal conductance fluctuations (UCF). However, at larger distances $\Delta\gamma$ is expected to decay further. The region of UCF extends roughly over the regime $10l < L < \zeta$.

In terms of Landauer's formula and the prior discussion on spatial correlations, there is a heuristic explanation for the existence of the UCF regime. Take the two-terminal version of Landauer's formula,

$$\gamma = \sum_{i,k}^{M_0} |\tau_{ik}|^2,$$

and consider the evaluation of $\Delta\gamma$. From the discussion in the previous section it is clear that due to interference effect, there can be strong correlation in the transmission coefficients $\{\tau_{ik}\}$, so they cannot be independent. However, Lee proposed (Lee, 1986) that the reflection coefficients, $\{\rho_{ik}\}$, should be roughly independent because they are dominated by the short paths near the input lead-sample interface. Since the reflection coefficients are related to the transmission coefficients due to the unitary requirement, we have from Eq. (10.58)

$$\sum_{i,k}^{M_0} |\tau_{ik}|^2 = M_0 - \sum_{i,k}^{M_0} |\rho_{ik}|^2. \qquad (10.65)$$

From the two-terminal Landauer's formula, it follows that

$$\Delta\gamma = \left[\langle \gamma^2 \rangle_c - \langle \gamma \rangle_c^2\right]^{1/2} \cong M_0 \left[\langle |\rho_{ik}|^4 \rangle_c - \langle |\rho_{ik}|^2 \rangle_c^2\right]^{1/2}, \qquad (10.66)$$

where the reflection coefficients are assumed to be independent so terms like $\langle |\rho_{ik}|^2 |\rho_{lm}|^2 \rangle_c = 0$ if $i \neq l$ or $k \neq m$. If $|\rho_{ik}|^2$ is considered to be the intensity of many interfering paths, then the argument of Section 10.2 applies, and the standard deviation of the distribution is equal to its mean in that case. That means

$$\Delta\gamma \cong M_0 \langle |\rho_{ik}|^2 \rangle_c = 1 - \frac{1}{M_0} \langle \sum_{i,k}^{M_0} |\tau_{ik}|^2 \rangle_c. \qquad (10.67)$$

Since the averaged total transmission varies as $1/L$, it follows from (10.67) that for a *sufficiently thick sample* the leading order term in $\Delta\gamma$ is independent of L, with the magnitude on the order of 1. This is precisely the universal conductance fluctuation.

The behavior of mesoscopic conductances can be summarized by the "phase" diagrams of Figures 10.8a and b. The two external control variables are L and p. It is seen that mesoscopic systems have an entirely different set of characteristics from the bulk systems. However, it is interesting to observe that it is precisely at the mesoscopic level that quantum and classical waves display very similar behavior due to the common underlying physics.

Problems and Solutions

10.1. Evaluate $S(\Delta\omega, \mathbf{r} - \mathbf{r}'|\omega)$ in 3D.
From the diffusive form of $S(\Delta\omega, \Delta\mathbf{k}|\omega)$, we get

$$S(\Delta\omega, \mathbf{r} - \mathbf{r}'|\omega)$$
$$= \frac{\rho_e(\omega)}{2\pi} \frac{d\omega}{d\kappa_0^2} \frac{d\omega}{d\kappa_e^2} \int_0^\infty (\Delta k)^2 \, d(\Delta k)$$
$$\times \int_{-1}^1 d\cos\theta \frac{\exp[i\,\Delta k|\mathbf{r} - \mathbf{r}'|\cos\theta]}{-i\,\Delta\omega + v_t l(\Delta k)^2/3}$$
$$= \frac{\rho_e(\omega)}{2\pi} \frac{d\omega}{d\kappa_0^2} \frac{d\omega}{d\kappa_e^2} \int_0^\infty d(\Delta k) \frac{(\Delta k) 2\sin[\Delta k|\mathbf{r} - \mathbf{r}'|]}{|\mathbf{r} - \mathbf{r}'|(v_t l/3)\left[(\Delta k)^2 - i\dfrac{3\,\Delta\omega}{v_t l}\right]}$$
$$= \frac{3\rho_e(\omega)(d\omega/d\kappa_0^2)(d\omega/d\kappa_e^2)}{2\pi v_t l|\mathbf{r} - \mathbf{r}'|} \int_{-\infty}^\infty dx \frac{x \sin x}{x^2 - i\left(\dfrac{3\,\Delta\omega|\mathbf{r} - \mathbf{r}'|^2}{v_t l}\right)}.$$

(P10.1)

The poles of the integrand are at

$$x = x_\pm = \pm(1 + i)\sqrt{\frac{3\,\Delta\omega}{2v_t l}}\,|\mathbf{r} - \mathbf{r}'|. \qquad (\text{P10.2})$$

By carrying out the integration in the complex plane, we get

$$S(\Delta\omega, \mathbf{r} - \mathbf{r}'|\omega) = \frac{3\rho_e(\omega)(d\omega/d\kappa_0^2)(d\omega/d\kappa_e^2)}{4v_t l|\mathbf{r} - \mathbf{r}'|}[\exp(ix_+) + \exp(-ix_-)]$$

$$= \frac{3\rho_e(\omega)(d\omega/d\kappa_0^2)(d\omega/d\kappa_e^2)}{2v_t l|\mathbf{r} - \mathbf{r}'|}$$

$$\times \exp\left[(1+i)\sqrt{\frac{3\Delta\omega}{2v_t l}}|\mathbf{r} - \mathbf{r}'|\right]. \quad \text{(P10.3)}$$

10.2. Show that the $G(\mathbf{k}, z, z')$ given by Eq. (10.37) solves Eq. (10.36). The Green's function to Eq. (10.36) must satisfy

$$\frac{d^2}{dz^2}G(\mathbf{k}, z_<, z_>) - k^2 G(\mathbf{k}, z_<, z_>) = \delta(z_< - z_>). \quad \text{(P10.4)}$$

Let us first take the $z_<, z_>$ derivatives of G:

$$\frac{dG(\mathbf{k}, z_<, z_>)}{dz_<} = -\frac{\cosh(kz_<)\sinh[k(L - z_>)]}{\sinh(kL)}. \quad \text{(P10.5)}$$

$$\frac{dG(\mathbf{k}, z_<, z_>)}{dz_>} = +\frac{\sinh(kz_<)\cosh[k(L - z_>)]}{\sinh(kL)}. \quad \text{(P10.6)}$$

By taking the second derivatives with respect to $z_<$ and $z_>$, one gets back exactly $k^2 G(\mathbf{k}, z_<, z_>)$ everywhere except at $z_< = z_>$. Therefore, Eq. (P10.4) is clearly satisfied everywhere $z_< \neq z_>$. At $z_< = z_>$, we have

$$\frac{dG(\mathbf{k}, z_<, z_>)}{dz_>} - \frac{dG(\mathbf{k}, z_<, z_>)}{dz_<} = 1. \quad \text{(P10.7)}$$

Therefore, there is a unit jump at $z_< = z_>$, which means the second derivative must give a delta function at $z_< = z_>$. Moreover, $G(\mathbf{k}, z_<, z_>)$ satisfies the boundary conditions that $G(\mathbf{k}, z_<, z_>) = 0$ at $z_< = 0, z_> = L$.

References

Buttiker, M., Imry, Y., Landauer, R., and Pinhas, S. (1985). *Phys. Rev.* **B31**, 6207.
Feller, W. (1957). "An Introduction to Probability Theory and Its Applications, Vol. 1. John Wiley & Sons, New York.
Goodman, J. W. (1985). "Statistical Optics," John Wiley & Sons, New York.

References

Landauer, R. (1957). *IBM J. Res. Dev.* **1**, 223.
Lee, P. A. (1986). *Physica* **140A**, 169.
Lee, P. A., Stone, A. D., and Fukuyama, H. (1987). *Phys. Rev.* **B35**, 1039.
Shapiro, B. (1986). *Phys. Rev. Lett.* **57**, 2168.
Sheng, P., and Zhang, Z. Q. (1991). *J. Phys. Cond. Matt.* **3**, 4257.
Spivak, B. A., and Zyuzin, A. Yu. (1988). *Solid State Commun.* **65**, 311.
Stephen, M. J., and Cwilich, G. (1987). *Phys. Rev. Lett.* **59**, 285.
van Wees, B. J., van Houten, H., Beenakker, C. W. J., Williamson, J. G., Kouwenhoven, L. P., van der Marel, D., and Foxon, C. T. (1988). *Phys. Rev. Lett.* **60**, 848.

Index

Accuracy, coherent-potential approximation, 84–85
Amplitude, wave
  diffusive wave spectroscopy, 162
  lattice structure, 26
  scattering, 62–63, 116
  speckle pattern, 302–303
Anderson model
  anisotropy, 234
  coherent-potential approximation, 69–73, 74
  configurational average, 142
  diffusion constant, 153–156
  Herbert-Jones-Thouless formula, 207
  ladder diagram, 154
  lattice structure, 31
  localization behavior, 244–258
    backscattering, 6
    Ioffe-Regel condition, 203, 251
    low frequency, 210
    mobility edge, 249
  mesoscopic conductance, 319, 321
  minimum length scale, 264
  optical theorem, 144
  scaling hypothesis, 227–228, 230
  vertex function, 140–141
  Ward identity, 150–151
Angular profile, 181–186, 188
Anisotropy, 132, 160, 234, *see also* Isotropy
Approximation, *see also* Coherent-potential approximation; Correction, Green's function
  Boltzmann, 197
  tight binding, 26
Average, configurational
  Anderson model vertex, 142
  backscattering, 186
  conductance, 218
  CPA condition, 69
  dissipation, 134
  Green's function, 49–50, 54, 87–88
  intensity transport, 121
  lattice structure, 30
  multiple collisions, 120
  reflection intensity, 181–182
  single versus average, 116
  speckle pattern, 177
  two Green's functions, 171–172

Backscattering, coherent, 177–192
  Anderson model, 6
  angular profile, 181–186, 188
  defined, 3
  diagrammatic representation, 193–195
  diffusion, 3
  discovery, 180
  dynamic implications, 193
  elastic scattering, 3
  electronic systems, 285–286
  features, 3–4
  Green's function, 193–195, 198
  inelastic scattering, 3–4, 283–284
  magnetoresistance, 287–290

329

Backscattering, coherent (*Cont.*)
  mesoscopic conductance, 319
  sample size, 4–7, 185–189
  scaling function, 277
  scaling theory, 234
  universal conductance fluctuation, 315
Bethe-Salpeter equation, 121–133
  flux, 273
  moments, 126
  transport velocity, 269
  Ward identity, 127, 131, 145
Boltzmann approximation, 197
  diffusion constant, 154, 159
  vertex function, 137
Boltzmann limit, 307
Boltzmann transport, 126, 159
Bra and ket notation, *see* Notation, Dirac's bra and ket
Bragg reflection, 28
Brewster angle, 206
Brillouin zone
  lattice Green's function, 32, 35
  localization, 203, 244
  magnetoresistance, 291
  pulse intensity evaluation, 119
Brownian motion
  diffusive transport, 2
  diffusive wave spectroscopy, 163
  pulse intensity, 120
Brownian particles, 161
Bruggeman's equation, 81

Calculation, scaling, 226–232
Causality
  Green's function, 26, 125
  speed of information, 3
Channels, definition, 121
Classical system, *see* System comparison
Clausius-Massotti relation, 84
Coefficient
  reflection, 316–317
  transmission
    Landauer's formula, 229
    mesoscopic sample, 315, 317, 321
    resonant effect, 204–205
Coherence, *see also* Backscattering, coherent
  effective medium, 50
  enhancement factor, 184, 187
  infinite scatterers, 68
  intermediate frequency, 85
  phase, 3
  pulse front, 116–117
  transport velocity, 269
Coherent backscattering effect, *see* Backscattering, coherent
Coherent-potential approximation
  accuracy, 84–85
  Anderson model, 69–73, 74
  classical wave case, 73–84
  diffusive regime transition, 115–116
  electromagnetic waves, 80–81, 83–84, 103–104
  generalized, 86–87, 160, 262
  intermediate frequency, 85–87, 262
  introduced, 68–69
  ladder diagrams, 139, 145
  lattice versus continuum, 71–72
  localization behavior, 250, 257, 261
  1D case, 84
  solvability, 115
  3D case, 71, 83
  2D case, 83
  vertex function, 134–145
  Ward identity, 75, 145–151
Colloid
  backscattering, 186
  generalized coherent-potential approximation, 87
Concentration, scatterer, 270
Conductance
  configuration dependence, 218
  Einstein relation, 217
  finite sample, 218
  finite-size scaling, 227, 229–230
  hopping conduction, 293–298
  localized regime, 219
  mesoscopic sample, 319–323
    measurement, 13
    transmission, 317
    tunneling, 315
  ordinary, 7
  percolation, 295–296
  quantized, 315–316
  scaling function, 271–272, 275
  scaling theory, 216–233
  universal fluctuation, 301–302, 315, 323

Index 331

Conduction
  heat, 2
  hopping, 293-298
  parallel, 295
Conductivity
  Anderson model, 155
  dissipation, 134, 302
  Einstein relation, 201, 242
  electronic sample, 301-302
  electron spin degeneracy, 201, 218
  intensity, 285
  measurement, 5
Conductor, as insulator, 6
Configurational average, *see* Average, configurational
Constant, *see* Dielectric constant; Diffusion constant; Discretization constant; Planck's constant
Continuum, *see* System comparison
Correction, to classical diffusion, 199-201, 242-256, 259-267
Correlation
  effective medium, 76
  frequency, 305
  length, 223, 250, 270
  long-range, 310-315
  operator notation, 54
  short-range, 310
  spatial, 306
  speckle pattern, 304-310
Correlation function, 310, 312
Coulomb interaction, 6
Coupling, spin-orbit, 292
CPA, *see* Coherent-potential approximation

Degeneracy, electron spin
  diffusion, 201
  disordered film, 286
  Landauer's formula, 318
  length scale, 218
  magnetoresistance, 289
  scaling hypothesis, 221
Delocalization, 205, *see also* Localization
Density, of states
  accumulated, 207
  Bethe-Salpeter equation, 127
  diffusive flux, 273-274
  disorder, 72
  frequency, 225

Green's function, 23-24
  infinite scatterers, 67
  Landauer's formula, 318
  lattice Green's function, 34-37
Density, photon, 166
Diagram, *see* Ladder diagram; Maximally crossed diagram; Phase, diagram
Dielectric constant, 18, 58, 60
Diffusion, 115-175, *see also* Dissipation; Transport
  backscattering, 6
  dimensionality, 10-11
  homogeneous, 168
  interference, 314
  light, 164, 166-171
  localization behavior, 250
  photon, 166-167
  regime transition, 115-116
  renormalized, 5, 193-213
  scaling function, 272
  spectroscopy application, 160-171
  static case, 182-183, 187
  suspended particles, 164-165
  transport, 2-3, 50, 68, 216
  wave versus classical, 177-178
Diffusion constant
  Anderson model, 153-156
  classical case, 151-153, 157-160
  classical versus quantum, 259-260
  dimensionality, 201-202
  Einstein relation, 12, 134, 155
  evaluation, 153-160
  frequency dependency, 151-153
  phase velocity, 11-12
  renormalized, 3-4, 199-202, 242
  spectral component, 120
  suspension, 161
  transport velocity, 269
Diffusive pole, 120, 133-134, 199, 306
Diffusive wave spectroscopy, *see* Spectroscopy, diffusive wave
Dimensionality
  coherent-potential approximation, 71, 83-84
  diffusion, 10-11
  diffusion constant, 201-202
  Green's function, 25
  localization, 203-211
  marginal, for localization, 8
  propagation, 8

332 Index

Dimensionality (Cont.)
  scattering, 8
  single scatterer, 63
Dirac's bra and ket, see Notation, Dirac's
    bra and ket
Discretization constant, 234
Disorder
  backscattering, 182
  density of states, 72
  effective, 228
  inaccuracy of Green's function, 49
  intensity, 139
  propagation, 204
Dispersion
  diffusion constant, 154
  intermediate frequency, 86
  isotropic, 251
  microstructure, 82–84
  spectral function, 126
  wavelength, 15
Dissipation
  conductivity, 134, 302
  inelastic scattering, 284
  tunneling, 302
Distance, transport, 115
DWS, see Spectroscopy, diffusive wave
Dyson equation, 55

Effective medium
  correlation, 76
  defined, 50
  description failure, 115
  infinite scatterers, 66–69
  intensity transport, 134
  microstructure, 76
  renormalized, 68
  scalar wave localization, 262
  scattering, 49–113
  scattering potential, 70
  symmetrical microstructure, 77–82
  transport velocity, 269
  unique determination, 76
Eigenfrequency, 206–207
Eigenfunction
  frequency, 20
  lattice structure, 26–28
  localization, 206
Eigenstate, sample size, 217

Eigenvalue
  localization, 216
  magnetoresistance, 288–289
Einstein relation
  conductance, 217
  conductivity, 201, 242
  diffusion constant, 12, 134, 155
  inelastic scattering, 285
Elastic scattering, see Scattering, elastic
Electrical conductivity, see Conductivity
Electromagnetic waves, 80–81, 83–84,
    103–104
Electronic localization, 283–300
Electronic system, see System comparison
Energy
  activation, 294
  conversion, 284
  lattice structure, 28
Energy, self, 124, 145
Enhancement, coherent, 184, 187
Equation, wave, see Green's function;
    Schrödinger equation
Equilibrium, thermal, 284
Ergodic hypothesis, 51
Evanescent wave, 28

Film, disordered, 284–293
Fluctuation
  decay, 314
  intensity, 177, 304, 310–315
  universal conductance, 301–302, 315,
    323–324
Flux
  conservation, 97
  diffusion, 166, 273–274, 311
Four terminal measurement, 316
Fraction, volume
  dispersion microstructure, 82, 83
  localized state, 264
  symmetrical microstructure, 77, 78
Frequency
  beating, 117
  correlation, 305
  density of states, 225
  diffusion constant, 151–153
  gap, 265–266
  Green's function, 20, 22
  high, 262

Index   333

intermediate, 85–87, 262
lattice structure, 26
Rayleigh scattering, 226
shift, 215–216
spatial, 20, 22
spectral, 118
variation, 306
Function, *see specific functions*

Gap, frequency, 265–266
Gaussian distribution
  conductance, 218–219
  diffusion, 163
  mesoscopic conductance, 321
  path length, 187
  speckle pattern, 303
  successive scattering, 205
GCPA, *see* Generalized coherent-potential approximation
Generalized coherent-potential approximation, *see also* Coherent-potential approximation
  diffusion constant, 160
  intermediate frequency, 86–87, 262
Green's function, *see also* Coherent-potential approximation
  advanced, 26, 117, 194
  backscattering, 193–195, 198
  causality, 26, 125
  configurational average, 49–50, 54, 87–88
  continuum case, 21
  correction term evaluation, 242–256, 259–267
  density of states, 23–24
  disorder inaccuracy, 49
  extension for strong scattering, 241–242
  infinite scatterers, 66–68
  inhomogeneous differential equations, 51
  initial condition, 19–20
  intensity fluctuation, 311–312
  lattice, 21, 31–37
  lattice versus continuum, 33
  matrix notation, 52
  operator notation, 52
  recursive, 230, 321
  represented graphically, 135
  retarded, 117, 194
  self-energy analog, 124

single scatterer, 55–56
symmetric microstructure, 79
uniform medium, 19–26
uniform versus random media, 24
usefulness, 19

Hankel function, 25, 63, 65
Herbert-Jones-Thouless formula, 206–211
Hermitian conjugate, 235
Homogeneity, 3, 19
  diffusion, 168
  sample, 215–216
  scaling theory, 233
Hopping conduction, 293–298

Impurity, scattering, 57–58, 60, 64
Incoherence
  background, 188–189
  propagation, 2
  scattering, 194
  speckle pattern, 307
Inelastic scattering
  absorption, 189
  backscattering, 3–4, 283–284
  diffusion, 177
  Einstein relation, 285
  electron-electron interaction, 285
  finite temperature, 283–284
  frequency, 2
  length, 283–285
  magnetoresistance, 289
  mesoscopic sample, 301
  probabilistic process, 283
  propagation direction, 2
  scattering time, 289
Inhomogeneity
  correlated, 38
  sample, 216
  scaling theory, 233
  scattering, 2, 54
Input and output, vertex function, 121–123
Insulator
  as conductor, 6
  disordered, 5
Intensity
  fluctuation, 177, 304, 310–315
  incident, 184

Intensity (Cont.)
  mean, 177
  pulse, 116–120
    configurationally averaged, 116, 120
    Ward identity, 148
    random medium, 116–120
    reflection, 181–182
    reflection coefficient, 316–317
    scattering, 139
    speckle pattern, 302–306, 310
    transmission coefficient, 204–205, 315, 317, 321
    transport, 50
      Bethe-Salpeter equation, 121, 133–134
      random medium, 117, 120
      renormalization, 177
      variation, 177
Interference, see also Speckle pattern
  backscattering, 177
  diffusion, 314
  reflection intensity, 181
Ioffe-Regel criterion
  classical case, 261
  frequency regimes, 263
  localization behavior, 251
  renormalized diffusion, 203
  scaling calculation, 232
  scaling hypothesis, 225
Isotropic scattering
  diffusion constant, 159–160
  ladder diagram, 160, 307
  optical theorem, 64, 65
  point scatterer, 129
  Rayleigh-like, 145
  renormalized diffusion constant, 199–200, 202
  speckle pattern, 307
Isotropy, see also Anisotropy; Isotropic scattering
  dispersion relation, 251
  medium, 16–17
  propagation direction, 265
  sample, 132
  scaling theory, 234

Junction, tunnel, 229, 302

Kramers-Kronig relation, 43, 207
Kronecker delta, 32, 142–143
Kubo-Greenwood formula, 154

Ladder diagram
  Anderson model, 154
  homogeneity, 145
  incoherent scattering, 194
  infinite, 139
  isotropic scattering, 160, 307
  maximally crossed diagram, 196–198
  speckle pattern, 307
Landauer-Buttiker formula, see Landauer's formula
Landauer's formula
  conductance, 302
  derivation, 315–319
  scaling theory, 229
Landau level, magnetic field, 288, 298–299
Laser beam, and spectroscopy, 165–167
Lattice
  Brillouin zone, 32, 35
  continuum comparison, 27–28, 39, 71–72
  continuum problems, 37–40
  correlated, 39–40
  energy, 28
  inhomogeneity, 38
  limits of lattice model, 37–39
  potential, 26
  randomness, 30–31
  simulation of continuum waves, 37
  spring-and-mass, 29
Legendre polynomial, 63
Length
  Anderson model, effect of a minimum scale, 264
  correlation, 223, 250, 270
  electronic versus classical, 301
  inelastic scattering, 283–285, 301
  localization, 5, 215, 250, 254, 257, 261, 266, 267
    Anderson model, 250, 254
    finite-sized scaling, 227–228
    frequency gap, 265–266
    scaling hypothesis, 220–226
    strong scattering, 243–244
  path, dependence of coherent backscattering, 185–189

Index

Light
  diffusion, 164, 166–171
  reflection, 182
Liquid, bubbly, 78
Local density of states, *see* Density, of states
Localization, *see also* Delocalization
  Anderson model, 244–258
    backscattering, 6
    electronic case, 203
    low frequency, 210
  backscattering, 3, 5, 6, 189
  classical, 259–268
  classical versus lattice, 39
  conductance, 219
  dimensionality, 203–211, 248, 251–257
  eigenfunction, 206
  electronic systems, 283–300
  finite temperature, 293–298
  frequency gap, 265–266
  Herbert–Jones–Thouless formula, 207, 209
  Ioffe-Regel criterion, 203
  length, 202, 250, 254, 257, 261, 266, 267
  localized states, 241–281
  mobility edges, 249, 261
  numerical simulation, 249, 254
  1D, 2D, and 3D cases
    Anderson model, 244–258
    classical waves, 259–268
  scalar, 264–265
  scalar versus quantum, 262
  scaling, 6–7, 215–239
  self-consistent theory, 241–244
  theory, 6, 241–244, 249, 251
  transport, 6
  weak, 189
Lorentz gauge, 103
Lyapunov exponent, 212

Magnetoresistance, 287–293, 299
  defined, 290–291
Mass, effective
  defined, 28
  lattice structure, 27
  scaling theory, effect of anisotropy, 219
Matrix
  notation for Green's function, 52
  scattering, 53

Maximally crossed diagram, 195–199
  diffusion constant, 199
  ladder diagram, 196–198
  localization behavior, 244, 246–248, 253
Maxwell equations, 80
Maxwell-Garnett theory, 84
MC diagram, *see* Maximally crossed diagram
Mean free path
  Boltzmann expression, 159
  dispersion microstructure, 82–84
  dispersion relation, 154
  effective medium, 50
  electromagnetic waves, 81, 83
  intensity transport, 134
  length scale, 2
  symmetric microstructure, 79, 80–81
  transport, 115, 159–160
Medium, *see also* Effective medium; Random medium
  disordered, 1–2, 49
  homogeneous effective, 2
  isotropic, 16–17
  renormalized, 67, 68
  single scatterer, 55–57, 61
  uniform, 19–26, 55–57, 61, 78
Mesoscopic phenomena, 1, 301
  conductance, 319–323
    measurement, 13
    transmission, 317
    tunneling, 315
  correlations, 304–315
  intensity distribution, 302–304
  intensity fluctuation, 310–315
  Landauer's formula, 315–319
  sample, 12, 50, 301–325
  speckle pattern, 302–310
Microstructure
  dispersion, 82–84
  inaccuracy of coherent-potential approximation, 85
  scalar wave localization, 263
  structural units, 76–77, 82
  symmetrical, 77–82
  transport velocity, 270
Mobility edge
  Anderson model, 249
  classical waves, 261
  defined, 7
  scaling hypothesis, 223–224

Mott's law, 296
Mott transition, 6

Neumann function, 93
Newton's equation, 100
Notation
    Dirac's bra and ket, 29–30, 54, 56, 60
    Green's function matrix, 52
    Green's function operator, 52
    vertex function, 121

Operator
    impurity potential, 52
    notation for Green's function, 52
Optical theorem
    Anderson model, 144
    flux conservation, 97–100
    infinite scatterers, 68
    single scatterer, 64–66
    symmetrical microstructure, 79
    Ward identity, 151
Output and input, vertex function, 121–123

Particle function, 75, 146
Path
    Green's function, 193
    hopping conduction, 294
    length, 182, 185–189
    mean free, *see* Mean free path
Percolation
    conductance, 295–296
    quantum model, 321
    threshold, 81–82, 84, 104–105
Phase
    diagram, transport regimes, 323
    relation, 179
    velocity, 11, 12
Phenomena, mesoscopic, *see* Mesoscopic phenomena
Phenomenology, 6, 215
Photon diffusion, 166–167
Planck's constant, 15
Pole, diffusive, 120, 133–134, 199, 306

Potential
    impurity operator, 52
    scattering, 70
Profile, angular, 181–186, 188
Propagation
    configurational average, 50
    dimensionality, 8
    disorder, 204
    incoherence, 2
    isotropy, 265
    local, 9
    sound in bubbly liquid, 78
    universal conductance fluctuation, 315
Pulse intensity, 116–120

Quantized conductance, 315–316
Quantum system, *see* System comparison

Random medium
    dispersion microstructure, 82–84
    microstructure, 76–77
    pulse intensity evolution, 116–120
    symmetric microstructure, 77–82
    transfer matrix, 205
Randomness
    Anderson model, 69
    frequency gap, 265–266
    lattice, 30–31
    scalar wave localization, 264–265
    scaling hypothesis, 220
Rayleigh frequency dependence, 67, 81
    localization, 209
Rayleigh scattering
    isotropy, 145
    scaling theory, 226, 233
    single scatterer, 64–66
    symmetrical microstructure, 80
Recursive Green's function method, 234–239
Reflection
    Bragg, 28
    diffusion wave spectroscopy, 169
    intensity, 181–182
    intensity coefficient, 316–317
    scattering direction, 66
Renormalization
    backscattering, 193–195

Index 337

diffusion, 5, 193-213
diffusion constant, 3-4, 199-202, 242
effective medium, 67, 68
Herbert-Jones-Thouless formula, 206-211
maximally crossed diagram, 195-199
Resistance, *see* Magnetoresistance
Resonance
  classical versus lattice, 39
  impurity state, 57-58
  leaky, 270
  natural, 20
  scattering, 153
  single scatterer, 57-58
  sphere, 264
  transmission, 204-205
  transport velocity, 269, 270

Sample
  anisotropic, 132
  backscattering, 4-7, 185-189
  bulk versus mesoscopic, 301
  dimensionality, 201-203
  electronic, 301-302
  finite, 218, 226-232, 271-273
  homogeneous, 215-216
  incoherent background, 188-189
  inhomogeneous, 216
  isotropic, 132
  localization length, 215
  macroscopic, 283, 285, 295, 301
  mesoscopic, 12, 50, 301-325
  microscopic, 283
  scaling, 216-218
  scaling hypothesis, 222, 226-232
  size limitation, 189
  time scale, 216
  weak scattering, 9
Scalar system, *see* System comparison
Scaling
  calculation, 226-232
  finite-size, 226-232
  meaning, 6-7
  sample size, 216-218
Scaling function, 226-232, 271-277
  asymptotic behavior, 276-277
  delocalized state, 273-274, 276-277
Scaling hypothesis
  Anderson model, 227-228, 230

configurational average, 218-219
Ioffe-Regel criterion, 225
mobility edge, 223-224
numerical simulation, 226-232
1D, 2D, and 3D cases, 220-222, 227-232
randomness, 220
Thouless argument, 217
Scaling theory
  conductance, 216-233
  correlation length, 250
  essential element, 217
  isotropy, 234
  limitations, 233-234
  localization, 6-8, 215-239
  quantum and classical, 233
  Rayleigh scattering, 233
  universality, 233
Scatterer, single
  continuum case, 58-66
  lattice case, 55-58
  optical theorem, 64-66
Scatterers, infinite, 66-69
Scattering, *see also* Backscattering, coherent;
    Inelastic scattering; Isotropic scattering
  amplitude, 116
    1D, 96
    2D, 94-95
    3D, 91-93
  anisotropic, 160
  backward, 66
  classical versus lattice, 39
  effective medium, 49-113
  elastic, 2, 3, 178
  electron-electron, 285, 287
  electron-phonon, 284, 286
  forward, 64, 66, 116
    dispersion microstructure, 82
    symmetrical microstructure, 77
  impurity, 57-58, 60, 64
  incoherent, 194
  inhomogeneous, 2
  intensity, 139
  long-wavelength, 160
  matrix, definition, 53
  nonisotropic, 199
  random, 302
  Rayleigh, 64-66, 80, 145
  scaling theory, 226, 233
  strong, 241-250
  successive, 205

Scattering (*Cont.*)
  time-reversed path, 180
  weak, 130, 132, 152, 222
  weak limit, 9
  weak versus strong, 224
Schrödinger's equation
  dispersion relation, 16–17
  lattice potential, 26
  time derivative, 17
Schrödinger wave, 28
Self-consistent theory, 241–244, 249, 251
Self-energy, 124, 145
Shift, frequency, 215–216
Siegert relation, 162, 173–174
Simulation, *see also* Calculation, scaling
  numerical, 249, 254
Size, sample, *see* Sample, size limitation
Spatial dimensionality, *see* Dimensionality
Speckle pattern
  correlations, frequency and spatial, 304–310
  defined, 177
  intensity distribution, 302–304
  intensity fluctuation, 310, 314
  isotropic scattering, 307
  mean intensity, 182
  mesoscopic sample, 301
Spectral component, 117–118
Spectral frequency, 118
Spectral function
  dispersion relation, 126
  intermediate frequency, 85–86, 262
  quasi-mode, 86
Spectroscopy, diffusive wave, 160
  measurement, 165–171
  reflected intensity, 182
  theory, 161–165
Speed, *see also* Velocity
  information, 3
  quantum versus classical, 17–18
  sound, 78
Spin degeneracy, electron, *see* Degeneracy, electron spin
State, localized and extended, 215–218
Structural unit, 76–77, 82, 263
Suspension
  backscattering, 186
  diffusion constant, 160–161
  particle diffusion, 164–165

Symmetry
  microstructure, 77–82
  particle-hole, 248
System comparison
  classical versus electronic, 301
  classical versus lattice, 39
  classical versus quantum, 12, 15–48, 146, 148, 259–260
  classical versus wave diffusion, 177–178
  continuum versus lattice, 27–28, 33, 39, 71–72
  scalar versus vector, 16
  scalar versus quantum, 262

Temperature
  disordered film, 284–287
  electrical conductivity, 5
  equilibrium, 284
  finite, 283–284, 293–298
  hopping conduction, 293–298
  mesoscopic sample, 12–13
Theory, *see also* Scaling theory
  phenomenological, 6, 215
Thouless argument, 217
Threshold, percolation, 81–82, 84, 296
Time
  inelastic scattering, 289
  proportional to path length, 182
Time reversal
  backscattering, 3
  inelastic scattering, 284
  invariance, 180
  maximally crossed diagram, 196–197
  scattering path, 180
Time scale, 216
T-matrix, 53
Trajectory, diffusive wave spectroscopy, 162
Transmission
  coefficient, 229
  intensity, 315, 317, 321
  scattering direction, 66
Transport, *see also* Diffusion
  backscattering, 193
  Boltzmann equation, 126
  Boltzmann expression, 159
  characteristics, 115
  diffusive, 2–3, 50, 68, 216
  distance, 115
  finite temperature, 293–298

Index 339

inside localization length, 202
intensity, 50
  Bethe-Salpeter equation, 121, 133-134
  random medium, 117, 120
magnetoresistance, 287
mean free path, 115, 159-160
mesoscopic sample, 301
nonclassical, 177
regimes in mesoscopic samples, 323
sample size, 5
transition behavior, 116
velocity, 262
Tunneling
  dissipation, 302
  Landauer's formula, 315
  transport, 294-296
Tunnel junction, 229, 302
Two terminal measurement, 315-316

Unit, structural, 76-77, 263
Universal conductance fluctuation, *see* Fluctuation, universal conductance

Velocity, *see also* Speed
  phase, 11-12
  time reversal, 180
  transport diffusion, 154, 159
    scalar waves, 260-262, 269-271
Vertex
  irreducible, 242
Vertex function
  Anderson model, 140-141
  coherent-potential approximation, 134-145
  diffusion constant, 157
  input and output channels, 121-123
  irreducible, 136, 140-142, 145, 242
  ladder diagram, 139, 145
  maximally crossed diagram, 197, 199
  reducible, 124, 134
  reducible versus irreducible, 139
  represented as a box, 122
Volume fraction, *see* Fraction, volume

Ward identity, 145-151
  Anderson model, 150-151
  Bethe-Salpeter equation, 127, 131, 145
  coherent-potential approximation, 75, 145-151
  diffusion constant, evaluation, 158
  generalized optical theorem, 151
  maximally crossed diagrams, evaluation, 198
  quantum versus classical, 146, 148
Wave
  classical versus quantum, 15-48
  quantum versus classical, 12
Wave diffusion, *see* Diffusion
Wave equation, *see* Green's function; Schrödinger equation
Wavelength
  dispersion relation, 15
  measure of scattering, 2
Wave localization, *see* Localization
Wave scattering, *see* Scattering
Wood's formula, 78
  derivation, 100-102

ISBN 0-12-639845-3